TECHNICALLY—WRITE!

COMMUNICATING
IN A TECHNOLOGICAL ERA

TECHNICALLY—WRITE!

COMMUNICATING IN A TECHNOLOGICAL ERA

SECOND EDITION

R. S. Blicq

RED RIVER COMMUNITY COLLEGE

Winnipeg, Manitoba, Canada

Prentice-Hall, Inc., Englewood Cliffs, New Jersey 07632

Library of Congress Cataloging in Publication Data

Blicq, R S

Technically—write!

Includes index.

1. Technical writing. I. Title.

T11.B62 1981 808'.0666021 80-22282

ISBN 0-13-898700-9

Printed in the United States of America

10 9 8 7

Editorial/production supervision by Diane A. Lange
Cover Design by Wanda Lubelska
Manufacturing buyer: Harry P. Baisley

The names of people and organizations are imaginary and no reference to real persons is implied or intended.

Prentice-Hall International, Inc., *London*
Prentice-Hall of Australia Pty. Limited, *Sydney*
Prentice-Hall of Canada, Ltd., *Toronto*
Prentice-Hall of India Private Limited, *New Delhi*
Prentice-Hall of Japan, Inc., *Tokyo*
Prentice-Hall of Southeast Asia Pte. Ltd., *Singapore*
Whitehall Books Limited, Wellington, *New Zealand*

CONTENTS

3 TECHNICAL CORRESPONDENCE

4 INFORMAL REPORTS DESCRIBING FACTS AND EVENTS

5 INFORMAL REPORTS DESCRIBING IDEAS AND CONCEPTS

6 FORMAL REPORTS

7 OTHER TECHNICAL DOCUMENTS

10 THE TECHNIQUE OF TECHNICAL WRITING

11 GLOSSARY OF TECHNICAL USAGE

PREFACE

This book presents all those aspects of technical communication that you, as a technician, technologist, engineer, scientist, or technical manager, are most likely to encounter in industry. It introduces you to the employees of two technically oriented companies, to the type of work they do, and to typical situations that call for them to communicate with clients, suppliers, and each other.

Since the first edition of *Technically—Write!* there have been some changes in management of the larger company (H. L. Winman and Associates—see Chapter 2). Its owner has stepped into semiretirement, and new faces and departments have been introduced to better reflect the company's involvement in current technology.

At the end of each major chapter there is a selection of detailed assignments of varying complexity, in which it is assumed you are employed by one of these companies and are engaged in projects that require you to write letters and reports, and sometimes to present your findings orally. A marking control chart (at the end of the book) offers a place for you to record your writing errors when assignments are returned to you, and to check whether you have corrected your faults in successive work. It is based on a Correction Key in "Writing to be Read," a Prentice-Hall Inc. textbook by Eleanor Newman Hutchens, and is printed here, in an altered form, with her permission.

The information on technical paper presentation (Chapter 8) was presented originally as a technical paper at the IRE (now IEEE) Sixth National Symposium on Engineering Writing and Speech, Washington, D.C. It was subsequently published by Electronic Design under the title "It's Your Turn at the Rostrum—Can You Deliver a Technical Paper?" and it is reprinted here through their courtesy.

R.S.B.

TECHNICALLY—WRITE!

COMMUNICATING
IN A TECHNOLOGICAL ERA

PEOPLE AS "COMMUNICATORS"

As communicators, people are lazy and inefficient. We are equipped with a highly sophisticated communication system, yet consistently fail to use it properly. (If our predecessors had been equally lazy and failed to develop their communication "senses" to the level required to ensure survival, many species would be extinct today; indeed, it's even questionable that we would exist in our present form.) This communication system comprises a transmitter and receiver combined into a single package controlled by a computer. It accepts multiple inputs and transmits in three mediums: action, speech, and writing.

We spend much of our wakeful hours communicating, half the time as a transmitter, half as a receiver. If, as a receiver, we mentally switch off or permit ourselves to change channels while someone else is transmitting, we contribute to information loss. Similarly, if as a transmitter we permit our narrative to become disorganized, unconvincing, or simply uninteresting, we encourage frequency drift. Our listeners detune their receivers and let their computers think about the lunch that's imminent, or wonder if they should take in a show tonight.

As long as a person transmits clearly, efficiently, and persuasively, the persons receiving keep their receivers "locked on" to the transmitting frequency (this applies to both written and spoken transmissions). Such conditions expedite the transfer of information, or "communication."

In the direct contact situation, in which one person is speaking directly to another, the receiver has the opportunity to ask the transmitter to clarify vaguely presented information. But in more formal speech situations, and in all forms of written communication, the receiver no longer has this advantage. He or she cannot stop a speaker who mumbles or uses unfamiliar terminology, and ask that parts of a talk be repeated or clarified; neither can the receiver ask a writer in another city to explain an incoherent passage of a business letter.

The results of failure to communicate efficiently soon become apparent. If people fail to make themselves clear in day-to-day communication, the consequences are likely to differ from those they anticipated, as Cam Collins has discovered to his chagrin.

Cam is a junior electrical engineer at Robertson Engineering Company, and his specialty is high voltage power generation. When he first read about a recent EHV DC power conference, he wanted urgently to attend. In a memorandum to Fred Stokes, the company's chief engineer, Cam described the conference in glowing terms which he hoped would convince Fred to approve his request. This is what he wrote:

> Fred:
>
> The EHV conference described in the attached brochure is just the thing we have been looking for. Only last week you and I discussed the shortage of good technical information in this area, and now here is a conference featuring papers on many of the topics we are interested in. The cost is only $75.00 for registration, which includes a visit to the Kettle Generating Station. Travel and accommodation will be about $250 extra. I'm informing you of this early so you can make a decision in time for me to arrange flight bookings and accommodation.
>
> Cam

Fred Stokes was equally enthusiastic and wrote back:

> Cam:
>
> Thanks for informing me of the EHV DC conference. I certainly don't want to miss it. Please make reservations for me as suggested in your memorandum.
>
> Fred

Cam has become the victim of his own carelessness: he has failed to communicate clearly exactly what he wanted. There may still be a chance to remedy the situation, but only at the expense of extra time and effort.

Don Ristowell, on the other hand, did not realize that he had missed a golden opportunity until it was much too late to do anything about it. His story stems from an incident that occurred several years ago, when he was a young engineering technician with a firm of mechanical engineers. The stenographers in the contracts department, whose job was to type up proposals for clients, constantly complained to Don that it was impossible to make the proposals look as neat as they would like.

"The right-hand side of the typing always looks so ragged," they would say. "If only we could make the margin straight—like the one on the left."

Don told them he would look at a typewriter he had at home to see if he could modify it. It was a heavy, old-fashioned machine, but it served his purpose. He reasoned that if the toothed bar that controls letter spacing at a fixed 10 or 12 characters to the inch could be made flexible, the length of each line could be adjusted by fractionally expanding or contracting the spacing between letters.

With the limited range of materials then available, Don could not find a practical way to stretch the toothed bar without destroying its rigidity. So he made a series of rigid toothed bars, each with a different number of teeth, which he mounted on a rotating axle that could be positioned immediately beneath the typewriter carriage. It was a clumsy contraption, too large and cumbersome for the average typewriter, but it proved his idea was feasible.

Don felt his employer should know about his idea—possibly the company could develop it into a marketable product, or even help him patent it. So the following day he stopped the Engineering Manager and blurted out his suggestion. This is the conversation that ensued:

Don	*Mr. White*
Mr. White! I've modified an old typewriter at home—to adjust the length of the typing line. . . .	
	Oh?
. . . The problem with most typewriters is the teeth are rigid, so the letter spacing is always the same. You can't get a straight right-hand margin. . . .	
	(Mr. White appeared to be listening politely, but internally he was growing impatient.)
. . . It would help them over in Contracts if we could. . . .	
	Ah! The Contracts Department asked you to modify their typewriters?
Well—uh—in a way. It's for their typists.	
	I don't remember giving you a work order. . . .
No. I did it on my own. (*He meant he did it at home, in his own time*)	
	You took the order directly from Contracts?
It was just an idea I had. . . .	
	There's no reason why the request should not have come through the proper channels. (*His tone was now cold and distant; already he was mentally composing a strong note to the Contracts Manager.*) You had better come into my office!

Don's simple little suggestion had become lost in a web of misunderstanding. By the time he was through explaining, he had given up trying to offer his idea to the company. It lay dormant for three years, until proportional-spacing typewriters first appeared on the scene. They served to remind him that perhaps there *had* been market potential in his modification.

If Cam Collins and Don Ristowell had paused to consider the needs of the persons who were to receive their information, they would never have launched precipitously into discourses that omitted essential facts. Cam had only to start his memorandum with a request ("May I have your approval to attend an EHV DC conference next month?"), and Don with a statement of purpose ("I have devised a gadget that I believe would make a marketable product. May I have a few moments to describe it to you?"), to command the attention of their department heads. Both Mr. Stokes and Mr. White could then have much more effectively appraised the information.

Such circumstances occur daily. They are frustrating to those who fail to communicate their ideas, and costly when the consequences are carried into business and industry.

Bill Carr recently devised and installed a monitor unit for the remote control panel at the microwave relay station where he is the resident engineering technician. As his modification greatly improved operating methods, head office asked him to submit an installation drawing and an accompanying description. Here is part of his description:

> Some difficulty was experienced in finding a suitable location for the monitor unit. Eventually it was mounted on a special bracket attached to the left-hand upright of the control panel, as shown on the attached drawing.

On the strength of Bill's explicit mounting description and detailed list of hardware, head office converted his description into an installation instruction, purchased materials, assembled 21 modification kits, and shipped them to the 21 other relay stations in the microwave link.

Within a week head office received reports from the 21 resident engineering technicians that it was impossible to mount the monitor unit as instructed, because of an adjoining control unit. No one at head office had remembered that Bill Carr was located at site 22, the last relay station in the microwave link, where there was no need for an additional control unit.

Bill had assumed that head office would be aware that the equipment layout at his station was unique. As he commented afterward: "No one said why I had to describe the modification, or told me what they planned to do with it."

In business and industry it is imperative that we communicate clearly and understand fully the implications of failing to do so. A poorly worded order that results in the wrong part being supplied to a job site, a weak report that fails to motivate the reader to take the urgent action needed to avert a costly equipment breakdown, and even an inadequate job application that fails to sell an employer on the right person for a prospective job, all increase the cost of doing business. Such mistakes and misunderstandings are wasteful of the country's labor and resources. Many of them can be prevented by more effective communication—communication that is receiver-oriented rather than transmitter-oriented, and that transmits messages using the most expeditious, economical, and efficient means at our command.

1

A TECHNICAL PERSON'S APPROACH TO WRITING

Almost every book on technical writing contains a statement that says in effect: "The key to effective writing is good organization." This rule is basically true, although many technical people who have tried to follow it too conscientiously find they have difficulty writing clearly. Frequently their trouble is caused by overorganization, or by trying to organize too early. Organizing that starts too early is self-defeating because it stifles a person's natural ability to write creatively.

Look at it this way: engineering technician Dan Skinner has a report to write on an investigation he completed seven weeks ago. He works for H. L. Winman and Associates, a consulting engineering firm we will meet in Chapter 2, and he has made several halfhearted attempts to get started. But each time has never seemed to be the right moment: maybe he was interrupted to resolve a circuit problem, or it was too near lunchtime, or a meeting was called, or, when nothing else interfered, he "just wasn't in the mood." And now he's up against the wire and he hasn't yet set pen to paper.

Unless Dan is one of those unusual persons who cannot produce except when under pressure, he is in danger of writing an inadequate, hastily prepared report that does not represent his true abilities. He knows he should not have left his report-writing project until the last moment, but he is human like the rest of us and constantly finds himself in situations like this. He does not realize that by leaving a writing task until it is too late to do a good job, and then frantically organizing the work, he is probably inhibiting his writing capabilities even more than necessary.

If Dan were to relax a little in the initial stages, instead of spending time trying to organize both himself and his writing task, he would find the physical process of writing reports a much more pleasant experience. But he must first change his whole approach to writing.

Every technical person, from student technician through potential scientist to practicing engineer, must recognize that he or she has the ability to write clearly and logically. (You may have to prove this to many responsible people in industry who continue to believe the old adage that "technical people just can't write.") This ability must be developed by continued practice: only by planning and writing all types of documents will you develop the confidence that is the prerequisite to good writing.

Planning the Writing Task

The word "planning" seems to contradict everything I have just said: it implies that report writers must start by organizing their material. However, I suggest they do so "creatively," to allow their latent writing ability to develop naturally. For example, I recommend that at first Dan Skinner do nothing about making an outline or taking any action that smacks of organization. All he has to do is work through several simple planning stages that will not be the chore he probably expects.

Normally, the first stage in planning any writing project is to gather information. This means assembling all the documents, results of tests, photographs, samples, specifications, and so on, that will be needed to write the report or that will be included with it. Dan should gather more information than he will probably need, because it is better to be selective and discard information than to look for additional facts and figures just when the writing is beginning to go well.

The next stage—and probably the most important—is for Dan Skinner to define his reader. He must conjure up an image of the actual person, or type of person, for whom he is writing. He must ask himself some pertinent questions: Who is this reader? What is his reading level? How much does he know about my project? Is he a technical person? Does he need to know all the details that I know? Where do his main interests lie? What will he do with my report? How will he use it? And who else is likely to read it? The answers to these questions will help Dan not only plan a good report, but also set the right tone when he is writing it.

When he has a reader in mind, Dan can start making notes. It is at this stage that he must "loosen up" enough to generate ideas. He needs to let his mind freewheel, so that it throws out ideas and pieces of information quickly and easily. He must not stop to question the relevance of this information—his role is purely to collect it. But first he must find a quiet place where he can work undisturbed; it's no use trying to be creative in a noisy, crowded office. (I will mention more about the need for establishing a good writing environment in the next section, when I discuss the practical aspects of settling down to write.)

Normally at this stage a technical person will take a blank sheet of paper and write down a set of arbitrary headings, such as "Introduction," "Initial Tests," and "Material Resources," and arrange them in logical order. (These seem to be standard-type headings in many reports.) But I want our report writer to be different. I want Dan Skinner to free his mind of the elementary organized headings, and even to refuse to divide his subject mentally into blocks of information. Then I want him to jot down a series of main topics that he will discuss, writing only brief headings rather than full sentences. He must do this in random order, making no attempt to force the topics into groups (although it is quite possible that grouping will occur naturally, since many interdependent topics are likely to occur to him in logical order). The topics that he knows best will spring most readily to mind, followed by a gradual slowdown as the flow of familiarity eases up.

When he stops his initial list, he should return to the top of it and examine each topic in turn to see if it will suggest less obvious topics. As additional topics come to mind, he must jot them down too, still in random order, and continue doing so until he finds he is straining to find new ideas. This should be a signal for him to stop before he becomes too objective.

Dan must not try to decide whether each topic is relevant during this spontaneous freewheeling session. If he does, he will immediately inhibit his creativeness because he will become too logical and organized. He must jot down all topics, regardless of their importance and eventual position in the final report.

At the end of this session Dan's list should look like Figure 1-1. He can now take a break, knowing that the first details of his report have been committed to paper. What he may not yet realize is that he has almost painlessly produced his first outline.

The fourth stage calls for Dan to examine his list of headings with a critical eye, dividing them into headings that bear directly on the subject and those that introduce topics of only marginal interest. His knowledge of the reader will help him decide whether each topic is really necessary so he can delete irrelevant topics, as has been done in Figure 1-2.

The headings that remain should be grouped into "topic areas" that will be discussed together. This he can do simply by coding related topics with the same symbol or letter. In Figure 1-2, letter (A) identifies one group of related topics, letter (B) another group, and so on.

Now, at last, Dan can take his first major organizational step, which is to arrange the groups of information in the most suitable order and at the same time sort out the order of the headings within each group. He must consider three factors: which order of presentation will be most interesting, which will be most logical, and which will be simplest to understand. The result will become his final writing plan or report outline (see Figure 1-3).

In summary, overorganizing a report, or organizing it too early in the writing process, has an inhibiting effect on writing. The key to good report writing is to organize material in a spontaneous, creative manner, allowing one's mind to freewheel through the initial planning stages until the topics have been collected, scrutinized for relevance, sorted, grouped, and written into a logical outline that will appeal to the reader. This method will not necessarily suit everyone. If you already have a workable method for planning and organizing, you should continue to use it. But if you do not, or if you have difficulty starting

Building OK -- needs strengthening
Elevators — too slow, too small
Talk with YoYo — elev. mfr (10% discount)
Waiting time too long — 70 sec
Shaft too small
How enlarge shaft?
Remove stairs?
Talk with fire inspector
Correspondence — other elev mfrs
Talk with Merrywell – Budget $500,000
Sent out questionnaire
Tenants' preferences—

Express elev
Executive elev
Prestige elev
Freight elev

No stop – 2nd floor
Faster service
No stop – ground flr

Freight elev – takes up too much space
Shaft only 35 x 8 ft
(when modified)
Big freight elev – omit basement
Tenants "OK" small freight elev ← YoYo "C" (8 ft)
YoYo – has office in Montrose
Basement level has loading dock
Service reputation – YoYo?
– others?

Figure 1-1. Initial list of topic headings, jotted down in random order.

Ⓐ Building OK -- needs strengthening

~~Elevators — too slow, too small~~

Ⓑ Talk with YoYo — elev. mfr (10% discount)

Ⓒ Waiting time too long — 70 sec

Ⓐ Shaft too small

Ⓐ How enlarge shaft?

Remove stairs? Ⓐ

~~Talk with fire inspector~~

Ⓑ Correspondence — other elev mfrs

Ⓒ Talk with Merrywell - Budget $500,000

Ⓒ Sent out questionnaire

Tenants' preferences—

Ⓒ {
Express elev
Executive elev ~~No stop 2nd floor~~
Prestige elev Faster service
Freight elev ~~No stop ground flr~~

Ⓓ Freight elev - takes up too much space

Ⓐ Shaft only 35 x 8 ft
(when modified)

~~Big freight elev — omit basement~~

Ⓒ Tenants "OK" small freight elev ← YoYo "C" (8 ft) Ⓓ

Ⓑ YoYo - has office in Montrose

~~Basement level has loading dock~~

Ⓑ Service reputation — YoYo?
— others?

Figure 1-2. The same list of topic headings, but with irrelevant topics deleted and remaining topics coded into subject groups (A—structural implications; B—elevator manufacturers; C—tenants' preferences; D—freight elevator).

Building Condition

- OK - needs strengthening (shaft area)
- Existing elev shaft too small
- Remove adjoining staircase
- Shaft size now 35 x 8 ft

Tenants' needs

- Sent out questionnaire
- Identified 5 major requests
- Requests we must meet:
  - Cut waiting time: <32 sec
  - Handle freight up to 7 ft 6 in.
- Requests we should try to meet:
  - Express elev to top 4 floors
  - Deluxe elev (for prestige)
  - Private elev (for executives)

Budget — Must be within $500,000

Elevator Manufacturers

- Researched 3
- Only YoYo company offers discount
- Only YoYo company has Montrose office

Figure 1-3. Topic headings rearranged into a writing outline. This was part of the writing plan for the formal report on elevator selection in Chapter 6; compare it with the final product on pages 198 to 200.

your writing tasks, try doing it this way. The stages are simple, as illustrated in Figure 1-4, and apply to any major writing project.

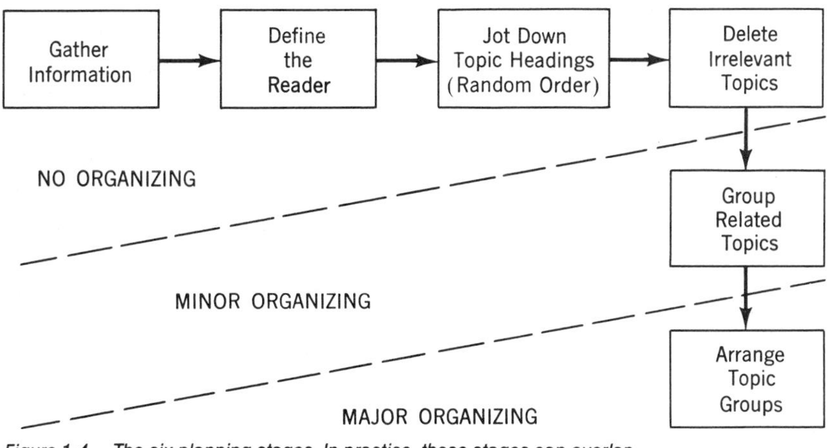

Figure 1-4. The six planning stages. In practice, these stages can overlap.

Writing the First Draft

As I sit at my desk, the heading "Writing the First Draft" at the top of a clean sheet of paper, I find that I am experiencing exactly the same problem that every writer encounters from time to time: an inability to find the right words—*any* words—that can be strung together to make coherent sentences and paragraphs. The ideas are there, circling around my skull, and the outline is there, so I cannot excuse myself by saying I have not prepared adequately. What, then, is wrong?

The answer is simple. In the distance I can hear the percolator in the kitchen murmuring quietly. Soon its tempo will increase to a rumbling crescendo, and then it will stop. Five minutes later my wife will appear at my study door bearing two cups of coffee and a plate of English biscuits on an old-fashioned carved tray, and I will be expected to pause for a ten-minute break. I cannot concentrate on writing when I know my continuity of thought is so soon to be broken.

Continuity is the key to getting one's writing done. In my case, this means writing at fairly long sittings during which I *know* I will not be disturbed. I must be out of reach of the telephone, visiting friends, children wanting to say good night, and even my wife, so that I can write continuously. Only when I have reached a logical break in the writing, or have temporarily exhausted an easy flow of words, can I afford to stop and enjoy that cup of coffee.

It is no easier to find a quiet place to write in the business world. The average technical person who tries to write a report in a large office cannot simply ignore the surroundings. A conversation taking place a few desks away will interfere with one's creative thought processes. And even a co-worker collecting money for the pool on that night's NHL game between the New York Rangers and the Toronto Maple Leafs will interrupt writing continuity.

The problem of finding a quiet place to write can be hard to solve. In business I recommend a short walk around the premises to find a hidden corner, or a small office which is temporarily unoccupied. Then perhaps, if you are lucky, you can just "disappear" for a while. For technical students, who frequently have to work on a tiny writing area in a crowded classroom, conditions are seldom ideal. Outlining in the classroom, followed by writing in the seclusion of a library cubicle, is a possible alternative.

When Dan Skinner has found a peaceful location out of reach of the telephone, his friends, and even his boss, he can take a fresh sheet of paper and start writing. This is exactly where a second difficulty may occur. Equipped with an outline on his left, a pad of paper in front of him, and a stack of sharpened pencils on his right, he may find that he does not know where to start. Or he may tackle the task enthusiastically, determined to write a really effective introduction, only to find that everything he writes sounds trite, unrealistic, or downright silly. Discouraged after many false starts, he sweeps the growing pile of discarded paper into the wastebasket.

I have frequently advised technical people who have encountered this "no start" block that the best place for them to start writing is at paragraph two, or even somewhere in the middle. For example, if Dan finds that a particular part of his project interests him more than other parts, he should write about that part first. His interest and familiarity with the subject will help him to get those first few words onto paper, and keep him going once he has started. The most important thing is to *start* writing, to put any first words down, even if they are not the right words or ideas, and to let them lead naturally into the next group of ideas.

This is where continuity becomes essential. A writer who has found a quiet place to work, and has allowed sufficient time to write a sizable chunk of text, simply must not interrupt the writing process to correct a minor point of construction. If you are to maintain continuity, you must not strive to write perfect grammar, find exactly the right word, insert perfect punctuation, or construct effective sentences and paragraphs of just the right length. That can be done later, during revision. The important thing is to keep building on that rough draft (no matter how "rough" it is), so that when you do stop for a break you know that you have some words on paper that can be worked up into a presentable document.

If Dan cannot find exactly the word he wants, he should jot down a similar word and draw a circle around it as a reminder to change it when the draft is finished. It's quite likely that when he does look back, the correct word will spring to mind. Similarly, if he is not sure how to spell certain words, he should resist the temptation to turn to the dictionary, for that would disrupt the natural flow of his writing. Again, he should draw a circle around the word as a reminder to consult his dictionary later.

I cannot stress too strongly that writers should not correct their work as they write. Writing and revising are two entirely separate functions, and they call for different approaches; they cannot be done simultaneously without the one lessening the effectiveness of the other. Writing calls for creativeness and total immersion in the subject so that the words tumble out in a constant flow. Revision calls for lucidity and logic, which will force a writer to reason and query the suitability of the words that have been written. The first requires exclusion of every thought but the subject; the second demands an objectivity that constantly challenges the material from the reader's point of view. Writers who try to correct their work as they write soon become frustrated, for creativity and objectivity are constantly fighting for control of their pens.

Most technical people find that, once they have written those first few difficult sentences, their inhibitions begin to fall away and they write more naturally. Their style begins to change from the dull, stereotyped writing we expect from the average technical person, to a relaxed narrative that sounds as though the writer is telling a co-worker about the subject. As the writer becomes interested in the topic, speed builds up and he or she has difficulty writing fast enough. The words may not be exactly those of the final report, but when the writer starts checking the first draft he or she will be surprised at the effectiveness of many of the sentences and paragraphs written earlier.

The length of each writing session will vary, depending on the writer's experience and the complexity of the topic. If a document is reasonably short, it should be written all at one sitting. If it is long, it should be divided into several medium-length sessions that suit the writer's particular staying power.

At the end of each session Dan should glance back over his work, note the words he has circled, and make a few necessary changes (Figure 1-5 is a page from a typical first draft). He must not yet attempt to rewrite paragraphs and sentences for better emphasis. Such major changes must be left until later, when enough time has elapsed for him to read his work objectively. Only then can he review his work as a complete document and see the relationship among its parts. Only then can he be completely critical.

Taking a Break

When the final paragraph of a long report has been written, I have to resist the temptation to start revising it immediately. There are sections that I know are weak, or passages with which I am not happy, and the desire to correct them is strong. But I know it is too soon. I must take my pages, staple them together, and set them aside while I tackle a task that is completely unrelated.

Immediate reading without a suitable waiting period encourages writers to look at their work through rose-tinted spectacles. Sentences which normally they would recognize as weak or too wordy appear to contain words of wisdom. Gross inaccuracies which under normal circumstances they would pounce upon go unnoticed. Paragraphs that may not be understood by a reader new to the subject, to their writers seem abundantly clear. Their familiarity with their work blinds them to its weaknesses.

The only remedy is to wait. If he can, Dan should arrange to have a double-

Tenants' Needs

To find out what the building's ~~tenants~~ most needed in elevator service, we asked each company to fill out a questionaire. From their answers we ~~identified~~ were able to identify 5 factors needing consideration:

1. A major problem seems to be the length of time a person must wait for an elevator. Every ~~tenant~~ said we must cut out lengthy waits. A survey was carried out to find out how long people had to wait (during rush hours). This averaged out at 70 sec, more than twice the 32 sec established by Johnson before people get impatient. From this we calculated we would need 3 or 4 passenger elevators. ref?

2. At first it seemed we would be forced to include a full-size freight elevator in our plan. Two companies (which?) both carry large but light displays up to their floors, but both later agreed they could hinge them, and to do this would mean they would need only 7 ft 6in. width (minimum). They also said they did not need a freight elevator all the time,

Figure 1-5. Part of the author's first draft. Note that he has not stopped to hunt up minor details. Several revisions were made between this first draft and the final product (see pages 198 and 199).

or triple-spaced draft typed during this time. Not only will this make his work easier to read, but he may also be fortunate enough to be assigned an efficient typist who corrects some of his punctuation and spelling errors.

Reading with a Plan

Always try to read from a typed draft, so you will see the same cold, clinical type your readers will see. There is a danger in reading your own handwriting, for you may "hear" an emphasis which your readers will not notice when they read the typewritten words.

Dan Skinner's first reading should take him straight through the draft without stopping to make corrections, so that he can gain an overall impression of the report. Subsequent readings should be slower and more critical, with changes written in as he goes along. As he reads he should check for clearness, correct tone and style, and technical and grammatical accuracy.

CHECKING FOR CLEARNESS

Checking for clearness means searching for passages that are vague or ambiguous. If the following paragraph remained uncorrected, it would confuse and annoy a reader:

Muddled Paragraph When the owners were contacted on April 15, the assistant manager, Mr. Pierson, informed the engineer that they were thinking of advertising Lot 36 for sale. He however reiterated his inability to make a definite decision by requesting this company to confirm their intentions with regard to buying the land within two months, when his boss, Mr. Davidson, general manager of the company, will have come back from a business tour in Europe. This will be June 8.

The only facts that a reader could be sure about after reading this paragraph are that the owners of the land were contacted on April 15 and that the general manager will be returning on June 8. The important information about the possible sale of Lot 36 is confusing. Probably the writer was trying to say something like this:

Revised Paragraph The engineer spoke to the owners on April 15 to inquire if Lot 36 was for sale. He was informed by Mr. Pierson, the assistant manager, that the company was thinking of selling the lot, but that no decision would be made until after June 8, when the general manager returns from a business tour in Europe. Mr. Pierson suggested that the engineer submit a formal request to purchase the land by that date.

The more complex the topic, the more important it is to write clear paragraphs. Although the paragraph below is quite technical, it would be generally understood even by nontechnical readers:

Clear Paragraph A sound survey confirmed that the high noise level was caused mainly by the radar equipment blower motors, with a lesser contribution from the air-conditioning equipment. Tests showed that with the radar equipment shut down the ambient noise level at the micro-

> phone positions dropped by 10 dB, whereas with the air-conditioning equipment shut down the noise level dropped by 2.5 dB. General clatter and impact noise caused by the movement of furniture and personnel also contributed to the noisy working conditions, but could not be measured other than as sudden sporadic peaks of 2 to 5 dB.

Here paragraph unity has added much to readability. The writer has made sure that:

> The topic is clearly stated in the first sentence (the topic sentence).
> The topic is developed adequately by the remaining sentences.
> No sentence contains information that does not substantiate the topic.

If any paragraph meets these basic requirements, its writer can feel reasonably sure that the message has been conveyed clearly.

Writers who know their subject thoroughly may find it difficult to identify paragraphs that contain ambiguities. A passage that is abundantly clear to them may be meaningless or offer alternative interpretations to a reader unfamiliar with the subject. For example:

> Our examination indicates that the receiver requires both repair and recalibration, whereas the transmitter needs recalibration only, and the modulator requires the same.

This sentence plants a question in the reader's mind: Does the modulator require both repair and recalibration, or only recalibration? The technician who wrote it knows, because he has been working on the equipment, but the reader will never know unless he or she cares to write or phone and ask. The technician could have clarified the message easily, simply by rearranging the information:

> Our examination indicates that the receiver requires both repair and recalibration, whereas the transmitter and modulator need recalibration only.

Sometimes ambiguities are so well buried that they are surprisingly difficult to identify, as in this excerpt from a chief draftsperson's report to a department head:

> The Drafting Section will need three Model D7 drawing boards. The current price is $975 and the supplier has indicated that his quotation is "firm" for three months. We should therefore budget accordingly.

The department head took the message at face value and inserted $975 for drawing boards in the budget. Two months later he received an invoice for $2925. Unable by then to return the three boards, he had to overshoot his budget by $1950. This financial mismanagement resulted from the chief draftsperson's omitting to state whether the price quoted applied to one drawing board or to three. If the word "each" had been inserted after $975, the message would have been clear.

Many ambiguities can be sorted out by simple deduction, although it really should not be the reader's job to interpret the author's intentions. Occasionally such ambiguities provide a humorous note, as in this extract from a field trip report:

> High grass and brush around the storage tanks impeded the technicians' progress and should be cleared before they grow too dense.

So that readers will not mistakenly think it is the technicians who are growing too dense, the two thoughts in this sentence should be separated:

> High grass and brush around the storage tanks impeded the technicians' progress. This undergrowth should be cleared before it grows too dense.

CHECKING FOR CORRECT TONE AND STYLE

How does a writer know when his or her work has the right tone? One of the most difficult aspects of technical writing is establishing a tone that is correct for the reader, suitable for the subject, and comfortable for the writer. If the writer tries to set a tone that does not feel natural, the reader will sense an artificiality, an unsureness that indicates the writer has not felt comfortable writing at that level.

A writer who knows a subject well and has identified the reader thoroughly will most likely write so confidently that the right tone will automatically be established. But a writer who does not know the subject well and has not taken the time to define the reader will produce writing that lacks confidence. And no matter how skillfully the work is edited, the hesitancy will show up in the final sentences and paragraphs.

Finding the Best Writing Level. To check that he has set the right tone, Dan Skinner must assess whether his writing is suitable for both the subject matter and the reader. If he is writing on a specific aspect of a very technical topic, and knows that his reader is an engineer with a thorough grounding in the subject, he can use all the technical terms and abbreviations that his reader will recognize. Conversely, if he is writing on the same topic for a nontechnical reader who has little or no knowledge of the subject, Dan must alter his approach. He may have to generalize rather than state specific details, explain technical terms that he would normally expect to be understood, and generally write in a more informative manner.

Writing on a technical subject for readers who do not have the same technical knowledge as yourself is not easy. You have to be much more objective when checking your work, and try to think in the same way as the nontechnical reader. You can use only those technical words which you know will be recognized, yet you must avoid oversimplified language that may irritate the reader.

When Dan Skinner wrote this excerpt from a modification report, he knew his readers would be electronics engineers at radar-equipped airfields:

> We modified the M.T.I. by installing a K-59 double-decade circuit. This brightened moving targets by 12% and reduced ground clutter by 23%.

Although this statement would be readily understood by the readers for whom it was intended, to any reader not familiar with radar terminology it would mean very little. So when Dan reported on the same subject to the airport manager, he wrote this:

> We modified the radar set's Moving Target Indicator by installing a special circuit known as the K-59. This increased the brightness of responses from aircraft and decreased returns from fixed objects on the ground.

For this reader Dan has included more descriptive details: "M.T.I." has become "Moving Target Indicator," and "moving targets" and "ground clutter" have become "responses from aircraft" and "returns from fixed objects on the ground." He has eliminated specific technical details because they might not be meaningful to the reader, and in their place has made a general statement that aircraft responses were "increased" and ground returns "decreased." He also knew that the airport manager would be familiar with terms such as *Moving Target Indicator, responses,* and *returns.*

Now suppose that Dan also had to write to the local Chamber of Commerce to describe improvements in the airport's traffic control system. This time his readers would be entirely nontechnical, so he would have to avoid using *any* technical terms:

> We have modified the airfield radar system to improve its performance, which has helped us to differentiate more clearly between low-flying aircraft and high objects on the ground.

Keeping to the Subject. Having established that he is writing at the correct level, Dan must now check that he has kept to the subject. He must take each paragraph and ask: Is this truly relevant? Is it direct and to the point?

If Dan prepared his outline using the method described earlier, and followed it closely as he wrote his report, he can be reasonably sure that most of his writing is relevant. To check that his subject development follows his planned theme, he should identify the topic sentences of some paragraphs and check them against the headings in his outline. If they follow the outline, he has kept to the main theme; if they tend to diverge from the outline or if he has difficulty in identifying them, he should read the paragraphs carefully to see whether they need to be rewritten, or possibly even eliminated.

Technical writing should always be as direct and specific as possible. The writer should convey just the right amount of information for the reader to understand the subject thoroughly—and no more. Technical writing, unlike literary writing, has no room for details that are not essential to the main theme. This is readily apparent in the following descriptions of the same equipment.

Literary Description

> The new cabinet has a rough-textured dove gray finish that reflects the sun's rays in varying hues. Contrary to most instruments of this type, its controls are grouped artistically in one corner, where the deep black of the knobs provides an interesting contrast with the soft gray and white background. A cover plate, hardly noticeable to the layman's inexperienced eye, conceals a cluster of unsightly adjustment screws that would otherwise mar the overall appearance of the cabinet and would nullify the esthetic appeal of its surprisingly effective design.

Technical Description The gray cabinet is extremely functional. All the controls used by the operator are grouped at the top right-hand corner, where they can be grasped easily with one hand. Subsidiary controls and adjustment screws used by the maintenance crews are grouped at the bottom left-hand corner, where they are hidden by a hinged cover plate.

Comparison of these examples shows how a technical description concentrates on details that are important to the reader (it tells *where* the controls are and why they have been so placed); it does not waste time being artistic. At the same time it observes the rules of good construction, using parallel structure to carry the reader easily through the description. Hence, it maintains an effective, businesslike tone.

Using Simple Words. A writer who uses unnecessary superlatives sets an unnaturally pompous tone. The engineer who writes that a design "contains ultrasophisticated circuitry" seems to be trying to justify the importance and complexity of his or her work rather than saying that the design has a very complex circuit. The supervisor who recommends that technician Janice Smith be "given an increase in remuneration" may be understood by the company controller but will only be considered pompous by Janice. If he had simply written that Janice should be "given a raise," he would have been understood by everyone. Unnecessary use of big words, when smaller, more generally recognized, and equally effective synonyms are available clouds technical writing and destroys the smooth flow that such writing demands.

Removing "Fat." During the reading stage Dan should be critical of sentences and paragraphs that seem to contain too many words. He should check that he has not inserted words of low information content, that is, phrases and expressions that add little or no information. Their removal, or replacement by simpler, more descriptive words, can tighten up a sentence and add to its clarity. Low information content words and phrases are often hard to identify, because the sentences in which they appear seem to be satisfactory. Consider this sentence:

> For your information, we have tested your spectrum analyzer and are of the opinion that it needs calibration.

The words of low information content are "for your information" and "are of the opinion that." The first can be deleted, and the second replaced by "consider," so that the sentence now reads:

> We have tested your spectrum analyzer and consider it needs calibration.

Now try to identify the low information content words in this sentence:

> If you require further information, please feel free to telephone Mr. Thompson at 489–9039.

The phrase "if you require" is not entirely wrong, although it could be replaced by the single word "for," but "please feel free to" is archaic and should be eliminated. The result:

For further information please telephone Mr. Thompson at 489-9039.

Inadvertent repetition of information can also contribute to excessive length. Although repetition can be an effective way to emphasize a point, in most cases its use is accidental; Dan may explain something in one paragraph, then several paragraphs later say the same thing in different words. By welding paragraphs that repeat previously mentioned facts into a single, cohesive block of information, he will help to clarify and shorten his report.

Many writers find it difficult to pay compliments or to apologize sincerely. Frequently they set an unnatural tone by trying to say too much. A simple, brief statement seems incomplete, so they "beef it up a bit" under the false impression that they are showing their true feelings. The resulting tone is so false, and the words are so forced that the reader immediately senses that the compliment or apology is insincere. Such writers have to learn that sincerity is enhanced by brevity: pay your compliment or make your apology, then forget about it.

For more information on this topic see the section on "Tone" in Chapter 3.

CHECKING FOR ACCURACY

Checking accuracy means examining one's work to ensure that the information is correct and that such technicalities as grammar, punctuation, and spelling have not been overlooked.

Nothing annoys readers more than to discover that they have been presented with information that is not absolutely accurate. They automatically assume that a writer knows the facts and has checked that they are correctly transcribed into the report. Errors may remain undetected for a long time, possibly until an inquiring reader starts conducting further tests and making calculations based on the author's results. Suspicion of an error in the original report can lead to tedious correspondence and wasted time until the inaccuracy is corrected. Worse, the readers' confidence in the writer, as well as the company, is downgraded.

There is no way to prevent some errors from occurring when quantities and details are being copied from one sheet of paper to another. These errors may be made either by the writer or by the typist. Therefore, Dan must personally check that facts, figures, equations, quantities, and extracts from other documents are all copied correctly. He may feel that it is the typist's duty to proofread the report, but the responsibility for checking the accuracy of the written work is ultimately his.

This is also the time for Dan to identify poor grammar, inadequate punctuation, and incorrect spelling. He must do this with care, because his knowledge of the subject may blind him to obvious errors. (How many of us have inadvertently written "their" when we intended to write "there," and "to" when we meant "too"?)

These examples illustrate a few of the ways that Dan can improve his report writing by careful reading and revision. Other methods are suggested in Chapter 10.

Revising One's Own Words

As Dan reads his work, he should make any corrections he finds necessary. Some corrections may require only minor changes to individual sentences; others may require complete revision of whole paragraphs, and even of complete sections. Whenever extensive revisions are made, he should have the draft retyped, then reread and revise it. This process should be repeated until a draft emerges which says exactly what he wants to say in as few words as possible.

As he reads, Dan must continually ask questions:

Can my readers understand me?

Will the person I am writing for be able to read my report all the way through without becoming lost?

What about other readers who might also see my report: will they understand it?

Is the focus right?

Is my report reader-oriented?

Are the important points clearly visible?

Have I summarized the main points in an opening statement which the reader will see right away?

Is my information correct?

Is it accurate?

Is it complete?

Is all of it necessary?

Is my language good?

Is it clear, definite, and unambiguous?

Are there any grammar, punctuation, or spelling errors?

Does every paragraph have a topic sentence (preferably at the start of the paragraph)?

Have I used any big "overblown" words where simpler words would do a better job?

Are there any low information content words and phrases?

Have I kept my report as short as possible while still meeting my readers' needs and covering the topic adequately?

By now Dan's draft should be in good shape. Any further reading and revising will be final polishing, and the amount will depend on the importance of the report. If his report is for limited or in-company distribution, a workmanlike job probably will suffice. But if it is to be distributed outside the company, or submitted to an important client, then Dan will spend as much time as necessary to ensure that it conveys a good image of both himself and H. L. Winman and Associates (his employer).

Reviewing the Final Draft

The final step occurs when Dan feels that his report is ready for final typing. Before charging across to the typing pool, he should ask himself three more questions:

1. Would I want to receive what I have written?
2. What reaction will it incur from the intended reader?
3. Is this the reaction I want?

Now he must be extremely self-critical, ready to doubt his ability to be truly objective. If his answers are at all hesitant, he should ask an independent reviewer to read his report. Ideally, this person will be technically and mentally equivalent to the eventual reader, but not too familiar with the project. The reviewer must be able to criticize constructively. He should read the report completely rather than scan it, and make notes describing possible weaknesses and ambiguities. He should then take time to discuss the report with Dan, who quite likely will discover that the reviewer has identified the very paragraphs that gave him most difficulty or did not entirely satisfy him.

Dan Skinner will now be able to issue his report with confidence, knowing that he has fashioned a good product. The approach described here will not have made report writing a simple task for him, but it will have assisted him through the difficult conceptual stages, and helped him to read and revise more efficiently. When Dan has to write his next report, he will be much less likely to put it off until it is so late that he has to do a rush job.

2

MEET "H.L. WINMAN AND ASSOCIATES"

Almost every project you read or write about in this book is based upon events affecting two companies: H. L. Winman and Associates, a Cleveland firm of consulting engineers, and its Canadian affiliate, Robertson Engineering Company. To help you identify yourself in this working environment, the following pages contain short histories and partial organization charts of the companies, plus brief biographical sketches of the men who have brought them to their current position, and the woman who most influences how their engineers, scientists, and technicians write.

Corporate Structure

H. L. Winman and Associates was founded by Harvey Winman early in 1935. Starting with a handful of engineers working in a small office in one of the older sections of Cleveland, Harvey gradually built up his company until now it has 760 employees: 380 at head office, 270 at the Canadian affiliate in Toronto, and 110 in smaller offices throughout the United States, Canada, and Mexico. Some are branches of H. L. Winman and Associates (there may be one in your area), while others are local engineering firms that Harvey has purchased. Figure 2-1 illustrates the company's two main locations.

Head Office: H. L. Winman and Associates

The head office of H. L. Winman and Associates is located in a modern building at 475 Reston Avenue, a major thoroughfare in the business district of Cleveland, Ohio. A partial organization chart in Figure 2-2 shows Harvey Winman as president, with nine department heads, who are responsible for the day-to-day operation of the company, reporting directly to him.

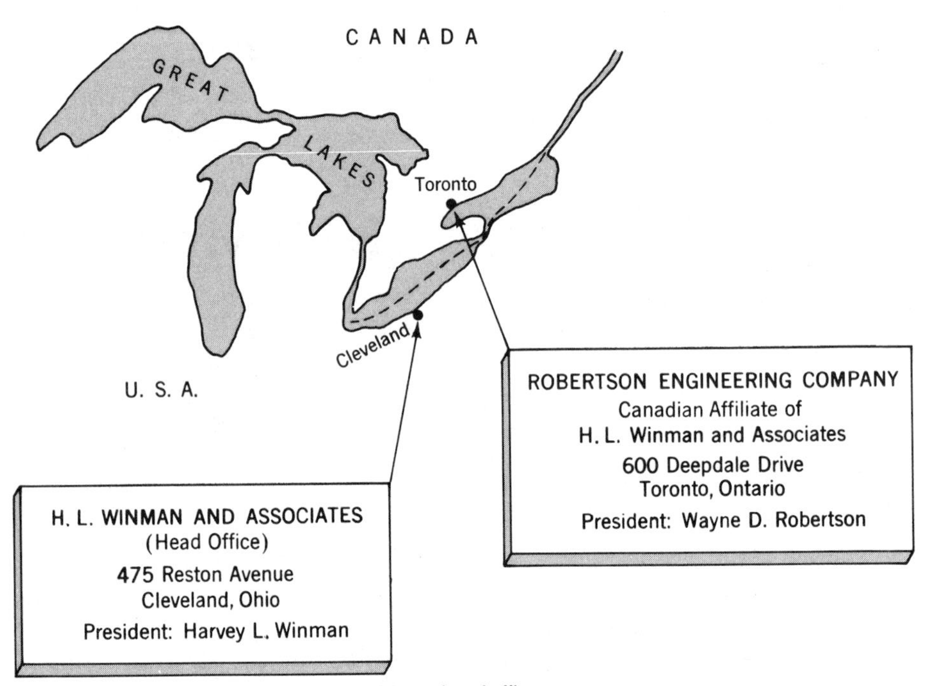

Figure 2-1. Locations of H. L. Winman and Associates' offices.

Over the years H. L. Winman and Associates has developed an excellent reputation among its clients for the management of large construction projects in remote areas of the United States and Canada. It has managed the construction of whole townsites, airfields, and dams for large hydroelectric power generating stations; engineered, designed, and supervised construction of major traffic intersections, shopping complexes, bridges, and industrial parks; and designed and supervised numerous large buildings such as hotels, schools, arenas, and manufacturing plants. More recently, the company has entered the systems engineering field. Some of its current studies involve problems in air pollution, protection of the environment, and the effects of oil exploration and pipeline construction on the Alaskan arctic tundra. To handle so many diverse tasks, Harvey has built up an extremely adaptable and flexible staff with representation from many scientific and technical disciplines.

Although he is past retirement age, Harvey maintains an active interest in his company's business. Short, with a round face topped by a bald head with a fringe of hair, he is often seen strolling from department to department. He stops frequently to talk to his staff (most of whom he knows by their first name) and to inquire about both the projects they are working on and their home lives. Well-liked and thoroughly respected by all, from senior management down to the newest clerk, he is a manager of the old school, a type seldom seen in today's brusque business atmosphere.

His chief executive is Vice-President Martin Dawes, to whom Harvey has handed operating control in anticipation of his coming retirement. Martin is a

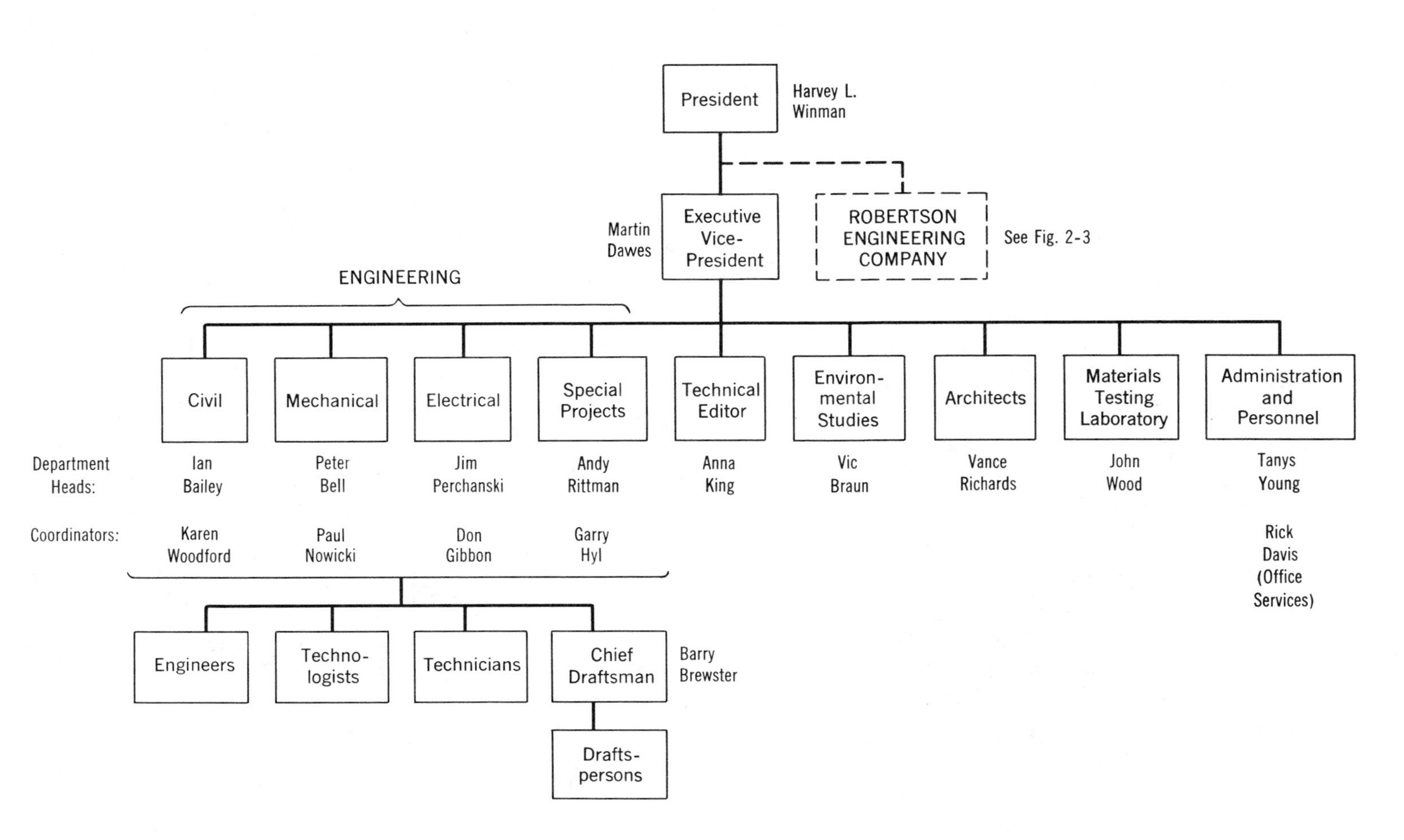

Figure 2-2. Partial organization chart—H. L. Winman and Associates.

Photo A. Harvey L. Winman. Photo: Anthony Simmonds.

civil engineer who developed a fine reputation for designing and constructing hydroelectric dams in remote, mountainous country. An ardent fisherman, he has often combined field projects with trips into the surrounding countryside, from which he has returned laden with salmon or trout. Before becoming vice-president, he was Head of the Civil Engineering Department for 15 years.

Both Harvey Winman and Martin Dawes write well, and they expect their staff to do likewise. They recognize that the written word is the means that most often conveys an image of their company to customers. So, to ensure that all letters, reports, and company publications are of top quality, they have introduced a technical editor into their management team.

The editor is Anna King. Born in Montreal of an English father and French mother, she grew up fully bilingual. Although her strength was languages, she wanted to be an engineer like her father and graduated from McGill University with a Bachelor of Science in Electrical Engineering. She worked for the South Georgia Power and Light Company for nine years, first in the relay department and later in development engineering, where her capabilities both as engineer and report writer soon became evident.

Eventually Anna wanted to return to ski country, so she accepted an offer from Harvey Winman to become H. L. Winman and Associates' technical editor. "Yours will be an exacting job," Harvey said on her first day. "You will be setting writing standards for the whole company. I don't want my engineers to think you are here simply to do their writing for them. Your role is to encourage them to become better writers."

Photo B. Technical Editor Anna King. Photo: George Tan.

Canadian Affiliate: Robertson Engineering Company

The Canadian affiliate of H. L. Winman and Associates was acquired by Harvey Winman in 1962, when he purchased an old, established firm of consulting engineers that held a controlling interest in the Robertson Engineering Company of Toronto, Ontario. At first Harvey considered selling the Canadian company, but when he realized that Wayne Robertson, its president, was running a very profitable business, he decided to retain it as an affiliate and to leave Wayne in full control.

Wayne formed Robertson Engineering Company when he left the Canadian Air Force in 1946, gathering together several ex-air force electronics specialists to set up a small equipment repair and maintenance service. He managed his business so well that his company was soon recognized for its competence and reliability. Consequently it grew steadily, with the emphasis gradually leaning toward research and development.

The company now has a well-established reputation for designing, developing, and custom-manufacturing sophisticated electronic, nucleonic, and electromechanical instruments. It also specializes in the installation and maintenance of complex communication systems across Western Canada. There are approximately 270 employees, of whom 115 are engineers, technologists, and technicians in the Engineering Department. The company occupies floors five and

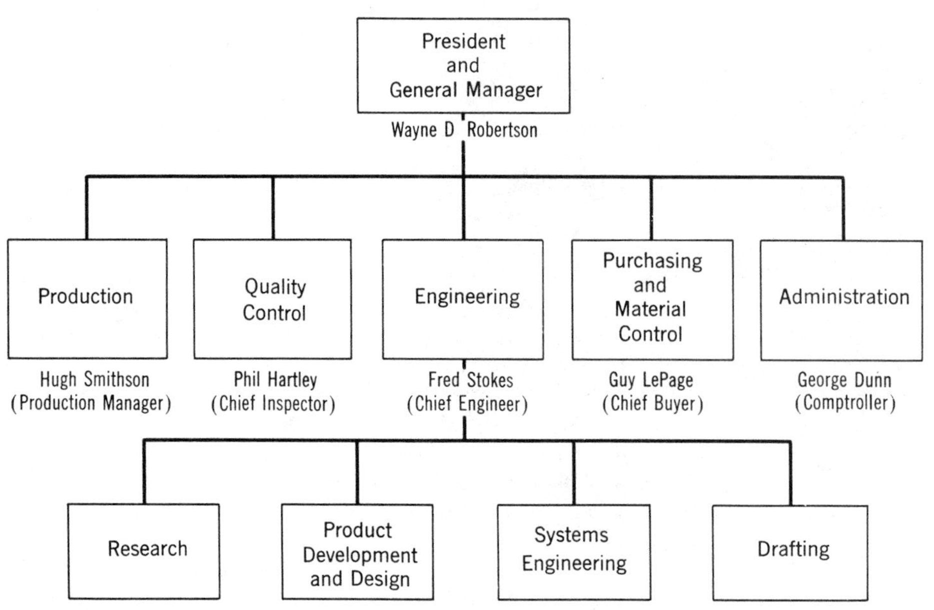

Figure 2-3. Partial organization chart showing Engineering Department of Robertson Engineering Company.

six of the Wilshire Building, a ten-story office and light manufacturing complex owned by the Wilshire Insurance Company. A partial organization chart in Figure 2-3 shows the structure of the company and the makeup of the Engineering Department.

Wayne Robertson is not only an experienced electronics specialist but also an astute businessman. Recognizing the need for a strong engineering capability supported by an effective sales force, he has taken care to build a particularly good management and engineering staff to carry out his company's primary function. The result has been recognition by Canadian businessmen of his company's technical capability and his personal managerial ability. His design engineers have developed a specialized range of electronic and nucleonic instruments for which a small but world-wide market exists. Consequently, Wayne travels frequently to South America, Australia, South Africa, and Europe, where he is able to talk knowledgeably of the technical capabilities of his company's equipment.

Like Harvey Winman, he recognizes the importance of communication in his business, and particularly stresses that the operating instructions and training manuals that accompany his instruments be abundantly clear. To this end,

Photo C. Wayne D. Robertson. Photo: Anthony Simmonds.

he employs a small technical writing group that can write manuals not only in English, but also in French, German, and Spanish.

The letters and reports that appear throughout this book are based on work done by these two companies. Similarly, many of the end-of-chapter assignments assume that you are employed by one of the companies and that the letter or report you are to write occurs naturally as part of your work.

3

TECHNICAL CORRESPONDENCE

Because the letter is our most common means of written communication, we frequently overlook its importance. We scribble down a few words and hope that someone will get the message; we bury the message under a mountain of unnecessary words; or we adopt an unnatural style that conceals our character and robs the message of humanity. Many of us are so accustomed to writing letters to friends and relatives that we find it difficult to transfer from the easygoing style of the personal letter to the more formal style of the business letter. This is where we err. Instead of letting our business letters be friendly, interesting, and persuasive, we force them to be stilted and dull, thinking that we are being businesslike.

Clarity and Conciseness

When we write personal letters to friends and relatives, we do so partly to be sociable and partly to convey a message. The length of our letters, and the amount of additional information we include, is unimportant. We can ramble along with very little organization, inserting interesting sidelights and comments as they occur to us. We assume our reader is pleased to have a letter from us, and so we use it mainly to portray what has happened since we last wrote.

But when we write a business letter we have to be much more direct. Our readers are busy men and women who are interested only in facts; information they do not need irks them. For these people we must keep strictly to the point. We must check that our letters are in focus, compact, brief, and clear, and that our facts are simply but effectively presented.

FOCUS

Letter writing is like photography: you have to center the image in the view finder and then focus the image sharply before you click the shutter. But first you have to decide exactly what you want to depict; otherwise viewers will wonder why they are looking at the picture.

Identify the Message. Before you pick up a pen to write a letter, identify clearly in your mind why you are writing and what your reader most needs to hear from you. Then focus sharply on this information by placing it right up front, where it will be seen immediately. Your opening words should capture the reader's attention.

To open a letter with background information rather than the main point makes a reader wonder why you are writing. This happened to Jim Connaught as he started to read this letter from Don McKelvey:

> Dear Mr. Connaught:
>
> I refer to our purchase order No. 21438 dated April 26, 19____, for a Whiteprint copier model XA21, which was installed on May 14. During tests following its installation your technician discovered that some components had been damaged in transit. He ordered replacements and in a letter dated May 20 informed me that they would be shipped to us on May 27 and that he would return here to install them shortly thereafter.
>
> It is now June 10, and I have neither received the parts nor heard from your technician. I would like to know when the replacement parts will be installed and when we can expect to use the copier.
>
> Sincerely,
> Don McKelvey

Jim had to read more than 70 words before he discovered what Don wanted him to do. He would have known immediately if Don had started with his main point:

> Dear Mr. Connaught:
>
> We are still unable to use the XA21 Whiteprint copier we purchased from you on April 26. Please inform me when I can expect it to be in service.

And Don would have found his explanation was a little shorter and easier to write:

> The copier was ordered on P.O. 21438 and installed on May 14. During tests, your technician discovered that some components had been damaged in transit. He ordered replacements and in a letter dated May 20 informed me that they would be shipped to us on May 27, and that he would return here to install them. To date, I have neither received the parts nor heard from your technician.
>
> Sincerely,

Focus the Message. Don knows he should always start his letters and reports with the main point, but often he doesn't. He says it makes him feel he is being too abrupt or "pushy."

If you feel the same way, you can borrow a technique used by Anna King, H. L. Winman and Associates' technical editor. Whenever Anna starts a letter, she first writes these six words:

> *I want to tell you that . . .*

and then she finishes the sentence with the main point, like this:

> I want to tell you that . . . the environmental data you submitted to us on October 8 will have to be substantiated if it is to be included in the TACMORE study.

Anna then deletes the first six words (the *I want to tell you that* phrase), to create an in-focus opening statement:

> The environmental data you submitted to us on October 8 will have to be substantiated if it is to be included in the TACMORE study.

Here are three more examples of opening statements, all of which originally contained the opening words *I want to tell you that*:

> Seven defective castings were received in shipment No. 308.
>
> We will complete the XRS modification on June 14, eight days earlier than originally scheduled.
>
> Your excellent paper "Export Engineering" arrived too late to be included in the Midwest Conference program.

If an opening statement seems too harsh or abrupt when the *I want to tell you that* phrase is removed, Anna inserts one or two additional words to soften its impact. For example, she would probably write "I regret that" in front of the third opening statement:

> I regret that your excellent paper . . . (etc.)

Avoid False Starts. Anna cautions H. L. Winman engineers against starting their letters with awkward expressions which can lead them into writing long, rambling sentences that say very little. For example:

> In answer to your inquiry of December 7, concerning erroneous read-outs you are experiencing with your Mark 1 Analyzer, and our subsequent telephone conversation of December 18, during which we tried to pinpoint the fault. . . .

When a sentence grows as long and complicated as this, there is always a danger that its writer may stop at this point and thus write an incomplete sentence. A much better opening sentence leads straight into the problem and the probable result, and then stops:

> (*I want to tell you that . . .*)
>
> I believe the problem with your Mark 1 Analyzer exists in the extrapolator circuit. Following your inquiry of December 7, in which you described the erroneous read-outs you are experiencing, we examined . . . (etc.)

Anna King equates such expressions with "spinning one's wheels" (see Figure 3-1). The problem is that we become accustomed to seeing them at the start of other people's letters, and so we use them ourselves rather than coming to grips right away with the real subject. If you can start a letter with an informative, meaningful opening statement, you are also using the pyramid technique recommended in Chapter 10 (page 300).

COMPACTNESS

Compactness means arrangement of information in the smallest possible space. When packing a small suitcase for a short trip, you have to be selective

WHEN YOU WRITE A LETTER . . .

Never start with a word that ends in "ing":

Referring . . .

Replying . . .

Never start with a phrase that ends with the preposition "to":

With reference to . . .

In answer to . . .

Pursuant to . . .

Never start with a redundant expression:

I am writing . . .

For your information . . .

This is to inform you . . .

The purpose of this letter is . . .

We have received your letter . . .

Enclosed please find . . .

Attached herewith . . .

IN OTHER WORDS . . .

DON'T SPIN YOUR WHEELS!

Figure 3-1. Anna King's warning to H. L. Winman Engineers.

and pick out only the items you will need for the environment you will encounter, choosing only one coat, one suit, one pair of shoes, and so on. In business writing you have to be equally selective, limiting your choice to the essential elements that will convey the message. The key word for compactness, both in packing a small case and in writing a business letter, is "one."

One Main Subject to a Letter. If you have to write to a company on two different subjects at the same time, write a separate letter for each subject. To cover both subjects in the same letter is to invite one subject to be overlooked, particularly if they demand action by different departments. The first department to receive your letter may inadvertently file it after replying to you, and the second department will never know of your inquiry.

Only when two subjects are related and are likely to be acted upon by the same person is it safe to write them in the same letter. In every other case, write separate letters.

One Idea to Each Paragraph. Let the first sentence of each paragraph introduce just one idea, then make sure that subsequent sentences in that paragraph develop it adequately and do not introduce any other ideas. In technical writing it is often wise to identify the first sentence of a paragraph as the "topic sentence," so that the reader knows immediately what the writer is trying to convey. This is particularly true of business letter writing, in which the first sentence of each paragraph should summarize the paragraph's contents. Remaining sentences must support the first sentence by providing additional information about the topic, as they do here:

> *We have tested your 12 Vancourt 60 Multimeters and find that 9 require repair and recalibration.* Only minor repairs will be necessary for 5 of these meters, which will be returned to you within one week. Of the 4 remaining meters, 3 require major repairs which will take approximately 20 days, and 1 is so badly damaged that repairs will cost over $60.00. Since this is more than the maximum repair cost imposed by you, no work will be done on this meter.

By reading the topic sentence (italicized in this example), the reader learns immediately what he or she most wants to know: the number of meters that need repair. The extent of the repairs can be determined from the supporting sentences.

One Thought to a Sentence. The intent here is to keep each sentence uncomplicated. Sometimes expert literary writers can successfully build sentences that develop more than one thought, but such sentences are confusing and out of place in the business world. Compare these two examples of the same information:

> *Complicated* There has been intermittent trouble with the vacuum pumps, although the flow valves and meters seem to be recording normal output, and the 18 cm pipe to the storage tanks has twice become clogged, causing backup in the system.
>
> *Clear* There has been intermittent trouble with the vacuum pumps, and twice the 18 cm pipe to the storage tank has become clogged and caused backup in the system. The flow valves and meters, however, seem to be recording normal output.

The first example is confusing because it jumps back and forth between trouble and satisfactory operation. The second example is clear because it uses two sentences to express the two different thoughts.

BREVITY

The key word for brevity is "short": short letters, short paragraphs, short sentences, and short words.

Short Letters. A short letter introduces its topic quickly, discusses it in sufficient depth, and then closes with a concluding statement. It is impossible to state arbitrarily how long a short letter should be. Length should be dictated by only one factor: the number of words a writer needs to convey a message adequately.

I know of a company in which the managing director has ruled that no letters issued by the organization should exceed one page. This is an effective means for forcing the staff to be brief, but it must be unbearably severe on some writers. There are occasions when one's letters have to extend to a second and even a third page. Limiting one's imagination and creativity at such times might reduce a convincing argument to a terse statement that fails to satisfy the reader's interest. Brevity at the expense of lucidity is false economy.

Very long letters can be shortened, however, by borrowing a technique from report writing. Instead of placing all your information in a letter, consider shaping it into a semiformal report similar to that shown in Figure 5-3 on pages 124–28. Then summarize the highlights (particularly the purpose and the outcome) into a short letter to be placed at the front of the report (p. 123), so that the report becomes *an attachment* to the letter. Or, if a letter contains supporting information such as a series of test results, a cost analysis, an excerpt from another document, or similar data that is not immediately relevant, place this data on a separate sheet and label it as an attachment. Always refer to each attachment in the body of the letter, preferably by drawing a main conclusion from it, as has been done here:

> During our investigation we took a series of noise readings to determine sound level variations at night, during the day, and on weekends. These readings (see attachment) show that a maximum of 55 dB was recorded on weekdays, and 49 dB on weekends. In both cases these peaks were recorded between 5 and 6 p.m.

Short Paragraphs. Novelists can afford to write long paragraphs because they assume that they will have their readers' attention, and that their readers have the time and patience to wend their way through leisurely description. But in business and industry readers are working against the clock: their time is limited and they want to read no more than is necessary. They need bite-size paragraphs of easy-to-digest information.

In the section on compactness I suggested that a paragraph should develop only one idea. If this paragraph is short, then readers will be able to grasp the idea, comprehend it, and proceed quickly to the next idea. Their progression from paragraph to paragraph will be rapid.

I am not suggesting that your letters should contain a series of small, evenly sized paragraphs, which would appear dull and stereotyped. Paragraphs should vary from quite short to medium-long to give the reader variety. How you can adjust paragraph and sentence length to suit both reader and topic, and also to place emphasis correctly, is covered in Chapter 10.

Short Sentences. If you present only one thought in each sentence, then your sentences are likely to be reasonably short. I hesitate to stipulate a maximum sentence length, but 25 words might be a realistic one. Even this figure can vary, depending on the technical level of the reader and complexity of the information. As a general rule, the more complex the subject, the shorter one's sentences should be.

But there is a danger in stringing together too many very short sentences. Each may be complete and clear in itself, but strung together they may create the effect of all starts and stops with inadequate subject development. Your objective should be to obtain rhythm by mixing short and medium-length sentences. Rhythm in sentence structure will lead the reader smoothly through each paragraph.

Short Words. Some engineers feel that the scientific environment in which they work, and the complex topics they have to write about, demand that they use long, complex words in their correspondence. They write "an error of considerable magnitude was perpetrated," rather than simply "we made a large error." In so doing, they make a reader's job unnecessarily difficult.

The scientific world encompasses many long and complex technical terms that have to be used in their original form. "Diphygenic" (to have two modes of development) is an example. If we surround these terms with simple words, we will make our correspondence more readable. This does not mean using only four- and five-letter words, unless we want our work to read like a third grade spelling text. The criterion is to use the right word in the right place, striking a happy balance between oversimplification and ponderousness.

CLARITY

A clear letter conveys information simply and effectively, so that the reader readily understands its message. To write clearly demands ingenuity and attention to detail. As a writer you must consider not only how you will write your letters, but also how you will present them.

Create a Good Visual Impression. Experienced writers know that a nicely laid out letter impresses a reader. Subconsciously it seems to be saying "my neat appearance demonstrates that I contain quality information that is logically and clearly organized."

It is a mistake to think that the person you are writing to is doing nothing but waiting for your letter to come in just so he can read it. The converse is more likely to be true. Your letter will be slipped into a pile of correspondence, with each piece crying out for attention. If the recipient is an astute reader, he or she will weed out the promotional literature (or a secretary may do it), so that your letter will demand attention among a number of similar-looking letters. If yours looks appealing, its reader will be attracted to it and place it among those to be read first.

The appearance of a letter tells much about the writer, as well as the company. If a letter is sloppily arranged, if it contains strikeovers, visible erasures, or spelling errors, then I imagine a careless individual working in a disorganized office. But if I am presented with a neat letter, tastefully placed in the middle of the page and carefully typed with a reasonably new ribbon, then I imagine a crisp, well-organized individual working for a forward-thinking company noted for the quality of its service. It is the latter company I want to deal with, and it is their correspondence that I will read first.

Many technical people feel that once they have scribbled out or dictated a letter their responsibility for its issue is over, apart from signing it. This is not true. It is the stenographer's job to type a neat-looking letter that presents a good image of the company. In signing the letter, the writer endorses this image. In effect, a writer has to act like quality control and set the standards of presentation he or she considers correct for conveying the company's image. If a letter fails to meet these standards, it should be rejected.

A good stenographer will recognize poor workmanship and never offer a letter of poor quality for signature; an inexperienced stenographer may need to learn what standards of workmanship are acceptable. But unless you know how you want your letters to look (see page 45), and explain your requirements clearly, it may be a long time before stenographers do the kind of work you want.

Develop the Subject Carefully. The key to effective subject development is to present the material logically, progressing gradually from a clear, understood point to one that is more complex. This means developing and consolidating each idea for the reader's full understanding before attempting to present the next idea. If you don't have a thorough knowledge of your intended readers, you may confuse them through inadequate development or annoy them by overstressing each point.

Break Up Long Paragraphs. Ideas that need thorough development may breed overly long paragraphs. You can overcome this by using subparagraphs, which can carry the idea of a main paragraph into a series of smaller paragraphs without disturbing the main theme. Subparagraphing offers a useful way to maintain continuity through a series of points that are only partly related, and to draw attention to specific items. If there are many subparagraphs, care must be taken that the mood and tone are carried correctly from the main paragraph to all of the subparagraphs. In the following example, subparagraphing has been handled properly:

> I have analyzed our present capabilities and estimate that we can increase our commercial business by at least $60,000 per month. To meet this objective we will have to:
>
> 1. Shift the emphasis from purely local customers to clients in major centers. To increase business from local customers alone will require an intensive sales effort for only a small increase in revenue, whereas a similar sales effort in a major center will attract a 30–40% increase in revenue.
> 2. Increase our staff and manufacturing facilities. The cost of additional personnel and new equipment will in turn have to be offset by an even larger increase in business. Properly administered, such a program should result in an ever-increasing workload.
> 3. Create a separate department for handling commercial business. If we remove the department from the existing production organization it will carry a lower overhead, which will result in products that are more competitively priced.

Notice how each subparagraph flows naturally from the lead-in words of the main paragraph, and the verbs create emphasis:

> . . . we will have to:
>
> 1. *Shift* the emphasis . . .

2. *Increase* our staff . . .
3. *Create* a separate . . .

This is known as *parallelism.* Without parallelism there would be no continuity, and the transition would jar the reader:

. . . we will have to:
1. *Shift* the emphasis . . .
2. *Increase* our staff . . .
3. The overhead could be reduced by *creating* a separate department for handling commercial business.

Insert Headings as Signposts. In longer letters, and particularly those discussing several aspects of a situation, a writer can help the reader by inserting headings that are indicators of specific information. Such headings must be informative, summarizing clearly what is covered in the paragraphs they precede (see Chapter 10). If, for instance, I had replaced the heading to this paragraph with the single word *Headings,* I would not have summarized what this paragraph is about. The same would be true in the next paragraph if I replaced the existing heading with the single word *Language.*

Use Good Language. It hardly seems necessary to tell you to use good language, but in this case I mean language that you know readers will understand. Use only those technical terms and abbreviations they will recognize immediately. If you are in doubt, define the term or abbreviation, or use a simple expression in its place.

SIMPLICITY

Keep your letters as simple as possible. Say exactly what you mean, removing any unnecessary or irrelevant information. Place complex technical details, drawings, and supporting data in an attachment, so that you deal only with the plain facts in the body of the letter.

Start your letters positively, particularly when replying to a letter:

Dear Ms. Novak:

We have investigated the problem of paint discoloration described in your letter of November 18 and have concluded that the paint you used may have exceeded its shelf life.

This letter assumes that Ms. Novak has an adequate filing system. There is no need to repeat much of the information contained in her original letter of inquiry, as has been done here:

Dear Ms. Novak:

With reference to your letter of November 18, in which you described discoloration of our paint color No. 177 used as a second coat on top of No. 134 primer, we have conducted an investigation into your problem. Our conclusion is that the paint you used may have exceeded its shelf life.

The middle part of this reply gives Ms. Novak information she already knows; hence it wastes both her and the writer's time.

The factors I have described so far are mostly manipulative details that can be learned. Armed with this knowledge and the basic letter formats illustrated later in this chapter, the inexperienced writer has some ground rules to follow in the practical aspects of business letter writing. Still to be acquired is the more difficult technique of letting one's character appear in letters without letting it become too obvious.

Sincerity and Tone

Sincerity and tone are intangible factors that defy close analysis. There is no quick and easy method that will make your letters sound sincere, nor is there a check list that will tell you when you have imparted the right tone. Both qualities are extensions of your own personality that cannot be taught. They can only be shaped and sharpened through knowledge of yourself and which of your attributes you most need to develop.

SINCERITY

At one time it was considered good manners not to permit one's personality to creep into business correspondence. Today, business letters are much less formal and, as a result, much more effective. Personal qualities that experienced letter writers develop to convey their messages efficiently are:

Enthusiasm
Humanity
Directness
Definiteness

These four qualities, together with knowledge of one's subject, form the five basic ingredients that new writers can use to inject sincerity into their correspondence.

Be Enthusiastic. Sincerity is the gift of making your readers feel that you are personally interested in them and their problems. You convey this by the words you use and the way you use them. A reader would be unlikely to believe you if you came straight out and said, "I am genuinely interested in your project." The secret is to be so involved in the subject, so interested by it, that you automatically convey the ring of enthusiasm that would appear in your voice if you were talking about it.

Be Human. Too many letters lack humanity. They are written from one company to another, without any indication that there is one human being at the firing end and another at the receiving end. The letters might just as well be from computer to computer.

Do not be afraid to use the personal pronouns "I," "you," "he," "she," "we," and "they." Let your reader believe you are personally involved by using "I" or "we," and that you know he or she is there by using "you." Contrary to what many of us were told in school, letters may be started in the first person. If you know the reader personally, or you have corresponded with each other before, or if your topic is informal, let a personal flavor appear in your letters by using "I":

> Dear Mr. Wicks:
>
> I read your report with interest and agree with all but one of your conclusions. . . .

If you do not know your reader personally and are writing formally as a representative of your company, then use the first person plural:

> Dear Mr. Wicks:
>
> We read your report with interest and agree with all but one of your conclusions. . . .

Be Direct. Very few people have acquired the knack of being naturally direct. Anna King tells H. L. Winman engineers to state their case immediately when they start a letter, but to do so carefully so they won't sound too abrupt. The following example demonstrates her approach.

A construction company wants to know if a supply of steel it has in stock can be used to build a bridge in northern Alaska, and has submitted a small sample to H. L. Winman and Associates for analysis. When tests are complete, Mechanical Engineering Coordinator Paul Nowicki writes this letter to the client:

> Dear Mr. Sparkes:
>
> We conducted extensive tests on the sample of steel you sent to us for analysis. A Charpy Impact Test told us that the steel has a low transition temperature, while a Tension Test demonstrated that the steel failed in a ductile manner, with a yield point of 44,000 psi (1.6236 MN/m^2).
>
> From these tests we determined the steel to be type G40.12. Although widely used as structural steel throughout southern Canada, it is less satisfactory for structures that are to be subjected to high loading in the far north. Its low temperature characteristics are only moderately good, which would make it unsuitable for a bridge at Peele Bay, Alaska.

Realizing that his indirect start forces Mr. Sparkes to wade through all the technical details before he finds the answer he wants, Paul heeds Anna's previous suggestions and rewrites the letter so it has a direct opening. But this time his directness makes him seem much too abrupt:

> You will not be able to use the type of steel you sent to us for analysis to build a bridge at Peele Bay, Alaska. We found your steel to be type G40.12, which has poor low temperature characteristics. A better type would be G40.8.

So finally Paul enlists Anna's help, and between them they write this naturally direct letter:

> Dear Mr. Sparkes:
>
> The sample of steel you sent to us for analysis is G40.12, a type that would not be suitable for the short span bridge you propose to build at Peele Bay, Alaska. A more suitable type would be G40.8, which has better low temperature characteristics than type G40.12.
>
> We analyzed your sample by conducting both a Charpy Impact Test and a Tension Test, details of which appear in attachment 1. These tests demonstrate that your sample has a low transition temperature and fails in a ductile manner.

This last version tells Mr. Sparkes immediately what he most wants to know: whether his supply of steel can be used for the bridge. It also helps him read the technical details that follow more intelligently.

A writer who fails to analyze exactly what the reader wants to know can easily be led into writing an indirect letter. The writer must ask him or herself, "What basic fact is most important to my reader?" and then introduce it as early as possible. If there are conditions or interpretations that need to be described, or additional factors that have strongly influenced the results of a project, they must be brought in afterwards. To introduce them before stating the main theme will weaken the effectiveness of the letter, and may even confuse or annoy the reader.

Be Definite. Being definite means deciding exactly what you want to say, then saying it. Persons who think better with pen in hand sometimes make decisions as they write, producing indecisive letters that are irritating to read. Their writers seem to examine and discard points without really grappling with the problem. By the time they have finished a letter, they have decided what they want to say, but it has been at the readers' expense.

Decision-making does not come easily to many people. Those of us who hesitate before making a decision, who evaluate its implications from all possible angles and weigh its pros and cons, may allow our indecisiveness to creep into our writing. We hedge a little, explain too much, or try to say how or why we reached a decision before we tell our reader what the decision is. This is particularly true when we have to tell readers something unfavorable or contrary to their expectations.

Reg Wasalusky, a field employee of H. L. Winman and Associates, has written from a project office in Alaska to ask for a raise. His supervisor, Andy Rittman, writes this letter in answer to Reg's request:

An Indefinite Letter

Dear Reg:

I could not answer your letter immediately because I wanted time to consider your request for a raise very carefully. We fully appreciate down here that you are working in rough climatic conditions and that much of your work takes place out of doors. On that count we feel that your request is justified.

On the other hand, we have to equate your position with those of all the other field staff, since to increase your salary when others are not similarly considered would be unfair. We have therefore also assessed your position and seniority, since we recognize that you have been with the company for some time and that you have a very good record.

As a result of our deliberations we find that it would not be possible to increase your salary at the moment. We would like to add, however, that your request will be reviewed again in three months, when a general salary updating program is to be inaugurated.

Andy Rittman

Andy and Reg are good friends, so Andy does not like to come right out and say "No." But in trying to demonstrate that he has not treated Reg's request lightly, he has used too many wishy-washy statements that have a hollow ring:

> "We fully appreciate" (para 1)
> "On the other hand" (para 2)
> "As a result of our deliberations" (para 3)
> "We find that it would not be possible" (para 3)

The letter is also misleading because at the end of both the first and second paragraphs Reg is led to believe he is being considered, but then in each case the next paragraph contradicts this impression. This "yes-no-yes-no" attitude has a disconcerting effect that may irritate Reg more than the denial of the request. So will Andy's sudden shift from the personal "I" to the company "we," which he adopts as soon as he starts explaining why the request is going to be turned down.

If Andy Rittman wants to be definite, he must be decisive. He must state his decision immediately, then bring in his reasons. If supporting information must be included, it must be brief and to the point. Compare this rewritten version with the original letter:

A Decisive Letter

Dear Reg:

I have considered your request for a salary increase but regret no increases can be granted at present. The company will be reviewing all departmental salaries in three months, when adjustments will be made for field staff who have seniority and who work under arduous conditions.

As soon as the salary updating program has been finalized, I will let you know if you are to receive an increase.

Andy

Reg may not like hearing "No," but at least he will know exactly where he stands and what is being done. If he knows he has seniority and a good working record, then he can reasonably anticipate receiving a raise in three months. (Note that Andy does not say anything that will make Reg assume his request will be granted; to do so would result in a very unhappy employee if a raise failed to come through.)

A writer will still sound indefinite, even though he approaches the subject directly and decisively, if he uses predominantly passive verbs. Passive verbs are weak, whereas active verbs are strong. Expressions such as "it was considered that" (I *consider*), "is an indication of" (*indicates*), "conducted an examination of" (*examined*), "it is recommended that" (*I recommend*), should be replaced by the much more direct expressions shown in parenthesis. For more information on this subject, see Chapter 10.

Know Your Facts. Insufficient background knowledge and inadequate preparation are the first steps to evasive, indefinite writing full of overblown words that avoid the issue. The big words act as camouflage to conceal uncertainty, but a discerning reader can easily penetrate them. Be absolutely convinced of your facts before you start to write, or you will find yourself filling in gaps with big words and many unnecessary adjectives and adverbs.

TONE

It is not easy to assess whether you are setting the right tone as you write. You will probably be doing so if you have successfully developed the personal qualities that contribute to sincerity; but you cannot assume this. To achieve the right tone, correspondence should be simple, dignified, but friendly. Approach your readers on a person-to-person basis, following the five suggestions below.

Know Your Reader. A writer cannot hope to set the right tone without having satisfactorily identified the reader. If the writer knows a subject well and the reader does not, the writer must take care not to be condescending by implying that the subject is so complex that the reader will be able to understand only a very simple description. Even if this is true, the careful writer will select just the right terminology to hold the reader's interest and perhaps offer a mild challenge. By letting readers feel that they are grasping some of the complexities of a subject (by using analogies within their range of knowledge), a writer can often present technical information without confusing or upsetting nontechnical readers.

Neither should inexperienced writers attempt to "write up" to the level of a technical reader whose experience and knowledge are far greater than their own. In trying to match the reader's intellect they may be tempted to insert long, important-sounding words that are incorrect, or even faintly amusing. They should write at their own level, using sensible technical terms and expressions that fit the subject.

Take It Easy with Apologies. When apologizing in writing do so quickly and cleanly, just as you would when apologizing to a person for stepping on his or her foot: "I'm sorry!" is all you need say. If you have been wrong, admit it and then consider the subject closed. To keep on apologizing, to say too much, will make your apologies seem insincere. These excerpts show the difference between an overstated apology and a brief one:

Too Wordy We were mortified to hear of the damage caused to your carpeting by our repair technician's carelessly placed soldering iron, and are most sincerely sorry that this should have happened. The person involved has been reprimanded and we can assure you that such an occurrence will not happen again. We trust that this accident will not affect our business relationship, particularly since this has never happened before and we extremely regret that it should have happened to you. Please accept our apologies and our guarantee of top-notch service in the future.

Brief I apologize for the inconvenience that my incorrectly worded telegram of February 20 has caused you. The telegram should have read . . . [and so on, but with no further apologies].

Keep Compliments Simple. Some people have a natural gift for paying compliments so that they always sound sincere. They have mastered an art that most of us find difficult: the ability to tread the narrow line between understatement and overstatement.

To pay a compliment effectively in writing is equally difficult. Most of us lean toward overstatement, forgetting that the best written compliments are stated very simply. To overstate a compliment, to gush, to overwhelm your reader with intensity, is to make the compliment seem insincere. This has happened in the following example:

> We have always received excellent service from your organization in the past, so it was only natural that we should turn to you again in the hour of our need. The assistance you provided in helping us to identify a new type of transducer was overwhelming, and we would like to extend our heartfelt thanks to all concerned for their help. The prompt attention you gave to help resolve our problems was very deeply appreciated.

A simple "thank you" would have been much more believable:

> We very much appreciated the prompt assistance you gave us in identifying a new type of transducer.

Criticize Effectively. A reader will usually accept criticism if it is presented "straight from the shoulder." Any attempt to soften the criticism by writing a lengthy preamble or by hiding it behind big words will reduce its effectiveness. To keep harping on the criticism once it has been stated, or to imply it without coming directly and clearly to the point, will lower the reader's opinion of you as a person who knows what he wants to say.

Implied criticism, or criticism that points a cautioning finger in the wrong direction, can so weaken a letter that the recipient will completely miss the message. On three separate occasions the manager of a radar maintenance section had ordered a very expensive high power transmitting tube called a magnetron, and each time it arrived damaged. Three times he wrote to the manufacturer, urging them to use a better packing method, but nothing was done. "How can I get them to do something about it?" the manager wailed. "They just ignore me!"

His letter said (in part):

> I regret that for the third time I must write to you to report still another of your magnetrons has been damaged in shipment. It is becoming increasingly obvious that the handling methods used by the railway express department are too rough for such delicate equipment, even though it is clearly marked as such on the outside container. I have spoken to the railway about this but their only comment is that the instrument must have been packed inadequately. To prevent damage to further shipments, I therefore suggest that you use a more resilient type of packing material.

The trouble with this letter is that its writer did not criticize directly. Instead he shifted the blame onto the railway express handlers, only *implying* that the manufacturer's packing methods should be improved. If he had placed the blame where it belonged, the manufacturer would have given his letter more serious attention:

> For the third time one of your magnetrons has been damaged in shipment. This clearly indicates that your packing methods are inadequate for delicate equipment that has to be shipped by rail express. In future, will you please use a more resilient type of packing material to protect the magnetron from rough handling.

Occasionally a writer has to both criticize and pay a compliment in the same letter. If this happens, always pay the compliment first, to make the criticism easier to accept. This is not the same as softening the criticism; it simply means that the writer recognizes the good work that has been done. Placing the criticism first can so reduce the effectiveness of the compliment that the reader may not even notice that it is there. Rebuked by the criticism, he (or she) may feel that it has been added only as a sop to make him feel better. Thus neither criticism nor compliment does its job effectively.

Avoid Words that Antagonize. Any statement that implies that a reader is wrong, has not tried to understand your point of view, or has failed to make himself or herself understood, will immediately place the reader on the defensive. Even though these factors may be true, they must be stated in words that will clear the air rather than electrify it. Tell readers gently if they are wrong, and demonstrate why; reiterate your point of view in clear-cut terms, to clarify any possibility of misunderstanding; or ask for further explanation of an ambiguous statement, refraining from pointing out that the writing is vague.

Business Letter Format

There are many opinions of what comprises the "correct" format for business letters. The two methods illustrated in Figures 3-2 and 3-3 are those most frequently used by technical business organizations. The former is a conservative format known as the Modified Block, and until recently it was the more common arrangement. The latter, a modern style known as the Full Block, is popular with organizations in which contemporary design plays an important role. Both are acceptable in today's business world. It is important, however, that a business organization be consistent and use only the one format for all correspondence.

Anna King has adopted the Full Block format for H. L. Winman and Associates' correspondence (see the letter reports in Chapter 4 and the cover letter preceding the second report at the end of Chapter 6). Anna is also aware that letter styles are continually changing. Some companies now omit the salutation and complimentary close (e.g. "Sincerely"), write dates military style (e.g. day-month-year: "27 January 1981"), omit all punctuation from names and addresses (e.g. "Ms Jayne K Tooke"), and use interoffice memorandums instead of letters for informal correspondence. If these trends become more firmly established, she will adapt H. L. Winman Associates' letter format to reflect the changes.

The memorandum is a flexible document normally written on a prepared form similar to that shown in Figure 3-4. Formats vary according to the preference of individual companies, although the basic information at the head of the form is generally similar. Examples of memorandums appear throughout Chapters 4 and 5.

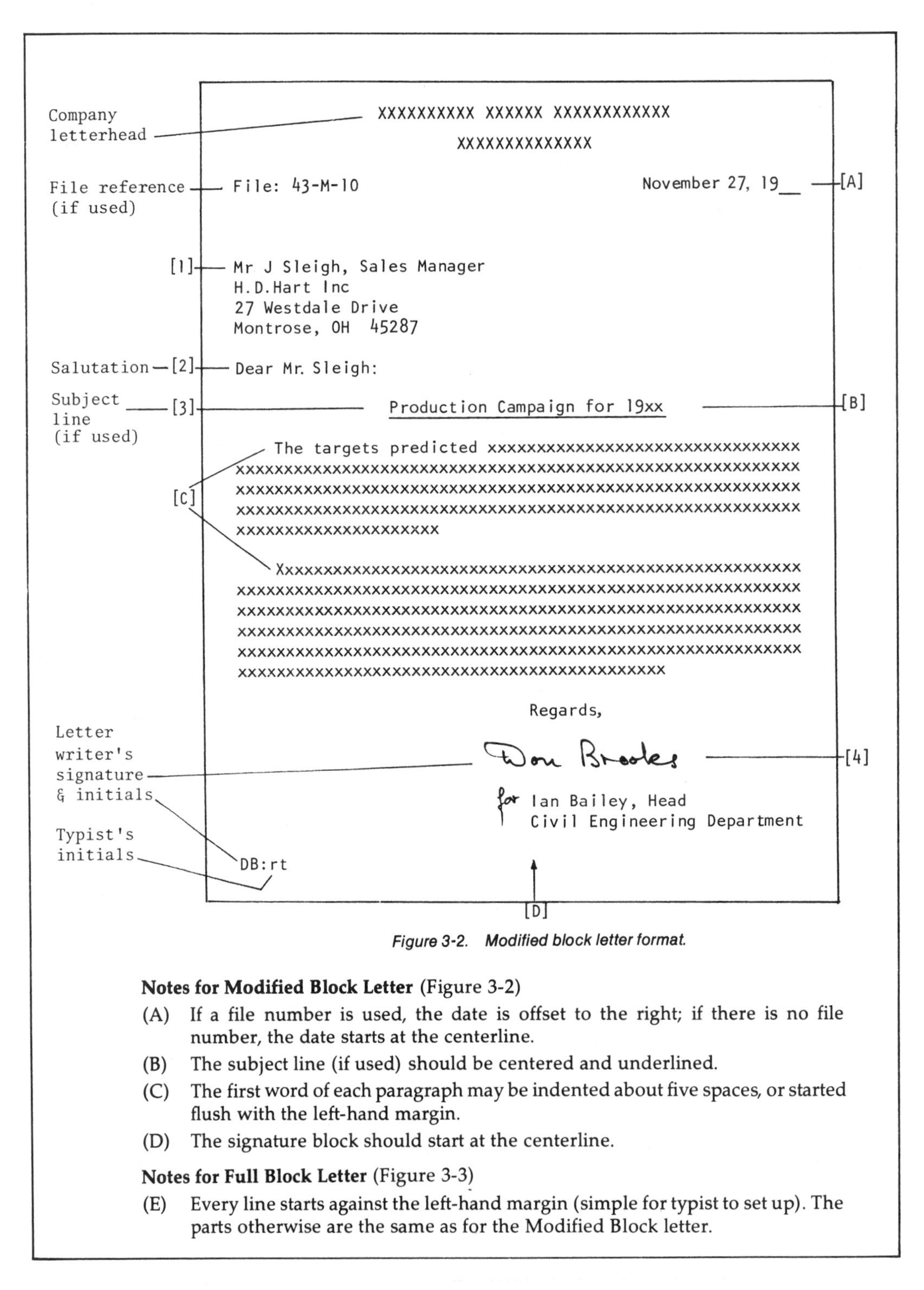

Figure 3-2. Modified block letter format.

Notes for Modified Block Letter (Figure 3-2)

(A) If a file number is used, the date is offset to the right; if there is no file number, the date starts at the centerline.

(B) The subject line (if used) should be centered and underlined.

(C) The first word of each paragraph may be indented about five spaces, or started flush with the left-hand margin.

(D) The signature block should start at the centerline.

Notes for Full Block Letter (Figure 3-3)

(E) Every line starts against the left-hand margin (simple for typist to set up). The parts otherwise are the same as for the Modified Block letter.

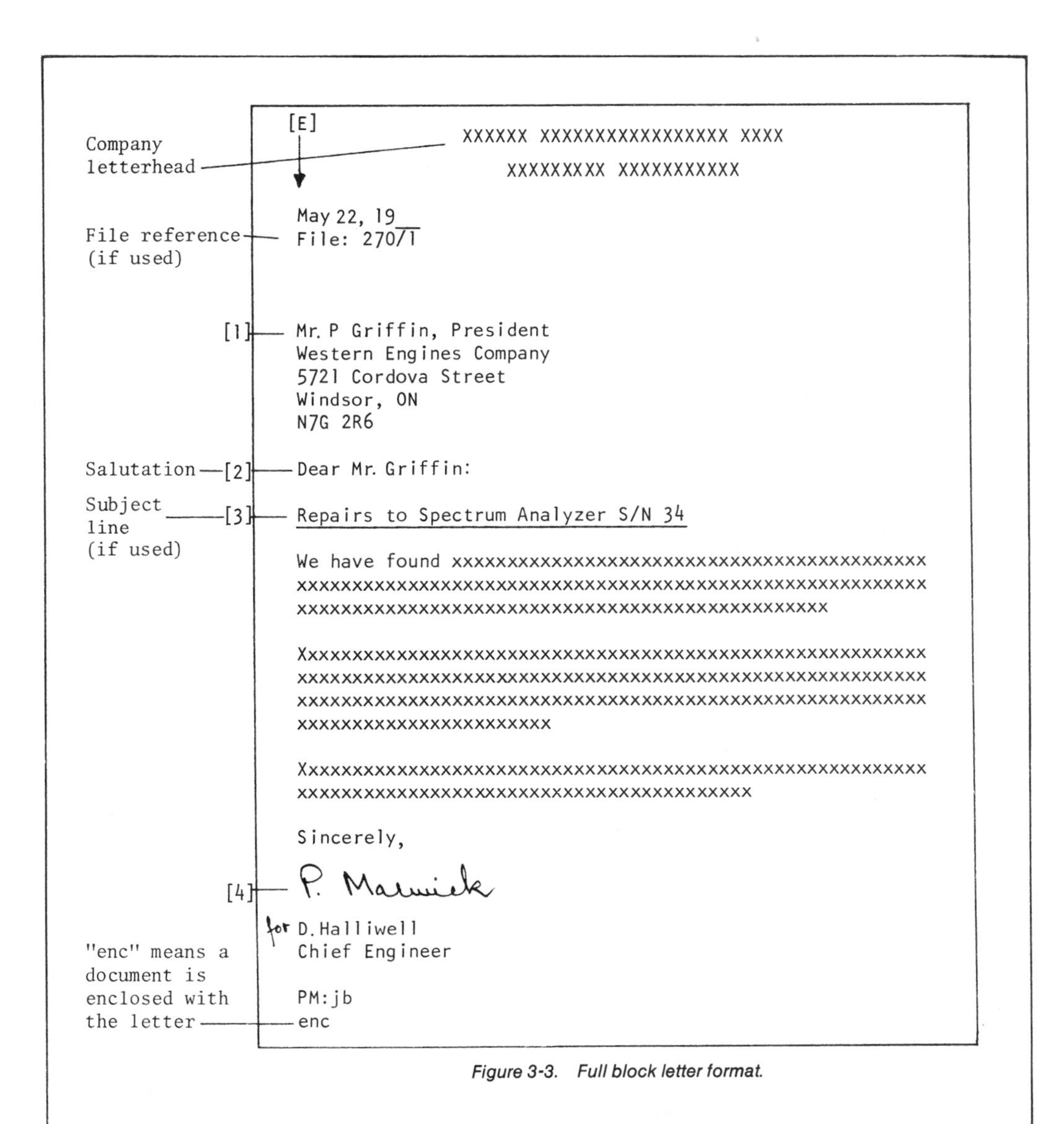

XXXXXX XXXXXXXXXXXXXXXXX XXXX
XXXXXXXXX XXXXXXXXXXX

May 22, 19__
File: 270/1

Mr. P Griffin, President
Western Engines Company
5721 Cordova Street
Windsor, ON
N7G 2R6

Dear Mr. Griffin:

Repairs to Spectrum Analyzer S/N 34

We have found xxxxxxxxxxxxxxxxxxxxxxxxxxxxxxxxxxxxxxxxxx
xxxxxxxxxxxxxxxxxxxxxxxxxxxxxxxxxxxxxxxxxxxxxxxxxxxxxxxx
xxxxxxxxxxxxxxxxxxxxxxxxxxxxxxxxxxxxxxxxxxxxxxx

Xxxxxxxxxxxxxxxxxxxxxxxxxxxxxxxxxxxxxxxxxxxxxxxxxxxxxxxx
xxxxxxxxxxxxxxxxxxxxxxxxxxxxxxxxxxxxxxxxxxxxxxxxxxxxxxxx
xxxxxxxxxxxxxxxxxxxxxxxxxxxxxxxxxxxxxxxxxxxxxxxxxxxxxxxx
xxxxxxxxxxxxxxxxxxxxxx

Xxxxxxxxxxxxxxxxxxxxxxxxxxxxxxxxxxxxxxxxxxxxxxxxxxxxxxxx
xxxxxxxxxxxxxxxxxxxxxxxxxxxxxxxxxxxxxxxx

Sincerely,

P. Marwick

for D.Halliwell
Chief Engineer

PM:jb
enc

Figure 3-3. Full block letter format.

Notes for Both Letter Styles

(1) In modern correspondence the name and title of the addressee are placed ahead of the company name. If the addressee's name is not known, a suitable title (e.g. Purchasing Agent) should be inserted. All but essential punctuation is eliminated from the address, salutation, and signature block.

(2) Today's trend toward informality encourages writers to use first names in the salutation: "Dear Jack" instead of "Dear Mr. Sleigh."

(3) Subject lines should be *informative* (not just "Production Plan" or "Spectrum Analyzer"); they may be preceded by *Subject:*, *Ref:*, or *Re:*.

(4) The letter writer may sign "for" the Department Head.

H L Winman and Associates

INTER - OFFICE MEMORANDUM

From: R Davis (1) Date: December 3, 19__

To: A Rittman Subject: Early Mailing of Pay Checks for Field Personnel (4)

(2) I need to know who you will have on field assignments during the week before Christmas so that I can mail their pay checks early. Please provide me with (3) a list of names, plus their anticipated mailing addresses, by December 8.

Checks will be mailed on Tuesday, December 14. I suggest you inform your field staff of the proposed early mailing.

RD (5)

Figure 3-4. Interoffice memorandum.

Notes for Memorandum (Figure 3-4)

(1) The informality of an interoffice memorandum means titles of individuals (such as Office Manager and Senior Project Engineer) may be omitted.

(2) No salutation or identification is necessary. The writer can jump straight into the subject.

(3) Paragraphs and sentences are developed properly. The informality of the memorandum is *not* an invitation to omit words so that sentences seem like extracts from telegrams.

(4) The subject line should offer the reader some information; a subject entry such as "Pay Checks" would be insufficient.

(5) The writer's initials are sufficient to finish the memorandum (although some organizations repeat the name in type beneath the initials). Some people prefer to write their initials beside their name, on the "From" line, instead of signing at the foot of the memo.

Letter of Application

Knowing how to write a good letter of application is just as important to the student seeking his or her first job as it is to the experienced engineer looking for a better position. Both have invested time, money, and personal effort into training that qualifies them to do specialized work, and both need to display their capabilities to the best advantage. Unfortunately, I hear too often of highly qualified persons disqualifying themselves by simply failing to present a good image in their letters of application.

Let's face it: a letter of application is a sales letter. It must convince a prospective employer that you are a commodity worth investigating, just as a land agent's letter must sell a prospective client the belief that the agent has a property worth looking into. The chief difference is that you, as an applicant, are trying to sell the most important commodity you own: yourself!

What happens when your letter of application first drops onto an employment manager's desk? As the manager reads it he or she forms an immediate impression, mentally classifying you as belonging to one of three groups:

1. "This applicant sounds interesting! I'd like to talk to her."
2. "This person *may* have something to offer. I *might* arrange an interview."
3. "This person has little to offer. Just a routine application—not worth following up."

You cannot afford to be classified in group 3, for then your dealings with this particular employer are already over. Even in group 2 you may be overlooked, particularly if you are facing competition from applicants whose letters have better reader appeal than yours. Only by writing a letter that creates enough interest for the reader to automatically classify you in group 1 can you be reasonably sure that you will reach the second stage of the hiring procedure. Your letter of application is the key that opens the door to the personal interview. If it fails to do its job properly, the door remains shut and your prospects of employment are negligible.

There are two situations that call for you to write a letter of application (excluding the many campus situations where a letter of application is not required, and those that simply call for the applicant to drop into the office and fill out an application form). Either you have seen an advertisement in a newspaper or technical magazine that invites applications from persons having your qualifications, or you know of an organization for which you would like to work but do not know whether they have any job openings. In the former case you can concentrate on displaying the talents required for the advertised position. In the latter situation your task is complicated because you do not know what the reader will most want to learn about you. But although the approach and tone are likely to differ (note the differences between the sample letters in Figure 3-5 and 3-6), both letters follow the same basic format.

THE PARTS

A letter of application has three parts:

The *initial contact,* in which the applicant creates interest in himself.

The *evidence,* in which he demonstrates why his application is worth considering.

The *close,* in which he holds open the door to further discussion.

These are not watertight compartments of information that stand alone; rather, they should carry a reader naturally and quickly through the letter so that he or she automatically classifies it in group 1.

The Initial Contact. A good opening arrests the attention of the recipient, making the reader want to read the remainder of the letter. A dull, uninteresting opening can so weaken the reader's attention that he or she may read the rest of the letter with only casual interest.

An interesting opening will avoid the routine "I would like to apply for a position as a laboratory technician. . . ," and replace it with a direct statement of intent:

> Please accept my application for the position of laboratory technician advertised in the Montrose Herald of May 17.

or

> I am applying for the position of. . . .

Many job applicants find the opening difficult to write because they fail to state in the first sentence why they are writing. It is relatively easy to do this when applying for a position known to be open, as in the two examples above, but can be more difficult when you do not know if a vacancy exists. Lack of knowledge can easily lead an applicant to an ineffective opening that fails to come to grips with the purpose of the letter. Colin Westdahl, the writer of the unsolicited letter of application in Figure 3-5, has clearly stated the purpose of his letter in the second sentence. But the writer below has only implied the reason for writing, and by not making a clear, specific statement of purpose has conveyed an indecisive, indefinite image:

A Weak, Indefinite Contact

> Dear Mr. Perchanski:
>
> While working in the HV Laboratory at the University of Montrose, I became very interested in HV DC power transmission. Now I understand that you have a project going on the same topic and would like to inquire if there are likely to be any job openings.

Writing a good initial contact demands imagination, plus the ability to distinguish between a really effective approach and one that is overstated to the point where it becomes silly, like this:

An Over-Written Contact

> Dear Mr. Perchanski:
>
> Did you know that I majored in high voltage power generation during my last term at Montrose Community College? And that I also worked in the University of Montrose high voltage laboratory during two summer vacations? Since you shortly will be involved in high voltage engineering projects, don't you think it would be a good idea if we were to get together?

Plainly, this applicant is trying too hard. He may be qualified for the job, but he should try proving it with a simpler approach.

The initial contact should command the reader's attention, state clearly why the letter is being written, and, if possible, demonstrate why the writer thinks he is suitable. All these requirements are met in the initial paragraph of Colin Westdahl's letter, and by this applicant:

42 Venables Street
Cleveland, Ohio 44103

Tel 777 1611

May 16, 19__

Mr. J. Perchanski, PE
Head, Electrical Engineering Department
H. L. Winman and Associates
475 Reston Avenue
Cleveland, OH 44104

Dear Mr. Perchanski:

In a radio interview last week you mentioned that your firm will shortly be setting up a project group to study HV DC power transmission for Montrose Hydro. If this project creates new openings in your Department will you please consider me for employment. I have had experience working in the University of Montrose high voltage laboratory during two summer vacations, and majored in high voltage power generation during my second year at Caravelle Junior College in Cleveland.

During periods of summer employment at the University I worked on project Mirron (a study of special applications of DC power transmission conducted for the United Nations) under the direction of Dr. David C. Crawford. Much of the time I was conducting tests of prototype systems, evaluating results, and preparing project reports. Then at Caravelle College I assisted in developing a HV DC Power Source for laboratory use. This work was supervised by Mr. Samuel Wolinsky, my instructor in advanced electrical technology. Both of these persons have agreed to vouch for my experience and capability.

If you think that a person having my qualifications could be of value to your firm, may I visit you for an interview? Any afternoon of the week starting May 27 would be particularly suitable as I will be writing my final examinations during the mornings and will be free every day after 1 p.m.

Sincerely,

Colin Westdahl

Colin Westdahl

Figure 3-5. Unsolicited letter of application.

A Good Contact

Dear Mr. Wood:

Will you please consider my application for the Programmer Analyst position advertised in Pro-Gram News of May 20, 19__? I am a graduate of the two-year Programmer-Analyst course at Montrose Community College and have three years experience as a programmer with Blackfriars Computers Incorporated of Montrose.

The Evidence. The second paragraph should present facts to support the writer's application. It should be written in firm, definite terms that emphasize the applicant's strong points, and offer specific information related to the position being sought. Most of all, it must never generalize. Vague statements such as "I have had many years of experience with the CanCel Chemical Corporation" make a reader suspect that the applicant is trying to conceal that he has only limited experience to draw on.

If an applicant has only limited work experience, it can all be placed in the evidence section, as Colin Westdahl has done in Figure 3-5. But if an applicant has a lot of evidence to include, then details of qualifications and work experience should be prepared on a separate sheet, known as a data sheet, biography, or resume, which is attached to the letter. The evidence section of the letter then contains only a brief synopsis of the more important points. This has been done by Alison Witney in her letter of application in Figure 3-6, which is supported by the biography in Figure 3-7.

The Close. Most applicants have difficulty in writing an effective closing paragraph, often saying too much because they do not know when they have said enough. The close should be short and should do the opposite of what its title suggests: it must open the door to the next step. Both sample letters do this successfully. By suggesting times when he can be available for an interview, Colin Westdahl (Figure 3-5) has opened the door even wider than usual. If he has not heard from Jim Perchanski by, say, May 28, he can telephone and inquire whether he is to visit H. L. Winman and Associates without feeling he is being too forward in taking the initiative.

THE PERSONAL BIOGRAPHY

Many names can be applied to the personal biography, ranging from the simple Data Sheet mentioned previously, through Resume and Biography of Experience, to the Curriculum Vitae used at professional levels. But each serves the same purpose: to provide details of a person's background, experience, and qualifications that are too cumbersome to insert in the body of a letter. They may also vary in format and tone, from the rather terse approach of the Data Sheet to the well-developed narrative often seen in the Resume and Curriculum Vitae.

Alison Witney's biographical details (Figure 3-7) attached to her letter of application offer a satisfactory middle-of-the-road treatment that is neither sketchy nor overdone. The numbered comments below refer to the circled figures beside her biography.

(1) Today's laws protect you from having to provide a prospective employer with personal details such as sex, age, color, and marital status. You can *choose* to

Alison V. Witney
210 - 1670 Fulham Boulevard
Amiento FL 32704

Tel: (305) 474 6318

March 23, 1980

Dr. Eugene Cartwright, Director
Animal Science Experimental Institute
Mount Ashburn University
Three Hills, Alabama 35107

Dear Dr. Cartwright:

I am applying for the position of Research Technician (Animal Sciences) advertised in the March 18, 1980 issue of the Amiento County Herald. I have been involved with animals and their care and treatment for many years, and shortly will receive my Diploma in Biological Science.

My interest in animals dates back to 1968, when I first learned to care for, groom, and ride horses. I now teach horse riding in my spare time. For the past three years my employer has been Dr. Alex Gavin, veterinary surgeon at the Amiento, Florida, Animal Treatment Center, where I assist in the medical treatment of small animals. It was my interest in horses, plus Dr. Gavin's influence, that led to my enrolment in the two-year Biological Sciences course at Amiento Technical College, from which I will graduate in early June. The attached biographical details provide further information on my education, employment background, and work experience.

I will be visiting your research station from April 21 to 23, as part of my college term research project. May I call on you then, while I am at Three Hills?

Regards,

Alison Witney

Alison V. Witney

Figure 3-6. Solicited letter of application (with reference to an attached biography).

BIOGRAPHICAL DETAILS

ALISON V. WITNEY

PERSONAL DATA

(1)

Address: 210 - 1670 Fulham Boulevard, Amiento FL 32704
Telephone: (305) 474 6318
Availability: June 15, 1980

EDUCATION AND TRAINING

(2)

Graduate of Morton Stanley High School, Corisand FL, 1975.
Expect to graduate with Diploma in Biological Science from Amiento Technical College, June 1980.

WORK EXPERIENCE

(3)

1977 to date — Animal Treatment Center, Amiento FL. One year full time, two years part time, as veterinary assistant. Responsible for reception, grooming, and exercising of animals, assisting veterinarian during operations, changing dressings, administering injections and anaesthetics, and performing administrative duties such as accounting and ordering of supplies.

1975 - 1977 — Remick Airlines, Orlando FL. Accounts clerk in air freight department; coordinating billings, preparing invoices, following-up lost shipments, assisting clients, and writing monthly reports. For nine months, assisted in payroll preparation.

1972 to date — Bar None Riding Stables, Corisand FL. Part time employment teaching the care and handling of horses, and basic riding techniques to young riders. Assisted in grooming, cleaning, feeding, and saddling-up.

ADDITIONAL INFORMATION

(4)

Winner of two educational awards: Morton Stanley Science Scholarship (1974), and Amiento Technical College Biology Scholarship (1979).
Member of YWCA since 1966, where I teach swimming and lifesaving.
Interests: teaching horseback riding and jumping, swimming, water skiing.

REFERENCES

The following persons have agreed to act as references on my behalf:

(5)

Dr. Alex Gavin
Veterinary Surgeon
Animal Treatment Center
2230 Wolverine Drive
Amiento FL 32704
Tel: 474 1260

Mr. Charles Devereaux
Owner--Manager
Bar None Riding Stables
2881 Westshore Drive
Corisand FL 32726
Tel: 632 2292

Figure 3-7. A biography or resume.

offer such information, but you cannot be asked for it. Some applicants think they should also include their height and weight, comments on their health and physical handicaps, and incidental items such as place of birth and ethnic origin. I do not believe such information is necessary. To include minor details seems to detract from the biography's purpose, which is *to offer evidence of the applicant's ability to do a good job.*

(2) There is no need to list all the schools you have attended. Simply state the name of your last school, the highest level you attained, and the year of graduation. Add to this the name of each college or university you have attended, the type of course you enrolled in, the diploma or degree you received, and the year that you graduated (or expect to graduate).

(3) Experience is usually presented in reverse order, with an applicant's most recent work experience appearing first and earliest experience last. Generally, say most about recent experience, providing that the length of employment has been long enough to warrant it, and earlier work that was similar to that of the position being sought.

(4) Employers are particularly interested in an applicant's activities and interests outside normal work. They want to know if the person is more than a routine employee who arrives at 8 a.m., works until 4:30 p.m., then drives home, eats supper, and watches television all evening. Information on your hobbies, interests, and participation in sports and community activities tells prospective employers that you recognize your role in society, are not too rigid or too narrow, and adapt well to your environment. Employers reason that such an applicant will make an interesting, active employee who will not only contribute much to the company, but also take part in social and sports functions.

(5) Try to draw your list of references from a cross-section of people you have worked for, been taught by, or served with on committees. Before including them in your list, check that all are willing to act as references. Almost everyone will be, but it is only courteous to ask them if they would mind.

THE OVERVIEW

The appearance of your letter can cause an employer to form a false impression of you. The image conveyed by a poorly presented letter may result in your being classified in group 2, or even group 3, when your qualifications warrant your being in group 1.

Tanys Young, H. L. Winman's Personnel Manager, has a letter of application on her desk from a young printer who technically is fully qualified to fill a vacancy in the company's print shop. But she has automatically classified him in group 3. Why? Because he has written so carelessly, scribbling a few lines onto grubby lined paper torn from a pad, that Tanys reasons he would be just as likely to carry his careless approach into his day-to-day work.

Finally, a word of warning: Try to avoid writing a letter of application that too closely resembles the stylized approach suggested by most textbooks (and I include this book!). The "textbook letter" is dangerous because its source is so obvious. Personnel managers like Tanys tell me they are more interested in an applicant who writes only an average letter but whose personality and enthusiasm show clearly through the writing, than an applicant whose letter is flawless but characterless. This by no means implies that you can be lax in your approach; rather, it means that you should try to fashion your letters of application as though *you* wrote them, without assistance from me or anyone else.

ACCEPTING OR DECLINING A JOB OFFER

Regardless of whether you are accepting or declining a job offer, write to the prospective employer as soon as you have made your decision. The employer can then inform other applicants that the position has been filled, or make an offer to another applicant. If you are accepting, write a brief letter stating so. Restate the important conditions of employment, such as the position you are accepting, your starting salary, and the date you will start work, to prevent any misunderstanding. For example:

> Dear Ms. Young:
>
> I shall be glad to accept the position of Engineering Technician II in your Test Laboratory, offered to me in your letter of May 10. Both the salary of $1230 per month and the employment starting date of July 1 are acceptable.
>
> Thank you for considering me for this position.
>
> Sincerely,

If you are declining the offer, do so briefly and politely, if possible stating your reason:

> Dear Ms. Young:
>
> Thank you for your letter of May 10 offering me employment as an Engineering Technician in your Test Laboratory. I regret that I must decline your offer, having already accepted employment elsewhere.
>
> Thank you for considering me.
>
> Sincerely,

Your tone and attitude in a letter declining a job offer are just as important as in the letter of acceptance. If the job you accept fails to work out satisfactorily, you may want to renegotiate with the employer whose offer you declined. It will help in reopening discussions if you turned down the previous offer gently.

Writing a letter of application and resume, and submitting them to a prospective employer, is only the first stage of the employment-seeking process. If your application creates interest, you are likely to enter into the second stage by being invited to attend an employment interview. How you should prepare for and present yourself at such an interview is discussed in Chapter 8.

Assignments

In addition to the letter-writing assignments printed below, the following projects in Chapter 4 relate to the material in this chapter:

Project 8, Part 2
Project 9, Part 2
Project 10, Part 2
Project 11, Part 1

Most projects include all the details you will need to write the assignment. You are encouraged, however, to introduce additional factors if you feel they will increase the depth or scope of your letter.

PROJECT NO. 1: A FAULTY HOME ENTERTAINMENT CENTER

Assume that you were recently in Waverly (1100 miles from home), where you visited friends Martin and Joan Tong. Martin gave Joan a PAM 78 "Home Entertainment Center" last Christmas, and you are impressed by its tone, appearance, and features. Martin tells you privately that he bought it from Craven's Discount Center at 1837 Kelly Street in Waverly, and offers to go with you if you are interested in buying one.

You are: but you are disappointed to discover that Craven's has sold all its PAM 78 entertainment centers, that no more are on order, and that no one else in Waverly carries them. However, Mr. Harry Craven, the store owner, has a suggestion: there is a demonstrator which he could sell to you at 5% off his regular discount price. You test it, and it seems okay. Martin suggests a 15% price reduction would be more normal for a demonstrator, but Mr. Craven won't budge; he adds, however, that he'll give it a good check-over if you'll leave it with him. You agree. Two days later you pick it up, pay $388.35, and receive Craven's invoice No. C7214 stamped "Paid."

But when you arrive home, you find that the tape section of the PAM 78 doesn't work. You also discover that there is no local service center for the PAM line, so you take the entertainment center to Modern TV and Radio at 280 Waltham Avenue. When you pick it up the following day, store manager Jim Williams hands you a printed circuit board (PCB) with several bent and twisted pins.

"There's your problem," Jim says. "Craven's in Waverly must have replaced this PCB—you can tell it's one of theirs because the name CRAVEN is stamped on it." He explains that whoever inserted the PCB didn't align the pins properly, and bent them by forcing it into its socket.

You pay $53.00 for the repair job on Modern TV and Radio's invoice No. 2656, and take both the PAM 78 and the ruined PCB with you.

Write to Harry Craven, tell him what has happened, and ask for a refund of $ . . . (you decide how much). You may assume you attach copies of the two invoices to your letter.

Note: The PAM 78 is made by VICOM in Korea. It contains an AM/FM stereo radio receiver, a record turntable, a cassette tapedeck, and two eight-inch speakers.

PROJECT NO. 2: TRACING A LOST SUITCASE

For the past three months you have been assessing the environmental impact of a proposed gas pipeline to be built some 450 miles (725 km) from your home office. You were based at Lake Winthrop, where you stayed at the Clock Inn Motel. Whenever you went on a field trip of more than two days' duration, you vacated your room and left unwanted baggage with the hotel receptionist.

The final field trip occurred on the last three days of the project, after which you flew home. Unfortunately, you forgot to pick up one of your cases at the motel. So you telephoned Marrianne Dubray, the receptionist at the Clock Inn, and told her what happened. You described your case to her, and she promised to ship it to you via Remick Airlines.

Today—one week later—your case arrives, but you realize that it is not your case. The size and color are right, but it has another person's initials engraved on it: KRZ.

So you telephone the motel and ask to speak to Marrianne. No luck: she is not there. You ask when she will be in.

"She won't," says a strange voice. "She quit."

You try describing the problem, but the voice is not very helpful. She can't even put you through to the manager, because he is away for three days. You ask her to check if there are any other bags or cases behind the desk or in the office, and she does so.

"There aren't any here," she says. "Why don't you write to the manager and tell him what happened? Maybe he will remember your case if you describe it for him."

Write to the manager of the Clock Inn, Lake Winthrop. Describe a case of your own.

PROJECT NO. 3: ORDERING AN AMMONIA PUMP

Your branch of H. L. Winman and Associates has a rather old blueprint machine known as a Diazo-Copy. It was made by the Vancourt Manufacturing Company, but the particular model is no longer in production and parts cannot be obtained locally.

Exactly two weeks ago the blueprint machine broke down, and you diagnosed the fault as an unrepairable ammonia pump. You immediately sent this telegram to the Vancourt Manufacturing Company, 430 Almeda Avenue, Dallas, Texas:

URGENTLY REQUIRE AMMONIA PUMP FOR DIAZO-COPY BLUEPRINTER. PLEASE SHIP AIR EXPRESS.

Exactly one week ago you received the pump but discovered that the Vancourt Manufacturing Company had sent you an air extraction pump instead of an ammonia pump. So you dug up an old, tattered operating manual for the Diazo-Copy and identified the ammonia pump's part number as 3701. Then you sent a second telegram to Vancourt:

AIR PUMP RECEIVED. STILL REQUIRE AMMONIA PUMP PART No. 3707. URGENT—SHIP AIR EXPRESS.

This morning you have received a heavy parcel on which the Vancourt

Manufacturing Company has prepaid $28.00 in air express charges. On opening it you notice that it is much larger and more complex than the pump you removed from the Diazo-Copy. With dismay you examine your second telegram and notice that you accidentally quoted part number 3707 instead of part number 3701. Unfortunately, the number you quoted also applied to an ammonia pump, but for a machine much newer and more complex than yours.

So you sit down and write a third telegram:

PLEASE SHIP BY AIR EXPRESS AMMONIA PUMP PART NO. 3701 FOR DIAZO-COPY BLUEPRINTER. LETTER OF EXPLANATION RE ORDERS FOR PREVIOUS PUMPS IN MAIL.

Now write the letter to Vancourt Manufacturing Company. Use real dates if you refer to the telegrams and pump deliveries. Offer to pay for air express charges for the second pump, and either ask what you are to do with the first two pumps, or say that you are returning them (at your own expense?).

PROJECT NO. 4: REQUEST FOR FREE PARTS

You are engaged in a lake level measurement program for the state of Minnesota. At a critical moment in the program your Hektik Model 370 Water Stage Manometer breaks down. You take it apart and identify that it needs a replacement spring and drive assembly. This is the third time that the fault has occurred in the last six months, and each time the thread on the drive shaft has stripped. You previously purchased spare spring and drive assemblies from the manufacturer (Hektik Company, 315 Pennsylvania Boulevard, Pennstown, Nebraska) on May 13 and August 22, at a cost of $178.15 each.

Today you send a telegram ordering a replacement spring and drive assembly against Purchase Order W7714.

Write to the Hektik Company to complain about the repeated failures (you may attribute the cause to any condition you wish, if you feel you need to point out the cause), and to request that this replacement part be supplied free of charge.

PROJECT NO. 5: CORRECTING A SHIPPING ERROR

You are the laboratory technician who discovered the loss of 3 millicuries (mCi) of radioactive iodine (131_I) in Project No. 3 of Chapter 4. Write to the manufacturer of the 131_I and ask for an adjustment of the missing 3 mCi (ask for either a credit note or an invoice for only 7 mCi). The manufacturer is Weatherdon Labs, 430 Freeling Street, West Grantham, New Jersey, 08617. Address your letter to Mr. Harold Klyman, Sales Manager, and prepare it for your lab supervisor's signature (she is Frances K. Eldon).

PROJECT NO. 6: ORDERING VIDEOTAPES FOR PREVIEW

Part 1. You need to find basic training information on computers for one of H. L. Winman and Associates' clients. During your research, you discover that Computronics Inc. of San Diego probably has just what you require. So this morning you telephone their training supervisor (Carl V. Gonzalez), who tells you:

1. Yes, they have just produced three videotapes, all on ¾-inch (15.8 mm) videocassettes.
2. Each videotape runs 20 minutes and is in color.
3. The three-part series is titled "Elements of Computer Control Systems."
4. The videotapes can be purchased or rented, but he doesn't have prices yet.
5. Computronics will let you preview the set.
6. The videotapes carry the identification numbers 572A, B, and C.
7. There is a preview charge of $25.00 per videotape (refundable if the tape is purchased).
8. He cannot send the videotapes on the basis of a telephone call. He needs a letter from H. L. Winman and Associates plus a check or purchase order as a deposit.

You obtain your project coordinator's approval to spend $75.00 to preview the three videotapes.

Write a letter to Carl V. Gonzales. In it:

Ask the purchase price of each videotape, both individually and as a set of three.
Refer to this morning's telephone conversation.
Refer to an attached check for $75.00 (check number 21606).
Ask to preview the three videotapes.
Ask him to ship the preview set very quickly.
Ask how long you may keep the videotapes.

You may introduce any other details you consider necessary. The address of Computronics Inc is 2148 El Monte Boulevard, San Diego, California, 92104. Prepare the letter for your Coordinator's signature (but you are to sign for him/her).

Part 2. One of the three videotapes you borrowed from Computronics Inc was destroyed when H. L. Winman and Associates' videotape recorder (VTR) malfunctioned. Write to Carl V. Gonzalez, training supervisor at Computronics Inc, to tell him what has happened. Ask him to quote you a price so that your company can pay for the damaged videotape. Use these details:

1. The damaged videotape is number 572B, "Elements of Computer Control Systems—Part 2."
2. Its purchase price is $245.00.
3. You paid a preview deposit of $25.00 on it.
4. You will be returning the other two preview videotapes.
5. You may mention that H. L. Winman and Associates will be buying the complete set of three videotapes (total cost $650.00).
6. Because you will be buying the set, you are hoping that the replacement price for the damaged videotape will be waived, or at least considerably less than the full purchase price.

If you would like to know how the videotape was damaged, read Project No. 5 of Chapter 4.

PROJECT NO. 7: REQUESTING A LONG LEAVE

You are a technician employed by H. L. Winman and Associates. Write a memorandum to your branch manager, Vern Rogers, asking for two years' leave without pay. Details are:

You want to take a computer analyst/programmer course at the local community college.

You would like to continue working part-time (Saturdays, and three months in the summer) to keep your head above water and your stomach full.

You know that in the past the company has not entertained such requests. It prefers to release employees completely, then rehire them as new employees if they choose to return after the course.

You would like the company to pay at least part of the tuition costs (assume that you plan to come back to the company for at least two years at the end of the course). There is no precedent for this; the company normally limits payment of tuition fees to night school and correspondence courses that are relevant to the person's normal duties.

The company uses a computer for many engineering applications.

You are a graduate engineering technician, and you have worked for H. L. Winman and Associates for 4½ years. You want leave rather than a complete release so that you will not lose seniority, your group insurance plan (if necessary, you'll pay the whole premium yourself while on leave), and the pension benefits that you have accrued.

You think the computer analyst/programmer course will be of value because it will help you analyze engineering problems and transcribe them onto the computer. You believe the company will also find your dual capabilities of value.

You did well (B average) in the computer segment of your engineering technician course.

PROJECT NO. 8: REQUEST TO ATTEND A COURSE

Assume that today is Monday (the third day of the current month), and that you are employed in the Engineering Department of H. L. Winman and Associates. For the past four weeks you have been on a field assignment to Duluth, where you have been conducting an engineering study for the RamSort Corporation. You have been assisted by two competent junior technicians (Ray Hicks and Len Dominie), and you are now three days ahead of schedule. The task is to be completed by the 28th of this month.

Today you have received a folder from the University of Minnesota in Minneapolis advertising a one-week course. Details are:

Course title:	Management for Engineering and the Sciences
Course dates:	Monday the 17th to Friday the 21st inclusive (of this month)
Type of course:	Maximum immersion: 8 A.M. to 5 P.M. daily plus 7 to 10 P.M. Wednesday evening; approximately 20 hours of home assignments
Cost:	$400; includes materials, books, and lunches, with a guest speaker from industry at each lunch

Registration:	No later than noon Wednesday the 12th; telephone registrations accepted
No. of participants:	Limited to 16

You are impressed by the technical standard of the course described in the folder and wish to attend. (Because of previous field assignments you missed a similar Extension Department evening course offered at your local university last winter. Your company sponsored four engineers to attend that course, for which the fee was $125 each.)

Write an interoffice memorandum to your project coordinator:

1. Tell him about the course; convince him it is a good one.
2. Ask if you can attend.
3. Ask if the company will pay the tuition fee, plus travel and lodging costs.
4. Convince him that you can be spared from the RamSort task for one week.
5. Tell him to reply in a hurry (because time is short).

Assume that your project coordinator can give technical approval for you to attend but must go to the department head for financial approval. Also assume that the project coordinator is a very busy person who is notorious for placing items in an *Action* tray, where they are quickly covered by other important items and often overlooked.

PROJECT NO. 9: LETTER OF THANKS

As a member of the Environmental Studies Department of H. L. Winman and Associates, you are assisting in a three-month technical/environmental project in the Delta Marsh at the southern tip of Lake Manitoba. At the end of your study you are invited by Delta Marsh Project Engineer Rudi Kane to spend a long weekend with him and his family at their summer cottage at Clear Lake in Riding Mountain National Park.

With Rudi, his wife Jean, and son and daughter Leon and Tracy, you swim, water-ski, golf, ride horseback along wild trails, and play tennis during the day, and then barbecue steaks beside the shore in the soft evening twilight—an idyllic existence.

On Monday evening you are to catch the 6:30 bus back to Delta Marsh in preparation for Tuesday's flight home. You have been water-skiing on the north side of the lake but have returned ahead of the Kane family to put your things together at the cottage (you have driven back with a neighbor).

You wait an hour and a half for their return, unaware that they have had car trouble. You want to say good-bye and thank them for having you for the weekend, but by 6:15 you can wait no longer and so trudge, with your suitcase, down to the bus depot.

Now you are back at the Delta Marsh, ready to leave for home tomorrow morning. Write to the Kanes to thank them for their hospitality. Address your letter to: c/o Wasagaming Post Office, Clear Lake, MB, ROJ 2HO.

PROJECT NO. 10: APPLICATION FOR EMPLOYMENT

This project assumes that you are seeking permanent employment at the end of a technical training program. You are to reply to any one of the following advertisements, using your present situation and actual background. If it is still early in your training program, you may update the time and assume that it is now two months before graduation date.

ROPER CORPORATION (OHIO DIVISION)
requires a
CHEMICAL TECHNICIAN

to join a project group conducting research and development into the organic polymers associated with the coating industry. The successful applicant will also assist in the development of control techniques for producing automated color tinting. Apply in writing, stating salary expected to:

Mr. P. Rassmusen
Personnel Manager
P. O. Box 1728
Montrose, Ohio 45287

(Advertisement in Montrose Herald, April 17)

H. L. WINMAN AND ASSOCIATES
475 Reston Avenue
Cleveland, Ohio 44104

CIVIL ENGINEERING TECHNICIANS

This established firm of consulting engineers needs three Engineering Technicians to assist in the construction supervision of several grade separation structures to be built over a two-year period. Graduation from a recognized Civil Engineering Technology course is a prerequisite. Successful applicants will be hired as term employees. Opportunities are excellent for transfer to permanent employment before the project ends.

Apply in writing to:

Mr. A. Rittman, Head
Special Projects

(Advertisement on College Notice Board)

We require a
CIVIL ENGINEERING TECHNICIAN

with an interest in building construction to prepare estimates, do quantity takeoffs, and assume responsibility for company sales and promotion activities.

Apply in writing to:

Personnel Department
PRECASCON CONCRETE COMPANY
227 Dryden Avenue

(Advertisement in your local newspaper, February 26)

ARCHITECTURAL DRAFTSPERSON
required by
a well-established firm of architects and planners

Applicants should have completed a Design and Drafting course at a community college or similar training institution, in which strength of materials and structural design were a curriculum requirement. Experience in preparation of reports and proposals will be considered in selection of successful applicant.

Write to:

The Corydon Agency
Room 604 – 300 Main St.

(Advertisement of March 28 on College Notice Board)

NORTH AMERICAN WILDLIFE INSTITUTE
requires
SCIENTISTS AND BIOLOGISTS
for its wildlife preservation projects in Florida,
New Mexico, Manitoba, Quebec, and Alaska.

Applicants should have a degree in the environmental sciences or an appropriate diploma from a junior college or technical institute. Persons anticipating graduation are particularly encouraged to apply.

Write to:

Dr. Frederick M. Hauser
Chairman, North American Wildlife Institute
Room 1620 – 385 East 47th Street
New York, N.Y. 10017

(Advertisement in last month's issue of American Wildlife*)*

ROBERTSON ENGINEERING COMPANY
invites applications from
ENGINEERS, TECHNOLOGISTS, AND TECHNICIANS
interested in Metrology

We operate a first-class Standards Laboratory that is a calibration center for precision electrical, electronic, and mechanical measuring instruments used in our military and commercial equipment maintenance programs.

Applicants for senior positions should hold a B.S. in Electrical Engineering. Laboratory Technicians should be graduates of a recognized electronics or mechanical technology course who have specialized in precision measurement.

Applications are invited from forthcoming graduates, who will work under the direction of our Standards Engineer. All applicants must be able to start work on July 1.

Apply in writing to:

Mr. F. Stokes
Chief Engineer
Robertson Engineering Company
600 Deepdale Drive
Toronto, Ontario M5W 4R9

(Advertisement in Toronto Globe and Mail, March 18)

PRODUCTION TECHNICIAN/DRAFTSPERSON

Duties:

Under the general direction of the Chief Engineer, to be responsible for detailed product design of prototype machines. Also will investigate and recommend improvements to existing product lines.

Qualifications:

Graduate of a recognized Mechanical Engineering Technician or Drafting course with emphasis on Mechanical Drawing. Experience with or knowledge of agricultural equipment would be an asset.

Please apply to:

Chief Engineer
Agricultural Manufacturing Industries
3720 Harvard Avenue

(Advertisement in your local newspaper, February 26)

MONTROSE PAPER COMPANY

Offers excellent opportunities for recent graduates to join an expanding manufacturing organization in the pulp and paper industry.

ENGINEERS AND ENGINEERING ASSISTANTS

Positions are available for mechanical engineers and technicians to assist in the design, installation, and testing of prototype production equipment. Previous experience in a manufacturing plant would be helpful. Innovative ability will be a decided asset.

ELECTRICAL TECHNICIAN

This person will assist the Plant Engineer in the maintenance of power distribution systems. Applicants should be graduates of a two-year course in Electrical Technology with good knowledge of automatic controls and machine application. Ability to read blueprints and working drawings is essential.

COMPUTER TECHNICIAN

This position will suit either a graduate of a Computer Technology course or an Electronics Technician who has specialized in Computer Electronics.

Duties will consist of installation, maintenance, and troubleshooting of existing and future computer equipment.

ENVIRONMENT SPECIALISTS

Persons selected will test air pollutants and water effluents from our paper mill and production plant, and assess their environmental impact. Applicants should be graduates of a recognized course in the environmental or biological sciences.

Salaries for the above positions will be commensurate with experience and qualifications. Excellent fringe benefit program available. Write in confidence to:

Manager of Industrial Relations
MONTROSE PAPER COMPANY
Montrose, Ohio 45287

(Advertisement in Montrose Herald, March 10)

TECHNICIAN

required by manufacturer of
automated car wash equipment

Duties: Assist plant engineer design and test mechanical and electrical components; supervise installations in cities across U.S. and Canada; troubleshoot problems at existing installations.

Applicants must be free to travel extensively. Excellent salary and promotion possibilities.

Apply in writing to:

Personnel Manager
DIAL-A-WASH Incorporated
2020 Waskeka Drive
Montrose, Ohio 45287

(Advertisement in Montrose Herald, April 23)

PROJECT NO. 11: APPLICATION FOR UNSOLICITED EMPLOYMENT

This project assumes that you are seeking permanent employment at the end of a technical training program but few job openings have been advertised in your field. Write to Robertson Engineering Company in Toronto, or to H. L. Winman Associates (for the attention of one of the department heads in Cleveland, or local branch manager Vern Rogers), applying for employment. Use your knowledge of the company and your real background. If it is still early in your training program, you may update the time and assume that it is now two months before graduation.

PROJECT NO. 12: APPLICATION FOR SUMMER EMPLOYMENT

Assume that you are looking for summer employment. Reply to any one of the following advertisements:

CITY OF MONTROSE, OHIO
TRAFFIC DIVISION

The city of Montrose is carrying out an Origination-Destination (OD) Study throughout June and July, with computation and analysis to be carried out in August.

Approximately 40 students will be required for the OD Study, and 20 for the computation and analysis. For the latter work, some knowledge of statistical analysis will be an advantage.

Students interested in this work are to apply in writing to Mr. G. Braithwaite, Room 310, Civic Center, City of Montrose, Ohio 45287.

(Notice on College Notice Board)

NORTHERN POWER COMPANY
BOX 1760, FAIRBANKS, ALASKA

Applications are invited from students in two-, three- or four-year technical programs who are seeking three months of well-paid summer employment. The work will include construction of living quarters and associated facilities for mining camps near Prudhoe Bay, Alaska. Terms of employment will be:

13 full weeks—June 10 to September 10
Excellent pay: $325 per week
Transportation: Paid
Accommodation: Paid
Health: Applicants should be in topnotch physical condition.

Students interested in this work are to write to Mr. G. R. Cardinal at the above address. Although previous experience in construction work is not essential, knowledge of general construction methods will be considered an asset for some positions.

(From Construction—North, *February 20)*

REMICK AIRLINES
offers
SUMMER EMPLOYMENT
to a limited number of U.S. and Canadian students.

Duties: To work as part-time dispatchers, baggage handlers, aircraft cleaners, cafeteria assistants, etc, during the peak summer period (June 15–September 10) at Remick Airlines terminals at:

Fort Wiwchar, Michigan
Weekaskasing Falls, Wisconsin
Lake Lawlong, Alberta

Good pay; ample free time; free transportation to destination and return.

Write, describing previous work experience (if any), to:

Carl Nickerson	or:	Mavis Chandler
Flight Operations Manager		Remick Airlines
Rocm 301		Winnipeg International Airport
Montrose Municipal Airport		Winnipeg, MB R2Y 3X7
Montrose, OH 45286		

(Notice on College Notice Board)

PROJECT NO. 13: UNSOLICITED SUMMER EMPLOYMENT APPLICATION

Assume that you are looking for summer employment but few summer jobs have been advertised. Write to H. L. Winman and Associates or Robertson Engineering Company, asking for a summer job. Use your knowledge of the companies, plus your actual background, to write an interesting letter.

Address your letter to the attention of George Dunn in Toronto or Tanys Young in Cleveland. If there is an H. L. Winman and Associates branch in your area, you may address your letter to Branch Manager Vern Rogers.

4

INFORMAL REPORTS DESCRIBING FACTS AND EVENTS

When you hear that someone has just finished writing a technical report, you may imagine a nicely bound formal document, tastefully typed and printed. In some cases you would be correct, but most of the time you would be wrong. Far fewer formal reports are issued than informal reports, which reach their readers as letters and memorandums.

THE MEMORANDUM REPORT

The memorandum report is the least formal of all technical reports. Normally an interoffice or interdepartmental communication, its length and tone can vary considerably. It can be very direct and to the point, it can develop its topic in great detail to present a convincing case, or it can lie anywhere in between.

Anna King is alone in the H. L. Winman office, working late one Tuesday evening, when the telephone rings. The caller is Bob Walton, a member of the electrical engineering staff who is on a field trip to Tangwell with a second staff member. He tells Anna they have had an automobile accident near Hadashville, and some of their equipment is damaged. He wants Jim Perchanski, his department head, to send out replacement items by air express.

Anna jots down notes while Bob talks. As she plans to be out of town the following day, she types out a report of the conversation and leaves it on Jim Perchanski's desk. She prepares it as a memorandum (Figure 4-1) which tells Jim Perchanski what has happened to two members of his staff, where they are now, how soon they will be able to move on, and that one of them is injured. it also tells him that equipment is damaged and replacements are needed.

Because she will not be available to answer questions the following morning, Anna takes care to describe the situation clearly (the paragraph numbers below are keyed to the memorandum):

H L Winman and Associates

INTER - OFFICE MEMORANDUM

From: Anna King   Date: November 18, 19__

To: Jim Perchanski   Subject: Accident Report and Request for Spare Parts   (1)

Bob Walton and Pete Crandell have been involved in a highway accident which will delay their inspection of the Sledgers Control project at Tangwell. They need replacement parts shipped to them tomorrow (Wednesday November 19).   (2)

Bob telephoned from Hadashville at 19:35 to report the accident, which occurred at 17:15 some two miles (3 km) north of Hadashville. Pete has been hospitalized with a fractured left knee and a suspected concussion. Bob was unhurt. The panel truck and some of their equipment were damaged.   (3)

Bob wants you to ship the following items to him by air express on the Wednesday evening flight to Montrose and to mark the shipment HOLD FOR PICKUP BY R WALTON NOV 20:   (4)

1	Spectrum Analyzer,	HK7741
1	Calibrator,	Vancourt Model 23R
24	Glass phials,	30 cm long x 5 cm dia

He has arranged to rent a van and will drive to Montrose to pick up the items Thursday morning. He will then drive on to Tangwell and will arrive there about 16:00 hr. He has informed Tangwell of the delay.   (5)

While at Hadashville Bob is staying at Hunter's Motel (tel 453 6671), where he is preparing an accident report for you.   (6)

Anna

Figure 4-1. A memorandum report.

(1) Anna knows that a subject line must be informative; it must tell what the memorandum is about and stress its importance to the reader. If she had simply written "Transcript of Telephone Call from R. Walton," she would not have captured Jim Perchanski's attention nearly as sharply.

(2) This brief summary gets right to the point by immediately telling Jim Perchanski in general terms what he most needs to know:

 Why the memo was written.
 What happened.
 What action has been taken.
 What action he has to take.

(3) In this paragraph Anna tells what she knows about the accident and its effects. It serves as background to the important facts that follow.

(4) Anna knows that Jim Perchanski must act quickly to ship the replacement items, so she uses a list to help him identify them. The indented list also is an attention-getter: if Anna had described the items in a paragraph like the following, they would not have been nearly as noticeable:

 He will need replacements for an HK7741 Spectrum Analyzer, a Vancourt 23R Calibrator, and 24 glass phials, each 30 cm long × 5 cm dia. He wants you to ship these items air express to the Remick Airlines terminal at Montrose, and to mark them . . . .

(5) It is important that instructions and movement details be explicit, otherwise the equipment and Bob Walton may not meet at Montrose. Anna has taken care throughout to identify specific days, and once even the date, to make sure that no misunderstandings occur. To state "tomorrow" or "the day after tomorrow" would be simple but might cause Jim Perchanski to assume a wrong date, since he will be reading the memorandum one day later than it was written.

(6) In this brief closing paragraph Anna indicates what further action is being taken and where Bob Walton can be contacted if more information is needed or if a message needs to be relayed to him.

When Jim Perchanski walks in on Wednesday morning, he will know immediately what has happened and what action he has to take. He does not need to ask questions because he has been placed fully in the picture.

THE LETTER REPORT

The letter report, although still basically informal, can vary in formality according to its purpose, the type of reader, and the subject being discussed. Some letter reports may be as informal as a memorandum report, particularly if they are conveying information between organizations whose members know each other well or have corresponded frequently. Others may be almost severely formal, although such reports are increasingly rare. Most often they are simple, dignified business letters that convey technical information from one company to another. As such, they serve their purpose by avoiding the rigidity of style and format demanded by the full formal report described in Chapter 6.

The letter in Figure 4-9 is a typical informal letter report written to a client by Bob Gray, an engineering assistant in the Mechanical Engineering Department of H. L. Winman and Associates. In the first paragraph he summarizes his findings to give the reader a quick understanding of the results of his investigation. Then in subsequent paragraphs he describes what he has found out, what methods can be used to alleviate the problem, and which methods he recommends. He has found the key to writing an effective letter report because he has organized his material into a logical, coherent order that will be readily understood.

Although there are many types of informal reports, most appear in the memorandum or letter form illustrated here. There are variations in format, content, and writing approach, but all follow the basic pattern shown in Figure 4-2: (1) a short introduction to the topic or problem; (2) a discussion of the data, situation, or problem and what has been done or could be done about it; and (3) a conclusion which sums up results and possibly recommends what should be done next.

	SUMMARY	A brief statement of the report's main features (often written last, but always placed *first*).
1.	**INTRODUCTION**	**Background:** information that "sets the scene."
2.	**DISCUSSION**	**Facts:** data, details; what has been and is being done.
3.	**CONCLUSIONS**	**Outcome:** results and effects; it may also suggest what needs to be done.

Figure 4-2. Basic report compartments.

These three compartments are preceded by a summary statement, which tells readers in as few words as possible what they *most* need to know about the topic (for more about summaries and summary statements, see Chapter 10). This pattern is evident in almost every report, whether it is a simple occurrence report, a field trip, inspection, or progress report, an investigation report similar to that in Figure 4-9, a feasibility study, or a technical proposal.

WRITING STYLE

The reports described in this chapter deal with specific information. Their purpose is *mainly to inform,* so they are written in a direct, informative style that is crisp and to the point. Their writers are usually describing events that have already occurred, so they write mostly in the past tense, which helps them to be consistent. They shift gear into the present or future tense only when they have to describe something which is presently occurring, outline what will happen in the future, or suggest what needs to be done. All three tenses occur in the following report from Engineering Technician Dan Skinner to Coordinator Don Gibbon:

SUMMARY The indoor/outdoor carpet we installed in station DMON-TV has corrected the noise problem but is "pilling" badly. I plan to examine the carpet with the manufacturer's representative, to find the cause and suggest a remedy.

BACK-GROUND *Past Tense*

The carpet was installed in the satellite studio control room during the night of January 8-9. Sound level readings taken at 12 different locations on January 10 showed that the ambient noise level had decreased an average 3.6 dB. Operating staff also noticed a marked decrease in clatter caused by impact noises.

FACTS *Mainly Past Tense* / *Present Tense*

At the station manager's request, I returned to the control room today and checked the carpet's condition. After only two weeks of use it has tight little balls of carpet material adhering to its surface. I called the manufacturer's representative, who said that the condition is not unusual and does not mean that the carpet is wearing quickly. He suggested that it may be caused by improper carpet-cleaning techniques and probably can be easily corrected. However, it is unsightly and our client is not pleased with the carpet's appearance.

OUTCOME *Future Tense*

The manufacturer's representative and I will return to the control room between midnight and 2:00 A.M. on January 31 to study the carpet-cleaning techniques used by maintenance staff. I will telephone our findings to you later in the day.

Dan has written the Background and most of the Facts paragraph mainly in the past tense because they deal with what has already been done. At the end of the Facts he has shifted into the present tense to report how the station manager feels *now.* Then for the Outcome he has jumped into the future tense to outline what he *plans* to do. This past-present-future arrangement is natural and logical; reader Don Gibbon will feel comfortable making the transitions from one tense to the next. Dan's Summary even follows the same pattern.

Occurrence Report

An occurrence report simply describes an event or incident that has occurred. Anna King's memo to Jim Perchanski in Figure 4-1 is an occurrence report. So is Bob Walton's report of the accident (Figure 4-3), which he is writing at Hunter's Motel in Hadashville. Although its topic is essentially the same as Anna's, its focus and emphasis are different.

Bob uses the three basic compartments in Figure 4-2 to organize his report (the letters below are keyed to the report in Figure 4-3):

(A) This is his summary: it takes a main piece of information from each compartment that follows.

(B) This is compartment 1, **Background.** By clearly describing the situation (*who? where? why? when?*) Bob is helping Jim more easily understand what happened. Notice how he:

Establishes where they were, how they happened to be there, their direction of travel, and who else was involved.

Itemizes vehicles, license numbers, and drivers' names in an easy-to-read list.

Mentions that he is enclosing a sketch (so Jim can look at it *before* he reads on).

H L Winman and Associates

INTER - OFFICE MEMORANDUM

From: Bob Walton
Date: November 19, 19__

To: Jim Perchanski
Subject: Report of Auto Accident, Hadashville, Ohio

Pete Crandell and I were involved in a multiple-vehicle accident on November 18, which resulted in injuries to Pete, damage to our panel truck and some equipment, and a two-day delay in our inspection of the Sledgers Control project. (A)

The accident occurred at 5:15 p.m. on highway 44, about two miles northeast of Hadashville. We were traveling north in company panel truck T2711, on our way to site RJ-17 at Tangwell. Pete was driving and we were approaching the intersection with highway 201.

Other vehicles involved in the accident were: (B)

Plymouth Fury, Lic AHJ735, driven by D.Varlick.
Ford truck, Lic T4851, driven by F.Zabetts.
Pontiac Catalina, Lic MD3720, driven by K.Schmitt.

Positions of these vehicles and our panel truck immediately before the accident are shown on the attached sketch.

As the Plymouth attempted a right turn into highway 201 it skidded into the Ford truck, which was standing at the intersection waiting to enter the highway. The impact caused the Plymouth's rear end to swing into our lane, where Pete could not avoid colliding with it. This in turn caused our panel truck to slide broadside into the southbound lane, where the Pontiac approaching from the opposite direction collided with its left side. (C)

Pete was taken to Hadashville hospital with a broken left knee and a suspected concussion; he is likely to be there for several days. The panel truck was extensively damaged and was towed to Art's Autobody, 1330 Kirby St., Hadashville; I have arranged to rent a replacement van from Budget. Some of our equipment was damaged or shaken out of calibration, so I telephoned the office on Tuesday evening and requested replacements (Anna King took the message and has a list for you). (D)

I also telephoned the duty engineer at site RJ-17 and told him that my inspection of the Sledgers Control project will start on Friday November 21, two days later than planned.

Bob

Figure 4-3. An occurrence report.

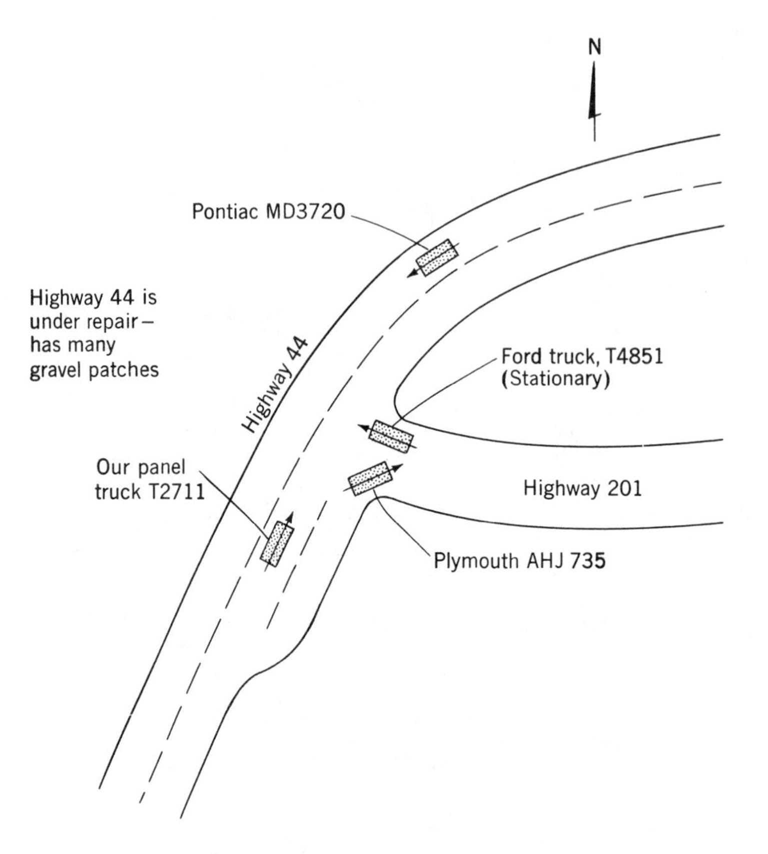

(C) Because his background information is complete, Bob's compartment 2, **Facts,** can be concise and to the point. He simply provides a chronological description of what happened from the time the Plymouth started to slide until all vehicles stopped moving.

(D) In compartment 3, **Outcome,** Bob describes the *results* of the accident (injuries, damage) and what he has done since (rented a van, requested replacement equipment). He closes on a strong point: What is being done about the project, which was his reason for passing through Hadashville.

Bob knows his role: to be an informative but objective (unbiased) reporter. No doubt he has an opinion of who is at fault, but to state it probably would have injected subjectivity into his report.

Field Trip Report

After returning from a field assignment, an engineer or technologist is expected to write a field trip report describing what has been done. He or she may have been absent only a few hours, inspecting cracks in a local water reservoir; or may have spent several days installing and testing a prototype pump at a power station in a nearby community; or may have been away for two months, overhauling communications equipment at a remote defense site. Regardless of the length and complexity of the assignment, many details will have to be remembered and transcribed into a logical, coherent, and factual report. On such assignments it is helpful to carry a pocket notebook in which to jot down daily occurrences. Without such a report to rely on, the engineer or technologist may write a disorganized report that omits many details and emphasizes the wrong parts of the project.

The simplest way to write a trip report is to answer three basic questions:

1. **Who went where?**
   Why and when did you go?
2. **What did you do?**
   What did you set out to do and were you able to do all of it?
   What problems did you run into? What additional work did you do?
3. **What remains to be done?**
   What work could you not complete, and who should now complete it? When and how should it be done?

The answers to these questions provide the information for the three compartments in Figure 4-2:

1. INTRODUCTION: Who went where? (*Background*)
2. DISCUSSION: What did you do? (*Facts*)
3. CONCLUSIONS: What remains to be done? (*Outcome*)

Short trip reports do not need headings. A brief narrative following the three-question pattern normally can carry the story:

Compartment 1 Dave Makepiece and I visited site RJ-17 from January 15 to 17 to install a prototype automatic alarm system for the Roper Corporation.

Compartment 2 The installation was completed without difficulty and no major problems were encountered. Installation instruction W27 was followed throughout with the exception of step 33, which called for connections to be made to the remote control panel. This step was omitted since the remote panel has been permanently disconnected.

Compartment 3 The prototype alarm will be field-evaluated for one month. It will be removed in mid-March by M. Tutanne, who will be visiting the site to discuss summer survey plans.

Long trip reports require headings to help their readers identify the compartments. Typical headings might be:

1. Assignment Details (Background).
2. Work Accomplished and Problems Encountered (Facts; best treated as two separate headings).
3. Suggested Follow-up or Follow-up Action Required (Outcome).

The whole report should be preceded by a short Summary that states very generally what was and was not achieved.

Anna King's memo at Figure 4-4 tells H. L. Winman and Associates'

H L Winman and Associates

INTER - OFFICE MEMORANDUM

From: Anna King

Date: February 20, 19__

To: Systems, Project, and Field Engineers

Subject: Guidelines for Writing Field Trip Reports

A standard format is to be used for all field trip reports. These notes suggest how you can organize your information under five main headings: Summary, Assignment Details, Work Accomplished, Problems, and Follow-up Action. The headings may be omitted from very short reports.

Summary

Make your summary a short opening statement that says what was and was not accomplished, and highlights any significant outcome.

Assignment Details

Under this main heading state the purpose of the trip and include any other information that the reader may want to know. If the information is too cumbersome to write as a single paragraph, use subheadings such as:

Purpose of Trip
Background
Project No./Authority
Personnel Involved
Date(s) of Trip

Work Accomplished

In this section describe all work that was done. Normally present it in chronological order unless more than one project is involved, in which case describe each project separately. Keep it short: don't describe at great length routine work that ran smoothly. Whenever possible, refer to your work instruction or specifications, and attach them to your report:

> The manual control was disconnected as described in steps 6 to 13 of modification instruction MI1403.

Go into more detail only if difficulty is encountered, or if work is necessary beyond that anticipated by the job specification:

> At the request of the site maintenance staff, the manual control was installed in the power house as a temporary replacement for a defective GG20 control. Parts removed from the panel, together with instructions for returning the panel to its original configuration, were left with Frank Mason, the senior power house engineer.

1

Figure 4-4. How to write a long trip report.

If parts of the assignment could not be completed, identify them and explain why the work was not done:

> Test No. 46 was omitted because the RamSort equipment was being overhauled.

Problems

In contrast to the conciseness recommended for the previous section, you should describe problems in detail. Knowledge of problems you encountered and how you overcame them can be invaluable to the engineering department, which may be able to prevent similar problems from occurring elsewhere.

A statement that does not tell the reader what the problem was or how it was overcome is virtually useless:

> Considerable time was spent in trying to mount the miniature control panel. Only by fabricating extra parts were we able to complete step 17.

If the information is to be used by the engineering department, it must be much more specific:

> Considerable time was spent in trying to mount the miniature control panel according to the instructions in step 17. We found that the main frame had additional equipment mounted on it, which prevented us from using most of the parts supplied. To overcome this problem we fabricated a small sheetmetal extension to the main frame and mounted it, with the miniature control panel, as shown in the attached drawing.

Follow-up Action

This section of the trip report ties up any loose ends by telling readers what still remains to be done. If any work has not been completed, draw attention to it here even though you may already have mentioned it under "Work Accomplished." Identify what needs to be done, if possible indicate how and when it should be done, and say whose responsibility it now becomes:

> The manual control mounted as a temporary replacement in the power house is to be removed when a new GG20 control is received. This will be done by F. Mason, who will use special instructions and spare parts left with him for this purpose.

In some cases you may direct follow-up action to someone else in your department:

> The manual control will be removed from the power house by R. Walton, who will be returning to the site on October 17.

If your report is very long, I suggest you insert frequent subheadings and use a paragraph numbering system to increase its readability.

Anna King

2

engineers how to organize their long trip eports, describes the type of information that normally would follow each heading, and includes excerpts and sample paragraphs.

With the exception of the Outcome section, trip reports should be written entirely in the past tense.

Occasional Progress Report

Progress reports keep management aware of what its project groups are doing. Even for a short-term project, management wants to hear how the project is progressing, especially if problems are affecting its schedule. Because delays can have a marked effect on costs, management needs to know about them early.

Jack Binscarth, one of Robertson Engineering Company's senior laboratory technicians, has been assigned to Cantor Petroleums in Edmonton to analyze oil samples. The job is expected to take five weeks, but problems develop that prevent Jack from completing the work on time. To let his chief know what is happening, he writes the brief progress report in Figure 4-5.

Occasional progress reports like this are similar to occurrence reports. They have an event or events to relate, and they do this best by adapting the standard Background-Facts-Outcome arrangement into a past-present-future pattern:

1. *Summary* A brief description of the overall situation.
2. *Progress* The work that has been done, the problems that have been encountered, and the effect these problems have had on progress.
3. *Situation Now* What is being done at present.
4. *Future Plans* What will be done to complete the project, and when it will be done.

Periodic Progress Report

If a project is to continue for several months, management normally will specify that progress reports be submitted at regular intervals.

A periodic progress report may be no more than a one-paragraph statement describing progress of a simple design task, or it may be a multipage document covering many facets of a large construction project. (There are also form-type progress reports, which call for simple entries of quantities consumed, yards of concrete poured, and so on, with cryptic comments.) Regardless of its size, the report should answer four main questions that the reader is likely to ask:

What are your plans/expectations?

What progress have you made?

Have you had any problems?

Will your project be completed on schedule?

robertson
engineering
company

MEMORANDUM

From: J Binscarth [at Cantor Petroleums]
Date: 14 October 19__

To: F Stokes, Chief Engineer
Head Office
Subject: Delay in Analysis of Oil Samples

My analysis of oil samples for Cantor Petroleums has been delayed by problems at the refinery. I now expect to complete the project on 25 October, nine days later than planned.

The first problem occurred on 23 September, when a strike of refinery personnel set the project back four working days. I had hoped to recover all of this lost time by working a partial overtime schedule, but failure of the refinery's spectrophotometer on 13 October again stopped my work. To date, I have analyzed 111 samples and have 21 more to do.

The spectrophotometer is being repaired by the manufacturer, who has promised to return it to the refinery on 19 October. Today I informed the Refinery Manager of the delay, and he agreed to an increase in the project price to offset the additional time. He will call you about this.

Providing there are no further delays I will analyze the remaining samples between 20 and 24 October, then submit my report to the client the following morning. This means I should be back in the office on 26 October.

Jack

Figure 4-5. An occasional progress report.

To answer these questions, a periodic progress report can readily use the standard Summary-Background-Facts-Outcome arrangement:

1. *Summary* — A brief overview of the project schedule, progress made, and plans (*answers the first question*).
2. *Background* — The situation at the start of the report period.
3. *Facts* — Progress made (*answers the second question*) and problems encountered (*answers the third question*).
4. *Outcome* — Plans/expectations for the next period (*answers the last question*).

Figure 4-6 shows how survey crew chief Pat Fraser used these four compartments to write an effective progress report (numbers are keyed to parts of the report):

(1) The **Summary** tells Civil Engineering Coordinator Karen Woodford how closely the survey project is adhering to schedule, and predicts future progress. This is the information she wants to read first.

(2) The **Background** section reminds Karen of the situation at the end of the previous reporting period and predicts what Pat expected to accomplish during this period. Background should always be stated briefly.

(3) The **Facts** (or **Discussion**) section is broken into two parts:
   Work done during the period (3A)
   Problems affecting the project (3B)
   Pat Fraser opens each paragraph of this compartment with a topic sentence (a summary statement) which states the main point of the paragraph in general terms:
   Dry, clear weather . . . enabled us to progress faster than anticipated.
   The electrical fault in the EDM equipment . . . recurred on May 23.
   I have had difficulty hiring reliable people to clear brush along the route.
   He then describes what happened in detail, using *facts* (exact dates and milepost numbers, for example) to support each topic sentence. To prevent the report from becoming too long, Pat attaches the survey results to it and simply refers to them in the narrative. (Because of their length, they have not been printed with Figure 4-6).

(4) In the **Outcome** paragraph Pat tells Karen what the crew expects to accomplish during the forthcoming period, and even suggests when they may eventually get back on schedule. This final statement clearly supports his opening paragraph, and so brings the report to a logical close.

Other factors you should consider when writing periodic reports are:

1. If a progress report is long, use headings such as these to help readers see your organization:
   Adherence to Schedule (*This is your* **Summary**).
   Progress During Period (*These are your* **Facts; Background** *information should go at the front of the Progress section.*)
   Problems Encountered
   Projection for Next Period (*This is the* **Outcome**.)
2. For lengthy progress or problems sections, start with a summarizing statement describing general progress, then write several subparagraphs each giving details of a particular aspect of the project:

   4. Interior construction work progressed rapidly but exterior work was hampered by heavy rain:

H L Winman and Associates

INTER - OFFICE MEMORANDUM

From: Pat Fraser, Survey Crew Chief   Date: May 31, 19__

To: Karen Woodford, Coordinator   Subject: Progress Report No. 4 --

Civil Engineering Department   Allardyce Survey Project

The Allardyce Route survey has progressed well during the May 16-31 period. The survey crew has regained two days, and now is only four days behind schedule. We expect to be back on schedule by June 30. (1)

Project plan AR-51 shows we should have surveyed mileposts 30 to 34 during this period. But, as stated in my May 15 report, we were six days behind schedule at the end of the previous period, having surveyed only as far as milepost 28. Consequently, we expected to survey only to milepost 32 by May 31. (2)

Dry, clear weather from May 18 to 23 enabled us to progress faster than anticipated. We reached milepost 31 on May 23, carried out a terrain analysis for the Catherine Lake diversion scheme on May 24 and 25, resumed surveying on May 26, and reached milepost 32 at 09:00 on May 29, two days earlier than expected. We then continued to within 300 meters of milepost 33, which we reached at 18:00 on May 31. Survey results are attached. (3A)

Two problems affected the project during this period:

1. The electrical fault in the EDM equipment, which delayed us several times early in the project, recurred on May 23. I had the unit repaired at Fort Wilson on May 24 and 25, while we conducted the terrain analysis, and it has since worked satisfactorily.
2. I have had difficulty hiring reliable people to clear brush along the route. Most remain with us for only a few days and then quit, so that I have had to waste time hiring replacements. This problem will continue until mid-June, when the college students we interviewed in March will join the crew.

(3B)

We plan to advance to milepost 37 by June 15, which should place us only two days behind schedule, and hope to make up the remaining two days during the June 16-30 period. (4)

P. Fraser

Figure 4-6. A periodic progress report.

4.1 In the east wing, all partitions were erected, 80% of the floor tiles were laid, and 20% of the light fixtures were installed.
4.2 In the west wing, all remaining floor tiles were laid, all light fixtures were installed, and 16 of the 24 benches were bolted down; 8 of the benches were also connected to the water supply and drains.
4.3 Landscaping started on September 16, but had to be abandoned from September 18 to 23 when heavy rains turned the soil into a quagmire. By the end of the month only the outer areas of the parking lot had been completed.

3. Be as brief as possible, particularly for routine work. Use specific information (*facts* rather than generalizations) in the report narrative, and place lengthy details in an attachment. If, for example, you are reporting an extensive analysis, in your progress section you might write:

   We analyzed 142 samples, 88 (62%) of which met specifications. Results of our analysis are shown in attachment 1.

   Attachment 1 would contain several pages of tabular data (numbers, quantities, measurements), which if included as part of the report narrative would have interfered with reading continuity.

4. Describe problems, difficulties, and unusual circumstances in depth. State clearly what the problem was, how it affected your project, what measures you took to overcome it, and whether the remedial measures were successful:

   Summary Statement — Juvenile vandalism has proved to be a petty but time-consuming problem.
   Facts — On September 3 (Labor Day) several youths scaled the fence around the materials compound and made off with about $170 worth of building supplies. On September 16 they managed to start up a front end loader, drove it into the excavation, then got it stuck in the mud and burned out the clutch while attempting to move it. From September 18 the night watch has been doubled and the site has been policed by a patrol dog.
   Outcome — There have been no further attempts at vandalism.

   Remember that management wants to hear about problems and how they were overcome. Such knowledge can be used to avert difficulties on future projects, or can indicate project trends. A series of problems encountered during a design project may indicate that insufficient time has been allowed for it, that more engineering skills are needed, or that funds budgeted for the project are inadequate. While the Progress section shows how much work is being done, the Problems section can act as a warning that progress may slow down unless immediate steps are taken.

5. Forewarn management of any situation which, although it may not yet affect your project, may become a future problem. With such knowledge, management may be able to help you prevent a costly work stoppage or equipment breakdown. Here's a typical situation:

   7.1 Unless the strike at Vulcan Steel Works ends shortly, it will soon curtail our construction program. Our present supply of reinforcing barmats will last until mid-October, by which time an alternative source of supply must be found. I have researched other suppliers, but have been unofficially warned by union representatives that my attempt to obtain steel from elsewhere may result in a walkout at other plants.

Where should such an entry appear in your progress report? The best position would be at the end of the Facts (Problems) section, immediately before or possibly part of the Outcome.

6. Don't be afraid to number your paragraphs and subparagraphs (see the examples in suggestions 2 and 5 above); this helps you to refer to a specific part of a previous report:

   The possibility of a shortage of steel mentioned in para 7.1 of my September report was averted when the strike at Vulcan Steel Works ended on October 6.

7. Maintain continuity between reports. If you introduce a problem that has not been resolved in one report, then you must refer to it in your next report, even though no change may have occurred or it has been solved only a day later (see the example in suggestion 6). You must never simply drop a problem because it no longer applies.
8. If management expects you to include project cost information in your progress report, insert it in three places:

   In the *Summary* (comment briefly on how closely you are adhering to projected costs).
   In the *Progress* section (give more details of costs, and particularly cost implications of problems).
   In the *Outcome* section (indicate future cost trends).

   Costs are usually closely linked with your adherence to schedule: the more you drop behind schedule, the more likely you will have to report a cost overrun.

Inspection Report

An inspection can range from a quick check of a small building to assess its suitability as a temporary storage center, to a full-scale examination of an airline's aircraft, avionic equipment, repair facilities, and maintenance methods. In both cases the inspectors will report their findings in an inspection report. The building inspectors' report will be brief: it will state that the building either is or is not suitable, and give reasons why. The airline inspector's report will be lengthy: it will describe in detail the condition of every aspect of the airline's operations and list every deficiency (condition that must be corrected). In both cases the inspectors' reports can follow the Summary-Background-Facts-Outcome arrangement.

For an inspection report, these four compartments are:

1. *Summary* — The main result(s) of the inspection (very brief); what the reader most wants to know.
2. *Background* — Why the inspection was necessary; what was being inspected; who was involved; where the inspection took place.
3. *Facts* — What the inspection revealed (the details). There are two parts to the Facts:
   A. CONDITIONS FOUND. A description of:
      Quality (condition) of an equipment or a facility, or of work done.
      Quantity of items examined, or of work done.
   B. DEFICIENCIES. A list of:
      Conditions that need to be corrected.
      Work that needs to be done (or redone).

4. *Outcome* A general statement of results, possibly with a recommendation.

Kevin Doherty's building inspection report in Figure 4-7 shows how these compartments helped him shape his report into a logical, easy-to-follow document. Note particularly how:

His **Summary** (1) tells the Production Manager the *one* thing he most wants to know: can they use the building?

Kevin has opened the **Conditions** section (3) with a summarizing general statement, and then supported it with facts.

He has presented the **Deficiencies** (3B) as a briefly stated list, which makes it easy to identify what has to be done.

The recommendation in his **Outcome** (4) supports his summary.

For a short inspection report like this, it's best to present all the Conditions first, and then list all the Deficiencies. But for a long report that covers many items, such an arrangement could become cumbersome. If Fran Hartley followed this sequence for an inspection at Remick Airlines, the organization of the Facts section would be like this:

A. CONDITIONS FOUND:
   1. Electrical Shop
   2. Avionics Calibration Center
   3. Flammable Materials Storage
   (etc . . .)

B. DEFICIENCIES:
   1. Electrical Shop
   2. Avionics Calibration Center
   3. Flammable Materials Storage
   (etc . . .)

The more departments that are inspected, the longer the report would become, and the further apart each department's Conditions and Deficiencies sections would be.

To overcome this difficulty, Fran should treat each department as a *separate* inspection and reorganize the report so that for each department the Deficiencies section immediately follows the Conditions section. The organization of the whole report then becomes:

SUMMARY
BACKGROUND
FACTS:
1. Electrical Shop:
   A. Conditions Found
   B. Deficiencies
2. Avionics Calibration Center:
   A. Conditions Found
   B. Deficiencies
3. Flammable Materials Storage:
   A. Conditions Found
   B. Deficiencies *(etc . . .)*

robertson
engineering
company

MEMORANDUM

From: Kevin Doherty
Date: 5 November 19__

To: Hugh Smithson, Production Manager
Subject: Inspection of Carter Building

The old Carter Building at the corner of River Avenue and 39th Street will make a suitable storage and assembly center for the Dennison contract. (1)

Guy LePage and I inspected the Carter Building on 3 November to assess its suitability for both storage and as a work area for 20 persons for 15 months. We were accompanied by Mr Ken Wiens of Wilshire Properties. (2)

We found the interior of the building to be spacious, to have good facilities, but to be unsightly. Our inspection showed that: (3)

1. There are 460 m^2 of usable floor space (see attached building plan, supplied by Mr Wiens); we need 350 m^2 for the project.
2. There are two offices, each 16 m^2, and a large unimpeded space ideal for partitioning into a storage area and four work stations.
3. The building is structurally sound and dry, but it is very dirty and smells strongly (the previous tenant was a fertilizer distributor).
4. There are numerous power outlets, newly installed with heavy-duty circuits, and the building has excellent overhead lighting. (3A)
5. Several wall surfaces are damaged and many contain obnoxious graffiti.
6. There is a new loading ramp on the north side of the building, suitable for semitrailers.
7. Washroom facilities are adequate for up to 30 persons, but one toilet and two washbasins are broken.

Before we rent the building, the rental agency will have to:

1. Clean it thoroughly.
2. Repair damaged walls, partitions, and toilet facilities. (3B)
3. Redecorate the interior.

Ken Wiens said his firm would be willing to do all this.

I recommend we rent the Carter Building from Wilshire Properties, with the provision that the deficiencies listed above are corrected. (4)

KD

Figure 4-7. A short informal inspection report.

OUTCOME:
Conclusions
Recommendations

Fran's inspection report now has a much more tightly knit, logical, coherent organization.

Investigation Report

The term "Investigation Report" covers several types of reports which can be grouped together. These include laboratory reports, project reports, and any other report in which the writer describes how he or she has had to perform tests, examine data, or conduct an investigation using tangible evidence. Basically, the writer starts with known data and then analyzes and examines it so that the reader can see how the investigation was conducted and the final results reached. The report may be issued as:

A letter, as in Figure 4-9.
An interoffice memorandum, as in Figure 5-2.
A semiformal report, as in Figure 5-3.

Sometimes an investigation report is written first as a memorandum from a project group leader to a department head. Then later its essential details are rewritten as a letter report from the department head to a customer or client. The investigation report in Figure 4-9 illustrates the informal yet businesslike tone of such a letter.

Although they are not always readily identifiable, there are standard parts to a well-written investigation report that help shape the narrative and guide the reader to a full understanding of its topic. They are shown in Figure 4-8, which is an expanded version of the basic compartments described at the start of this chapter. These parts are easy to recognize in long investigation reports, where headings act as signposts introducing each parcel of information. They are more difficult to identify in short reports which use a continuous narrative. In the investigation report in Figure 4-9, the parts are identified by circled numbers:

(1) This is the **Summary.**
(2) The **Background** is only one sentence; the reader knows the circumstances, so details are omitted.
(3) The **Discussion** starts here, with a very brief reference to the **Method.**
(4) These are the **Findings.**
(5) This is the first **Idea** and its **Evaluation** or **Analysis.**
(6) These are **Ideas** 2 and 3, each followed by an **Evaluation.**
(7) The **Outcome** draws **Conclusions** and makes a **Recommendation.**
(There are no **Attachments.**)

Every investigation report will differ, depending on the topic you are investigating and the results you obtain. Sometimes you will use all of the

SUMMARY	A brief statement of the situation or problem and what should be done about it.	
INTRODUCTION	**Background** of the situation or problem.	
DISCUSSION	The **Facts** comprising:	
	METHOD	How the investigation was tackled.
	FINDINGS	What the investigation revealed.
	IDEAS	Different ways the situation can be improved or the problem resolved.
	ANALYSIS	Evaluation of each idea.
CONCLUSIONS	The **Outcome**, or result of the investigation; a summing-up.	
RECOMMENDATION	A positive statement advocating action.*	
ATTACHMENTS	Evidence: detailed facts, figures, and statistics which support the Discussion.*	

**Included only when appropriate.*

Figure 4-8. Compartments for an investigation report.

standard parts, sometimes you will use only some of them, and occasionally you will need to devise and insert additional parts of your own. Typical standard parts are described below.

Summary. This is a very brief description of the whole report, stated in as few words as possible. It gives busy readers a quick understanding of the investigation from which they can learn the results and assess whether they should read the whole report.

Background. Events that led up to the investigation, knowledge of which will help the reader place the report in the proper perspective. This part is often referred to as the Introduction.

Investigation Details. A complete narrative description of the investigation, carefully organized so that the reader can easily follow your line of reasoning. This is the Discussion. In a long report the parts should contain information similar to that described here, and appear roughly in this order:

Start with Known Facts. Introduce readers to all the technical data and tangible evidence available at the start of the investigation or used during the investigation.

Introduce Guiding Factors. These are the requirements or limitations that controlled the direction of the investigation. They may be as diverse as a major specification stipulating definite results that must be attained, or a minor limitation such as price, size, weight, complexity, or operating speed of a recommended prototype or modification.

Outline Investigation Method. Tell readers the planned approach for the investigation so that they can understand why certain steps were taken.

H L Winman and Associates

PROFESSIONAL CONSULTING ENGINEERS
475 Reston Avenue-Cleveland, Ohio, 44104

October 21, 19__

Mr. D. R. Carlisle
Technical Operations Manager
Television Station DCMO-TV
PO Box 890
Montrose, OH 45287

Dear Mr. Carlisle:

Reduction in Noise Level - Satellite Studio Control Room

We have investigated the high background noise reported in the control room of Station DCMO-TV's satellite studio at 21 Union Road, and have traced it to the building's air-conditioning equipment. The noise could be reduced to an acceptable level by soundproofing the air-conditioning ducts and blowers, and by carpeting the control room. (1)

(2)

Our investigation was authorized by your letter of August 28, 19__. Tests conducted with a noise meter at various locations in the control room established that the average ambient noise level is 36 dB, with occasional peaks of 39 dB near the west wall. This is approximately 8 to 10 dB higher than the noise levels measured in the control room of your No. 1 studio on Westover Road. (3)

The unusually high noise level is caused by the air-conditioning equipment, which is located in an annex adjacent to the west wall of the control room. Air-conditioner rumble and blower fan noise are carried easily into the control room because the short air ducts permit little noise dissipation between the equipment and the work area. The flat hardboard surface of the west wall also acts as a sounding board and amplifies the noise. (4)

We have considered three methods that could be used to reduce the ambient noise to an acceptable level:

1. The most effective method would be to move the air-conditioning equipment to a remote location. This, however, would require major structural alterations which would make the approach impractical except as a last resort. (5)

Figure 4-9. An investigation report in letter form.

2. A noise reduction of approximately 6 dB could be effected by replacing the existing blower fan assembly with a Model TL-1 blower manufactured by the Quietaire Corporation of Detroit, and by lining the ducts with Agrafoam, a new soundproofing product developed by the automobile industry in West Germany.

3. A further reduction of 1.5 dB could be achieved by replacing the vinyl floor tiles with indoor/outdoor carpet, a practice that has been successful in Air Traffic Control centers. Mounting carpet on the whole of the control room's west wall, and possibly along the first five feet of the north wall, would also reduce the noise by an additional 1 dB.

(6)

Since relocation of the air-conditioning equipment would be unduly expensive, we recommend that methods 2 and 3 be adopted to reduce overall ambient noise by approximately 8.5 dB for a total cost of $8080.00. We suggest that the modifications be made progressively so that noise reductions can be measured on completion of each step. This will permit either or both of the final steps to be eliminated if earlier modifications achieve better results than anticipated. The four steps will be:

(7)

Modification	Anticipated Ambient Noise Reduction	Approximate Cost
(a) Replace blower fan assembly	2.5 dB	$2560.00
(b) Line ducts with Agrafoam	3.5 dB	$2600.00
(c) Install floor carpet	1.5 dB	$1580.00
(d) Install wall carpet	1.0 dB	$1340.00

We believe that these modifications will provide the quieter working environment desired for your operating staff.

Sincerely

RW Gray

for Peter Bell, Head
Mechanical Engineering Department

RWG:bls

Describe Equipment Used. If tests were performed that called for special instruments, describe the test setup and, if possible, include a sketch of it. In long reports this may have to be done several times, to keep details of each test setup close to its description.

Narrate Investigation Steps. Many investigation reports lend themselves to a chronological presentation of information. In most cases a step-by-step description will be both logical and easy to follow, particularly if the investigation has progressed naturally through a series of steps involving the introduction to a problem, development of a means to overcome it, application of corrective measures, and tabulation of results. But when an investigation has comprised an analysis of processes, equipment, or methods, chronological presentation cannot be used. In such cases divide the information into suitable subject areas (e.g. equipment, processes, application) and examine the advantages and disadvantages of each. For more information on the chronological and subject method of report development, refer to the "Discussion" section in Chapter 6.

Discuss Test Results. Analyze the results of each test and discuss whether or not they satisfy the needs of the investigation. (It is not sufficient simply to tabulate the test results and assume that the readers will infer their meaning and implications.) Such an analysis after a test or series of tests can demonstrate that a specific trend influenced or altered the course of the investigation.

Develop Ideas and Concepts. At some point in your narrative you may want to introduce ideas and concepts which have evolved as a result of your investigation. You may introduce them periodically throughout the report to demonstrate what you had in mind before taking the next investigative step, or you may group them together near the end. They should develop naturally from the flow of the report and should be persuasive: they may have to influence your readers' acceptance of a new technique or an unusual method which you want to recommend.

Analyze Ideas. If you have developed alternative ideas (methods) for resolving a problem, evaluate them to assess which is best, using factors such as effectiveness, cost, and simplicity as your evaluation criteria. (Ensure that your readers understand what criteria you are using, and why they are important, before they read your evaluation.) You may evaluate each idea immediately it is presented, as has been done in Figure 4-9, or you may present all the ideas first and then evaluate all of them, as Figure 4-8 suggests.

Conclusions. In a brief summing-up, draw the main conclusions that have evolved from your investigation. You may also recommend what steps need to be taken next, providing that they develop naturally from your Discussion and Conclusions.

Evidence. Detailed data such as calculations, cost analyses, specifications, drawings, and photographs that support the facts presented in the Investigation Details section would interrupt reading continuity if included with it. These pages of evidence are numbered consecutively and named Attachments or, in a long report, Appendices, and placed at the end of the report.

The process of an engineering investigation, and the report that evolves from it, can be illustrated as a flow diagram (see Figure 4-10).

As an example, suppose that Harvey Winman, as president of his company, decides to purchase an executive jet aircraft because he travels a lot, but cannot decide which aircraft would be most useful. He might assign two of his

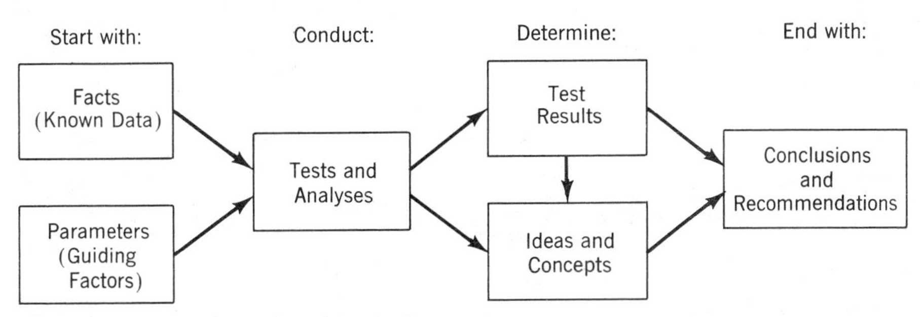

Figure 4-10. Basic flow diagram—investigation report.

engineers to investigate a selected range of aircraft and to recommend the most suitable model. Starting with facts (specifications of the types of aircraft Harvey is known to favor) and parameters (probable route mileage, maximum operating costs, maximum acceptable purchase price), they would analyze the different aircraft and determine which offers the most advantages under the proposed operating conditions. One of them might also conceive the idea of leasing an executive aircraft which could be rented from a local aviation company, complete with crew, on an as-required basis. Armed with all this information the engineers would prepare a report of their findings, ending with a conclusion that identifies the most suitable aircraft to purchase but which also points out the economic advantage of leasing. The report they submit to Harvey Winman would be an investigation report that starts with known information but ends by developing an idea.

Some companies preface their investigation reports with a standard title and summary page similar to the H. L. Winman and Associates' design illustrated in Figure 4-11. This page saves a reader the trouble of searching for the summary and the report's identification details. Subsequent pages, which have not been included with the example, contain the report narrative, starting with the Background (Introduction). Reports written in this way tend to adopt a slightly more formal tone than memorandum and letter reports and are less likely to be in the first person. Sometimes they are called Form Reports, although I prefer to identify them as Semiformal Investigation Reports.

Assignments
Occurrence Reports

PROJECT NO. 1: ONE OF OUR CRATES IS MISSING

You are employed by the local branch of H. L. Winman and Associates. When you arrive for work this morning, branch manager Vern Rogers calls you into his office.

"We've had a call from Hugh Smithson (he's production manager for Robertson Engineering Company in Toronto)," Vern says. "Twelve cartons of special instruments they shipped three days ago were in a semitrailer which rolled near Market Drayton."

H L Winman and Associates

INVESTIGATION REPORT

Report No. 70/26 File Ref 53-Civ-26 Date 20 March 19__

Prepared for City of Montrose, Ohio

Authority City of Montrose letter Hwy/69/38; 7 Nov 19__

Report Prepared by [signature] Approved by M. Dawes [signature]

SUBJECT OR TITLE

INVESTIGATION OF STORMWATER DRAINAGE PROBLEM:

Proposed Interchange at Intersection of Highways 6 and 54

SUMMARY OF INVESTIGATION

The proposed interchange to be constructed at the intersection of Highways 6 and 54, on the northern perimeter of Montrose, Ohio, incorporates an underpass which will depress part of Highway 54 and some of its approach roads below the average surface level of the surrounding area. A special method for draining the stormwater from the depressed roads will have to be developed.

Two methods were investigated that could contend with the anticipated peak runoff. The standard method of direct pumping would be feasible but would demand installation of four heavy duty pumps, plus enlargement of the three-quarter mile (1.21 km) drainage ditch between the interchange and Lake McKing. An alternative method of storage-pumping would allow the runoff to collect quickly in a deep storage pond which would be excavated beside the interchange; after each storm was over the pond would be pumped slowly into the existing drainage ditch to Lake McKing.

Although both methods would be equally effective, the storage-pumping method is recommended because it would be the most economical to construct. Construction cost of a storage-pumping stormwater drainage system would be $784,000, whereas that of a direct pumping system would be $967,000.

Figure 4-11. Title and summary page for a semiformal investigation report.

He pulls out a map and points to Market Drayton, some 40 highway miles (64 km) from your office. "The insurance company wants someone to look over the damage with one of their adjusters to confirm how much can be repaired or salvaged."

You drive to Market Drayton with Mike Shoemacher of Mansask Insurance Corporation. He takes you to a warehouse on the other side of town, where the smashed crates tell their own story of the violence of the accident. Very little of the delicate instruments could have survived such an impact.

You examine the crates one at a time. Broken glass, tangled wire, and chipped and splintered instrument cases are jumbled together. It is a sad sight to see, particularly when you remember the pressure of this special job and the overtime the people at Robertson Engineering put in to meet the customer's deadline.

As you check each container, the adjuster notes the numbers in his book: 7, 4, 12, 11, 5, 3, 1, 9, 8 and 2 are totally beyond repair and obviously have no salvage value. Crate number 6 surprisingly is hardly marked: somehow its movements must have been cushioned. You open it carefully and check the instruments.

"This one seems okay," you say. "I think we could do something with it."

The adjuster adds up the totals. "Not very good for us," he says. "Ten out of eleven means a heavy claim."

"Twelve," you say. "There were twelve crates."

The adjuster checks his figures again while you count crates. There are 10 smashed ones and only one good one. And the smashed ones are not so smashed that one would be unrecognizable and so not be counted.

"One is missing," you say.

"Number 10," says Mike. And the two of you check the numbers against the crates.

Plainly, crate number 10 is missing. You and Mike check all the other items removed from the semitrailer, but number 10 is not there.

"Looks as if it has been stolen," Mike comments. "Probably before the accident. I don't think there would have been time afterward: the police were on the scene almost right away."

When you return to your office, you tell Vern Rogers what has happened.

"I'll telephone Hugh Smithson in Toronto," he says. "In the meantime, will you write a report and send it to him special delivery? And give me a copy."

Information you may need to write this report:

1. The shipping company was Merryhurst Express Lines, Hamilton, Ontario.
2. The waybill number was C7218.
3. Robertson Engineering Company's invoice number was R2261.
4. The 12 cartons were being shipped to Melwood Test Labs, Circular Route No. 1, Harmonsville.
5. Mansask Insurance Corporation's local address is room 414–2800 Western Avenue, in the same town as your office.
6. The semitrailer rolled at Berryman's Corner, one mile (1.6 km) southeast of Market Drayton.

7. The crates are being held at Drayton Storage, 313 Crane Street, Market Drayton.

PROJECT NO. 2: OUR SURVEYOR'S TRANSIT HAS BEEN DAMAGED

You are the team leader of a four-man H. L. Winman crew en route to an assignment at Minnowin Point in Alaska. On Sunday evening you take a cab to your local airport to catch Remick Airlines Flight 679 to Edmonton, Canada, the first stop on your route. As one of the crew's suitcases is lifted by an airline employee from the weigh scale to the conveyor belt, its shipping tag tangles with the strap of the case containing a surveyor's transit, which has been placed carelessly on the counter with its strap hanging over the edge. The moving suitcase pulls the transit to the floor, then drags it beside the conveyor belt and bumps it against obstacles along the way. You try to rescue the transit before it disappears through the swinging doors to the baggage room, but it crashes through them and disappears from view. When you retrieve the transit you note that its carrying case is badly battered.

You consider that the transit cannot be used in its present condition and decide to leave it with a Remick Airlines' representative to be picked up by your company early in the week. You obtain a receipt for it and board the aircraft to Edmonton.

While in flight you write a report of the incident for your department coordinator, asking him to pick up the damaged transit and ship it to the manufacturer's service office. You also ask him to borrow or rent a replacement transit from the local manufacturer's representative, and to ship it on Monday night's flight to Edmonton, for routing to you at Minnowin Point. You hand your report to the flight crew and ask them to send it by messenger to your office on their return flight to your home city.

Additional information for the report:

Type of Transit: Fennel "Tropl" $5\frac{1}{4}$ in. with optical plummet, Model A0150

Manufacturer: Otto Fennel Co., Kassel, West Germany

North American Service Office: 35 Kringle Street, Peterborough, Ontario, Canada

U.S. Sales Representative: Engineers' Supply Company, 409 Cumberland Avenue, Cleveland, Ohio

Which member of your crew placed the transit carelessly on the counter?: Doug Rickerson

PROJECT NO. 3: RADIOACTIVE CONTAMINATION IN THE LAB

You are a laboratory technician for H. L. Winman and Associates, and your lab supervisor is Frances K. Eldon. This morning you are measuring a new vial of radioactive iodine (131_{I}) in a dose calibrator, when you:

1. Find the vial contains 7 mCi (millicuries) of 131_{I}.
2. Sense that "7" is an unusual quantity.

3. Check the manufacturer's shipping record.
4. Discover that the vial contained 10 mCi when shipped.
5. Realize that 3 mCi is missing.
6. Remember that you unpacked the shipment yesterday afternoon at 4:45 P.M., just before you went home.
7. Remember, too, thinking that the vial top felt a little loose, so tightened it before storing the vial yesterday afternoon.
8. Realize that the 131_I has probably spilled into the packing material.
9. Remove the packing materials from the ordinary garbage bin (where you put them yesterday afternoon), and place them in the contaminated garbage area.

You inform Frances of the occurrence, and she asks: "Have you monitored the lab for radioactive contamination?"

You say you haven't, and Frances adds: "Do it right now. And check yourself too."

You find contamination in four places:

Your own thyroid (you detect iodine).
The dose calibrator (you measured the vial with it).
The counters (where you did the measurements).
The ordinary garbage bin (where you put the packing materials initially).

Frances tells you to decontaminate the hot areas (which you do), and then to monitor again for contamination (you find none in the lab, but traces of iodine are still present in your thyroid).

Frances tells you to write a report of the occurrence and send it to Dr. Raymond B. Bolton, the local radiation officer, with a carbon copy for herself.

Information you may need:

The manufacturer of the 131_I was Weatherdon Labs of West Grantham, New Jersey
The 131_I lot number was 16024; the shipping certificate was No. XL233, dated seven days ago; and H. L. Winman's purchase order number was W4467, dated 14 days ago.
Dr. Bolton's office is in room 212 of the Cordon Building, 2860 Vermont Avenue.

PROJECT NO. 4: INTERRUPTED TEMPERATURE TESTS

H. L. Winman and Associates has been carrying out a series of extreme cold and heat tests on electronic and mechanical switches for Terrapin Control Systems of Denver, Colorado. The tests have been running for four months and will last another two months. The schedule is tight because of initial problems with measuring equipment, which delayed the start by nine days and used up any spare time the project had available.

Currently, you are testing the switches for continuous periods of from 8 to 14 hours. The tests have two parts:

1. For the first 6 hours each day you increase or decrease temperature in 2°C increments until a predetermined high or low temperature is reached. At each 2° increment you test the switches and record how they perform.
2. For the remaining 2 to 8 hours, you bake or deep-freeze the switches at the

preselected temperature. No monitoring is necessary during this period (although the switches are tested at room temperature the following day).

To avoid having a technician stay throughout part 2, which on some evenings runs as late as 10:00 P.M., you have installed electrical timers in the circuits of the oven and freezer chamber. At the end of part 2 each afternoon, the timers are set to switch off at the end of the prescribed bake and deep-freeze periods.

This morning when you remove batches 87H and 84C from the oven and freezer chambers you notice that, instead of being close to room temperature, the oven is still very hot and the freezer is still very cold. You check the electrical timers, but both are "off." Then you notice that the electric clock on the lab wall reads only 3:39; your wristwatch reads 8:47—a difference of 5 hours and 8 minutes. You phone the local power utility:

"Was there a power cut last night?" you ask.

"Where do you live?" a voice replies.

"I'm calling from my office," you say, and quote the address.

"Yes, there was," the voice answers. "We had a transformer blowout at Penns Vale. It affected everyone in your area."

You ask when the power cut started and ended. The voice asks you to wait a minute.

"The transformer blew out at 9:18 last night," the voice eventually announces. "And we restored power to buildings in your area at 2:26 A.M."

You thank the voice, and consult your log for the previous day's tests:

You started part 1 at 9:55 A.M.

At 3:55 P.M. you started part 2, and set the timers to run for 8 hours (they were to swtich off at 11:55 P.M.).

You consider what has happened:

The continuous bake and deep-freeze was interrupted part way through.

The oven temperature dropped, and the freezer temperature rose, for 5 hours and 8 minutes (but to what temperature?).

The power was restored and the oven temperature again increased, and the freezer temperature decreased (but to what temperature?).

The electric timers switched off at 5:03 A.M. (after their eight hours *total* running time).

You consider the implications of the power cut:

The batches have had uncontrolled, nonstandard testing and will have to be discarded.

Yesterday's tests will have to be run again (on two new batches).

The cost is:

Labor:	14 hours (7 hours per batch) = $210.00.
Materials:	Two complete batches at $42.00 each.
Time:	One day extra to be added to the program schedule.

Write an occurrence report to your project coordinator (J. H. Grayson).

Tell him what has happened, describe the implications, and possibly suggest what might be done to prevent a recurrence.

PROJECT NO. 5: A VIDEOTAPE RECORDER HAS BEEN DAMAGED

Yesterday afternoon you were previewing some videotapes when you were called away to take a long-distance telephone call. You hurried away, leaving the videotape machine running. Here are the details:

1. You were using a Vancourt Model XB-75 videotape recorder (VTR) and a Vancourt Model GL-27 color playback monitor, both company-owned.
2. You were viewing three videotapes, Computronics No. 572A, B, C, all part of a series titled "Elements of Computer Control Systems."
3. You were working alone in the A/V room.
4. You had placed the second videocassette (No. 572B), Part 2 of the series, onto the machine, and you were taking notes when you were called to the telephone.
5. It was 4:20 P.M. when you were called away.
6. It was 4:40 P.M. when you returned to the A/V room.
7. Company quitting time is 4:30 P.M.; no one was around.
8. The TV monitor screen was black.
9. The preformed plastic cover had been placed over the VTR.
10. You switched off the lights, shut and locked the door, and went home.

This morning when you unlocked the door of the A/V room, you:

11. Noticed a heavy smell; an acrid, hot smell; a pungent smell.
12. Switched on the lights, but noticed nothing unusual.
13. Felt warmth near the VTR.
14. Noticed that the plastic cover over the VTR was caved inward; it was hot and pliable.
15. Immediately pulled the VTR and monitor plugs out of the wall sockets.

You had the VTR and cassette checked by Vern Corley, the A/V technician. He reported that:

16. The VTR was badly damaged internally; repair cost: about $900.00.
17. The videocassette was partly melted internally; not repairable.
18. The VTR autostop had failed to function at the end of the tape.
19. The red light which normally indicates that the VTR is "on" had burned out, probably the previous day.

Vern explained that he had switched off the monitor and covered the VTR at 4:30 the previous evening. He hadn't switched off the VTR because he thought it was already off (no red light was on).

Other factors affecting the incident are:

A new VTR would cost $1600.00

The videotape is valued at $245.00, and your company has paid a $25.00 deposit to borrow it.

How you borrowed the videotape is described in Part 1 of Project No. 6 in Chapter 3.

Write a report of the incident to your project coordinator. Describe what happened and outline what the costs to the company are likely to be.

Trip Reports and Progress Reports

PROJECT NO. 6: SUBSTANDARD SPLIT-BOLT CONNECTORS

H. L. Winman and Associates is management consultant for Montrose Hydro during construction and start-up of a power transmission line between Wabagoon Falls and Montrose, Ohio. Coordinator for the project is Don Gibbon, and you are an engineering assistant in his project group. The contractor is Welland and Welland Inc, also of Montrose.

The supervisor at site M9 has called in to report that he has found three faulty KS-28 split-bolt connectors (SBCs) in the site's stock and asks if this is a common occurrence. Don Gibbon decides to investigate whether the fault has occurred elsewhere and sends you to check on the connectors at site M17, one mile east of Wabagoon Falls. Your authority to proceed on the assignment is Travel Order W67.

At M17 you test all the split-bolt connectors on site and find four that are faulty (three in stock and one that has been installed). You also notice that there appear to be two types of connectors, one a slightly darker gray than the other.

On returning to the office you write a report of your trip for Don Gibbon. In addition to informing him of your findings, you suggest that all the SBCs be tested, including those in the contractor's stock. You also try to draw a conclusion from the difference in shading.

Additional information:

1. The method you used to test the SBCs is specified in company procedure PR27-7.
2. There are 18 installation sites.
3. All the faulty connectors are in the darker gray group.
4. The fault is visible as a fine hairline crack when the connector is placed in tension.

PROJECT NO. 7: INSTALLING EDP EQUIPMENT

Robertson Engineering Company is converting to Electronic Data Processing (EDP) equipment for use in inventory control, accounting, and other departmental operations. Coordination of the installation phase has been assigned to you. The installation contractor is the Milward Corporation of Detroit.

The EDP equipment is scheduled to be fully installed and ready for hand-over by the contractor on the last working day of the current month. The changeover date from manual accounting and inventory control to the EDP system has been set for the last working day of next month.

You already know that the contractor will not meet his target date. Today he is eleven working days behind schedule.

He has had problems:

1. There has been a ten-day delay in delivery of some major components for the system (they were shipped to Montreal in error).
2. The main control panel was damaged in transit; the contractor wasted five days trying to repair it; a replacement ordered by Telex five days ago is due at Robertson Engineering one week from today.
3. The contractor's crew normally comprises four men and a supervisor. But one man quit four days ago (a replacement being flown in from Detroit is due here tomorrow), and another is sick (absent six days to date, due back in about a week).

Work completed so far: the main frame and input console have been installed, wiring is 70% complete, and the Vancourt 61 work station is in position except for the readout display (it's one of the units diverted to Montreal).

Write a progress report to George Dunn, Robertson Engineering's comptroller. Advise him of the problems and delays. Forecast a new installation completion and handover date, and a revised system changeover date.

PROJECT NO. 8: METRICATION CONVERSION FOR MULTIPLE INDUSTRIES

Part 1. H. L. Winman and Associates has a contract with Multiple Industries Inc to design and install dual-purpose scales on their measurement instruments, control equipment, and recording charts. The contract anticipates conversion from U.S. to metric methods of measurement, and calls for an eleven-week conversion schedule. The original plan was for a three-person conversion team to work at the head office of Multiple Industries' Control Systems Division for four weeks, and then to visit the seven satellite plants during the next seven weeks. It was estimated that conversion could be done in four days at each satellite plant and that the fifth day would be for travel between plants (see schedule).

You are the conversion team leader. Today is the fifth day (a Friday) of the seventh week, and you are packing up late in the afternoon at Gatsano satellite plant. You are somewhat dispirited, because you are one week behind schedule and there is little chance of making up the lost time. By now you should have finished at the Meridian plant and be traveling toward the Sioux plant.

The trouble started at head office, where conversion took five weeks instead of four weeks. You weren't too worried, because you figured you could make up the lost time on the road. You discussed the situation with technicians Theresa Cookson and Wilf Dormer, who are working on the project with you, and they agreed to work the fifth day each week and travel at night and on weekends. But the two satellite plants have required five days each instead of four, so that no time has been gained.

The satellite plants you have completed are Axcell and Gatsano. You still

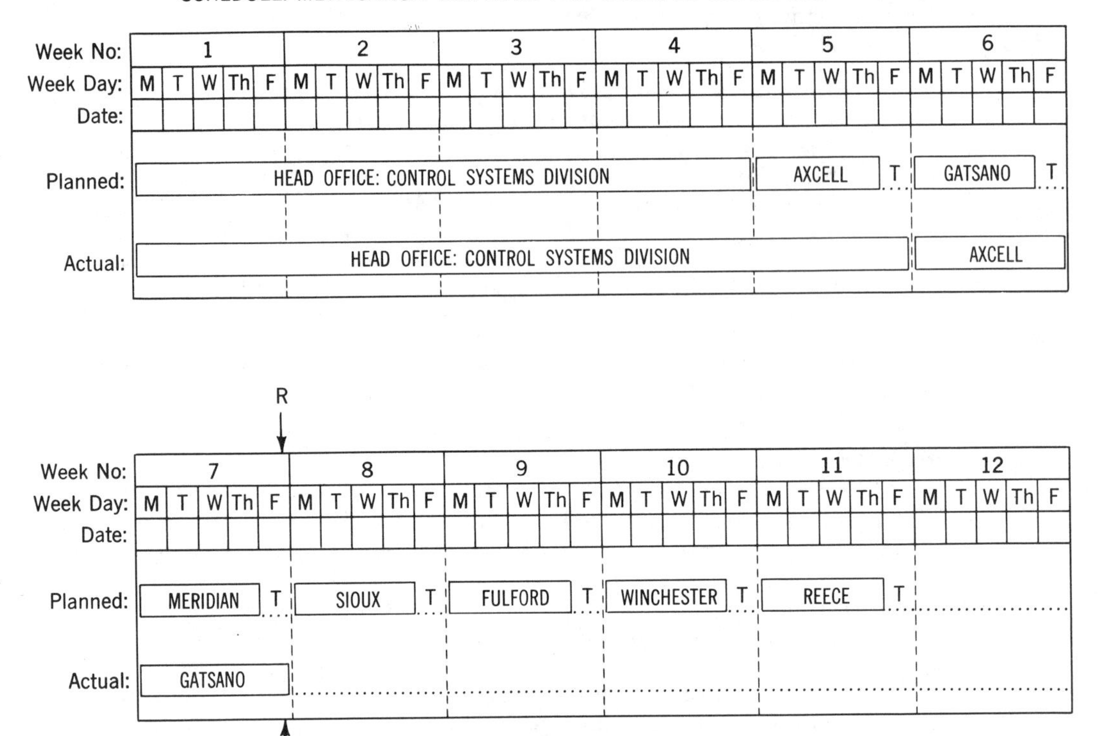

have to visit Meridian, Sioux, Fulford, Winchester, and Reece (see schedule). If each requires one full week, as is likely, you will be finished five weeks from today and the whole contract will have taken 12 weeks (an overrun of one week).

You decide to write a progress report today to your department head (he is Jim Perchanski) to tell him that you anticipate being one week late in finishing the conversion project. In your report, tell Jim what has happened, forewarn him of the overrun, and make any recommendation you consider necessary.

Here are some points you may want to include:

1. The extra time was caused by:

   Unavailability of equipment when you needed it. The satellite plants were supposed to have the equipment ready for your arrival, but you wasted considerable time rounding up equipment at both Axcell and Gatsano.

   Lack of interest in conversion to metrication on the part of many people, both at the Control Systems Division head office and at the satellite plants, some of whom said openly that "it's all a waste of time."

   More equipment (134 pieces more) at the head office than Multiple Industries had quoted in their contract.

2. You tried to arrange to work overtime (in the evenings and on weekends), but security arrangements prohibit you from entering Multiple Industries' plants outside normal work hours (8:00 A.M. to 4:30 P.M.).
3. You are writing to each satellite plant five days before your expected arrival, to say that you are coming and to ask that equipment be ready (you have already written to Meridian; you will write to Sioux the day you arrive at Meridian; and so on).
4. Because Multiple Industries' employees are taking metrication so lightly, you think a training/familiarization program in metric (SI) units is necessary, particularly at the satellite plants.

Before writing your report, calculate real dates backward and forward from today (Friday), enter them onto the schedule, and predict your revised completion date.

Part 2. It is late the following day (Saturday). You have arrived at Meridian, Nebraska, and have checked into the Belmont Motel. Write a letter to the production superintendent at Multiple Industries' Sioux plant to warn him of your arrival in another week and to ask him to have all control equipment, measurement instruments, and recording charts available for you and the Metric Conversion team to work on. His name is Jack K. Laporte, and the plant's address is 156 Wellborne Avenue, Sioux Valley, Iowa 51806.

PROJECT NO. 9: POSITIONING A PROCESSOR/CONTROLLER

Part 1. H. L. Winman and Associates has been hired by the Royston Gas Transmission Company to manage the installation of Vancourt Model 301 Processor/Controllers at its 38 pumping stations. You have been assigned by Jim Perchanski, head of the Electrical Engineering Department, to visit one of the Royston pumping stations to:

1. Select a suitable location for the processor/controller.
2. Check that the floor of the building is able to withstand the weight of the processor/controller.
3. Estimate the cost of bringing 208 V, 30 amp power to the chosen location.

In a pre-trip briefing, Jim Perchanski tells you that:

You are to visit station R22 at Long Acres.

You will be there for one day.

You are authorized to travel under project R674.

Station R22 is a "standard" site: the buildings at the 37 other stations have been constructed using the same plan; equipment layout is identical at all stations.

The results of your visit to station R22 will be used to fix the location of the processor/controller at all stations and to identify an approximate cost for the job at each station.

A Model 301 processor/controller weighs 280 lb (127 kg) and measures 38 in. (0.96 m) wide, 32 in. (0.81 m) deep, and 64 in. (1.62 m) high. When installed, it must have a minimum 8 in. (20 cm) clearance on its back and sides for heat dissipation.

You arrive at station R22 at 9:15 this morning and are greeted by Gary

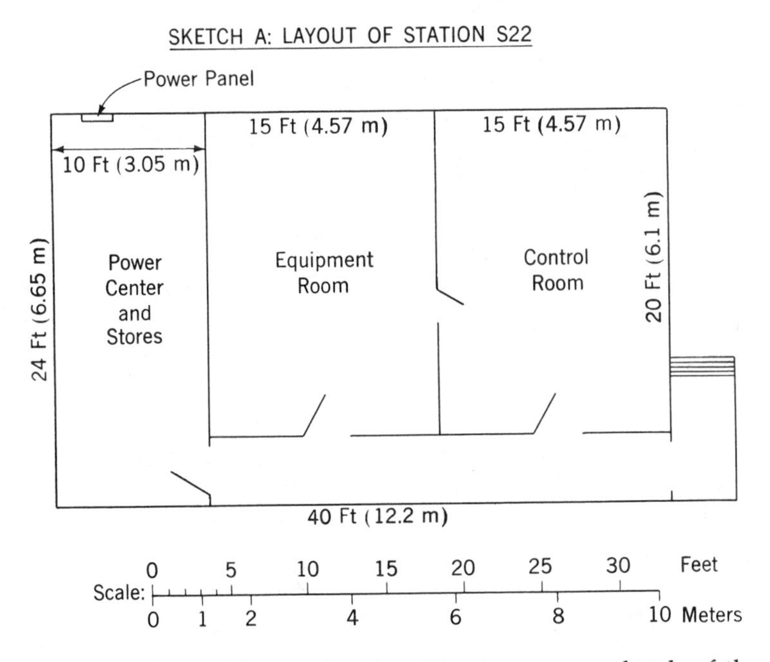

Manders, the resident technician. He gives you a sketch of the station layout (sketch A) and says: "Except for little things, the layout is the same at every station."

You prepare a sketch of the control room and draw each piece of fixed equipment in its correct position (see sketch B). You then examine the floor structure and condition, and estimate it to be strong enough to accept the processor/controller anywhere in the control room. You also check:

1. Width of doors: all are 36 in. (0.92 m) wide, except for an access door between the equipment room and the control room which is only 24 in. (0.62 m) wide.
2. Condition of loading dock: it's not good; some of the timbers are in such bad condition that you doubt whether it will support the processor/controller.
3. Position of electric power panel: it's shown on sketch A.
4. Amount of open space required in front of control stations: Gary Manders says he needs a minimum 30 in. (0.77 m).

Now you can plan where the processor/controller should go, and draw in the position you have chosen on sketch B (not forgetting to allow 8 in. (20 cm) clearance between it and the wall. Then you can draw in the cable run that will have to be installed between the power panel and the processor/controller, and calculate its length and cost using these parameters:

The cable run must travel only along walls and must not cross any doorways.

Its length can be calculated by measuring its route length and then adding 20% for variations in height, etc, to obtain its total length.

Its installation cost can be calculated at \$6.60 per foot (\$21.65 per meter), which includes materials and labor, over its total length.

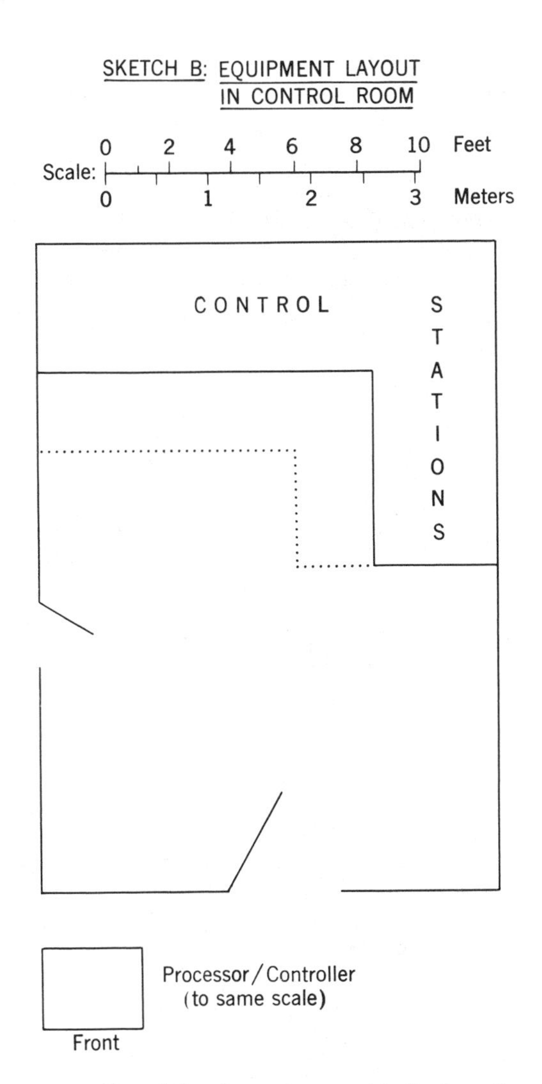

Gary Manders agrees with the location you have chosen for the processor/controller. When you tell him of your concern that the loading dock may not be able to take its weight, he says: "Oh, you don't have to worry about that. A requisition has already been approved to have the dock repaired, not only here but at other stations too. It should be done within the next three weeks." You decide, however, to mention in your report that loading docks need to be repaired before the equipment is brought in.

You have also noticed stains on the ceiling, and discoloration of the insulation around the cable runs that lead up to the antenna on the roof. You borrow a ladder and check the roofing material, which you find to be soft and spongy, and badly deteriorated around the chimney and where the cable run passes through the roof.

You discuss the condition of the roof with Gary Manders, who says that he has mentioned it in his last two progress reports to Royston's head office but so far nothing has been done about it. He also says that the technicians at the two adjacent stations have similar problems, although theirs are not quite as serious as his. He does not have any materials on site to make repairs. You consider that three 50-meter rolls of roofing material, 4 liters of adhesive, and 4 liters of bitumastic compound to seal joints around the chimney and the cable run should be sufficient to do the job. Gary says the work could be done by station personnel, if adequate instructions are sent with the materials.

As you fly back to your office, you write a memo-form trip report to Jim Perchanski, with a copy to Gary Manders.

Part 2. Jim Perchanski asks you to prepare the report and cost estimate he will send to Royston Gas Transmission Company (the address is 2720 Girton Avenue, Vienna, Oklahoma, 73406). In the letter you are to:

1. Say that a standard location has been selected (describe it, and attach a sketch).
2. State the length of cable required and the connection cost, both per site and for all 38 sites (see note 2 below).
3. Mention that there are two problems which must be corrected before the processor/controllers can be brought on to the sites (comment on what is being done to repair the loading docks and what needs to be done to repair the roof at some sites).

Additional notes:

1. The project is identified by Royston Gas Transmission Company as RG674.
2. If you have previously written the trip report (Part 1), use the cable length and price you have already calculated; if you have not written the trip report, use a total cable length of 48 ft (14.63 m) and an installation cost of $316.80 per station.
3. Address the letter to Wayne Campbell, Transmission Line Superintendent. You may sign it for Jim Perchanski.

PROJECT NO. 10: MANUALS AND A DRAWING FOR DIAMIN CORPORATION

Part 1. When you came to work this morning, there was a message asking you and drafting technologist Janice Kempson to meet with Vern Rogers at 8:45; he is manager of the H. L. Winman and Associates branch in your area. He asked the two of you to drive over to Diamin Corporation for a 10:00 A.M. appointment, and gave you these details:

Diamin makes mining and drilling equipment.

Diamin has made a new electronically controlled industrial drill, for which it wants H. L. Winman and Associates to prepare an instruction manual, a parts list, and an isometric exploded drawing.

The drill is called the "Electro-Max."

To date, only one drill has been built, and it is a prototype pre-manufactured unit.

Copies of the manual, drawing, and parts list are to accompany the production models.

You and Janice are to examine the drill and work out how long it will take to write the manual, make the drawing, and build up the parts list. Vern Rogers will use your estimates to calculate a contract price which he can quote to Diamin Corporation.

The person you are to meet at Diamin is Production Manager Gil Wishton.

When you and Janice drive over to Diamin, which is at 2760 Westmorland Drive, the following occurs:

When Gil Wishton meets you, he apologizes because he cannot show you the Electro-Max right away. One of his salesmen is demonstrating it to a potential buyer: Northwest Mining Company. Gil expects him back "about eleven," so you and Janice go out for coffee and return at 10:45.

By 11:15, the salesman still hasn't returned, so Gil gives you some Electro-Max manufacturing drawings to look at.

At noon, you go to lunch. When you return at 1:00, Gill promises to have the Electro-Max there by 2:00. He keeps apologizing and saying the salesman will be back "at any moment," so you keep waiting. At 2 P.M. you phone Vern Rogers to explain the delay (he expected you back by noon).

At 3:45, the salesman returns with the Electro-Max.

You and Janice examine the drill, compare it with the manufacturing drawings, and ask suitable questions. From your examination you and Janice determine that:

The drill will be difficult to describe, because it has so many interlocking parts. You estimate total writing time to be 85 hours.

The isometric exploded drawing will require about 50 hours' drawing time. Janice is enthusiastic and says it would be a challenging unit to draw. The parts list will require 20 hours' writing time.

The instruction manual and drawing could be done at the same time. You and Janice would need to have the unit beside you for the first five days that you are working on the job.

The parts list will have to be prepared from the drawing, so it cannot be started until the drawing is finished.

Gil tells you that:

Diamin Corporation can let you have the drill for five working days (but no more). But he can't let you have it until 14 days from today, because the salesman has promised to let Northwest Mining Company field-test it until then.

The drill can be taken to your office for the five days (you and Janice won't have to work at Diamin).

Diamin will pick up and deliver the drill.

The Electro-Max weighs 62 lb (28 kg) as it stands, 84 lb (38 kg) in its shipping case.

He wants the cost estimate one week from today.

He must have the instruction manual, drawing, and parts list six weeks from today.

H. L. Winman and Associates is to supply three typed copies of the instruction manual and parts list, and an ink original on mylar film, plus two blueline prints, of the drawing.

The drawing probably will be large. When reduced photographically, it must fit onto an 8½ in. × 11 in. page (21.6 × 28 cm).

It is 5:15 when you leave Diamin Corporation—too late to go back to the office. But because you know that Vern will want your report quickly, you decide to write it at home and take it in to work with you tomorrow morning. Before doing so, you make a plan of the three parts of the project (see sketch) to ensure that it can be done within the requested time.

Calculate the correct date for two weeks from today, and then enter it and subsequent dates on the sketch so that you can predict the project completion date. (Note: the "Actual" line on the plan is for you to plot progress later, as you do the job.)

Write the trip report as a memorandum addressed to Vern Rogers.

CONTRACT SCHEDULE: "ELECTRO-MAX" MANUALS & DRAWINGS FOR DIAMIN CORPORATION

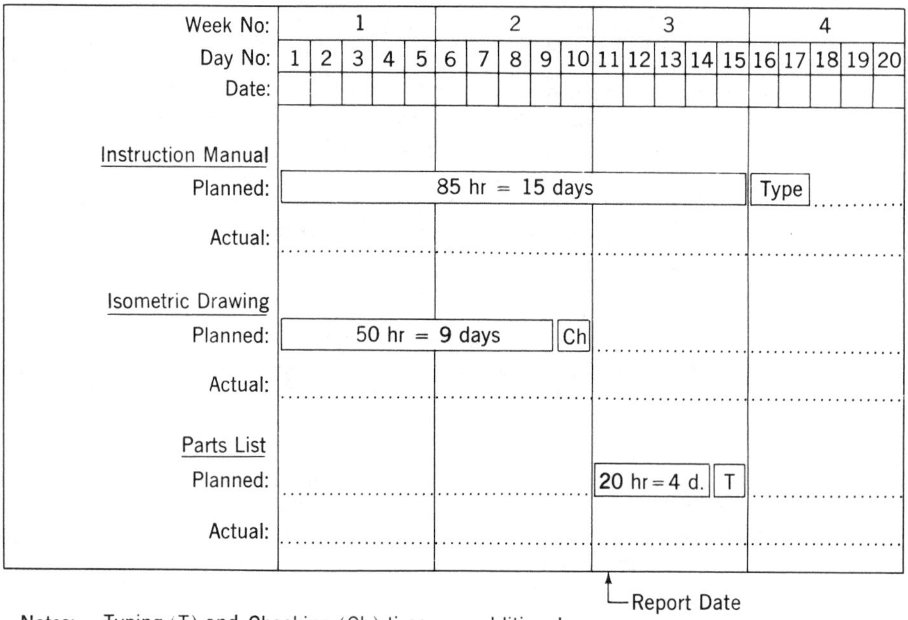

Week No:	1					2					3					4				
Day No:	1	2	3	4	5	6	7	8	9	10	11	12	13	14	15	16	17	18	19	20
Date:																				
Instruction Manual																				
Planned:	85 hr = 15 days															Type				
Actual:																				
Isometric Drawing																				
Planned:	50 hr = 9 days									Ch										
Actual:																				
Parts List																				
Planned:											20 hr = 4 d.				T					
Actual:																				

Report Date (start of Week 3)

Notes: Typing (T) and Checking (Ch) time are additional:

Instruction Manual:	2 days
Isometric Drawing:	1 day
Parts List:	1 day

Part 2. Two days later, Vern asks you to write the cost proposal to Diamin Corporation's production manager, Gil Wishton. Vern gives you these figures to quote:

The instruction manual will cost $2800.00.

The parts list will cost $430.00.

The drawing will cost $1040.00.

The cost for all three will be $4270.00, federal and state taxes extra.

Vern suggests you include specific details, such as project start date, delivery date of the finished manuscripts and drawing, exactly what you will provide to Diamin Corporation, and other factors you consider necessary to avoid a misunderstanding (details of Diamin Corporation's requirements are in Part 1).

You are to write the cost proposal as a letter, which Vern Rogers will sign.

Part 3. Diamin Corporation awarded the contract to H. L. Winman and Associates, with work due to start on the date shown on your schedule (see sketch).

This morning your branch manager (Vern Rogers) asks you to write him a brief progress report describing work done to date, any problems you have run into, and an estimate of whether you will complete the project on schedule. He says he will be seeing Diamin's production manager (Gil Wishton) tomorrow morning on another matter but would like to have progress information available if Gil asks for it.

You use these details to write your progress report:

As this is the morning of day 11, the midpoint of the contract period, theoretically the job should be more than half complete; it isn't.

The Electro-Max drill didn't arrive from Diamin Corporation until the morning of the third day, instead of the first day as promised by Gil Wishton. Because you couldn't work until day 3, Gil let you have the drill for days 3, 4, 5, 6, and 7 of the contract period, instead of days 1 through 5. This immediately put you two days behind schedule.

Janice Kempson, who is doing the isometric drawing, had a flu virus and was absent for days 9, 10, and 11 of the contract period. She called you this morning to say that she thinks she will be back at work tomorrow (day 12).

Work cannot start on the parts list until Janice's isometric drawing is complete.

Although the late arrival of the drill should have allowed you to overshoot the original project completion date by two days, you know that Vern Rogers wants you to stick to the contract schedule (he feels it will help gain future contracts for H. L. Winman and Associates).

To visualize the situation better, draw in the actual situation and predict your completion dates for each part of the contract (don't forget to "block out" any holidays which occur during the period). You can describe how much work has been done either as a number or as a percentage of the workdays completed. This can be done both for the total job (which has 32 workdays), and for each part of the job (including typing and checking, the instruction manual requires 17 workdays, the drawing requires 10 workdays, and the parts list requires 5 workdays).

Investigation Report

PROJECT NO. 11: COPING WITH A NOISY NEIGHBOR

Part 1. You are a technologist in the Cleveland office of H. L. Winman and Associates. This morning Department Head Peter Bell calls you into his office.

"I've had a letter from the area manager of Mansask Insurance Corporation," he explains. "His staff are complaining it's too noisy in the office—that they get headaches from it and go home tired."

Peter hands you the letter, dated yesterday:

MANSASK INSURANCE CORPORATION
210 – 381 Fort Street
Montrose, Ohio 45287

Mr. P. Bell, P.E.
H. L. Winman and Associates
475 Reston Avenue
Cleveland OH 44104

Dear Mr. Bell:

My staff complain that for the past three months the noise level in our office has been too high. They claim that it is affecting their work and causing fatigue.

Will you please look into this problem for me to see if their complaints are justified. If they are, will you suggest what can be done to remedy the problem, recommend the most suitable method, and include a cost estimate.

Yours truly,

David L. Wizowaty
Area Manager

"I also talked to him briefly on the phone," Peter continues. "He figures the staff are exaggerating a bit, that the noise isn't nearly as bad as they make out. Just the same, he's worried: staff turnover has been much higher recently, and it's costing him a fortune to train the replacements he has to hire."

Peter asks you to take on the project and assigns it project number C62.

At 4:00 P.M. the same day you visit Mansask Insurance Corporation and talk to David Wizowaty. You notice a background hum, which you consider to be caused by motors in the electric typewriters and accounting machines. You are still there when the office staff quits at 4:30. After they go, you notice you can still hear the hum, but at a lower level.

You walk around the office with Wizowaty, who plagues you with questions: "What do you think?" he asks. "Seems like the same noise level you get in any business office. Eh?"

It's apparent he is hoping for a good report from you, which he can use to prove to his staff that their complaints are imaginary.

"I can't tell without taking readings," you hedge. "Noise is a pretty tricky thing. What some people think is too noisy, others hardly notice."

But you do notice that the hum gets significantly louder near the east wall of the office. Then suddenly it stops; or, rather, dies away. The time is 4:45.

"What's on the other side of this wall?" you ask.

"Oh, that's Adanac Novelties," Wizowaty replies. "They distribute cheap imports—that sort of thing."

"Have you talked to them about the hum?"

"Yeah! I asked Frank Doherty about it—he's the manager next door. Pretty hostile, he was."

"And what time do they quit work?" you ask.

"Right now," Wizowaty replies. "You can always tell, because they switch their machines off."

You arrange with Wizowaty to take sound-level measurements one week from today. You want to find out how much of the noise in the insurance office is generated by normal office activity and how much by the machines next door.

You consider a visit to Adanac is essential, since you think that the machines next door offer the only real problem. You want to know the sound levels on both sides of the wall between the two companies, to assess the extent of soundproofing you probably will recommend.

Write to Frank Doherty, Manager of Adanac Novelties, to ask permission to carry out sound-level measurements one week from today. His business is at room 208, 381 Fort Street.

Part 2. It is now one week later. Taking a GenRad 1551-C Sound-Level Meter with you (see Figure 7-1 in Chapter 7), you return to 381 Fort Street. You plan to measure sound levels at various locations in the Mansask Insurance Corporation offices under four circumstances:

When both businesses are empty.
When only Adanac Novelties is working (4:30–4:45).
When only Mansask Insurance is working (8:00–8:15).
When both businesses are working.

You also plan to take readings in Adanac Novelties' office.

As Mr. Doherty has not replied to your letter, yesterday afternoon you telephoned him to ask if you could come in today to take the measurements. He said he was "terribly busy" and that it's "damned inconvenient," but he somewhat reluctantly agreed.

You record the measurements you take (see table), and compare them to the general ratings for office noise, which are:

Quiet office: 30–40 dB
Average office: 40–55 dB
Noisy office: 55–75 dB

You note that the sound level in the Mansask office increases as you move toward the dividing wall between the two offices (see drawing).

You also notice there seem to be two components of noise in Mansask's offices, some being transmitted through the air and some being transmitted through the structure (from Adanac's machines, through the floor). A hand placed on the walls or floor feels the vibration. Floors in both offices are tiled.

Before leaving Mansask you tell David Wizowaty that there seems to be a noise problem, but it can be corrected. You warn him, however, that it may prove expensive. He says he will have a job justifying the costs to his head office.

Average Sound Levels				
Second Floor, 381 Fort Street, Montrose, Ohio				
Mansask Insurance Corporation	*Both Offices Working*	*Only Adanac Working*	*Only Mansask Working*	*No One* Working*
A	74 dB	73 dB	48 dB	27 dB
B	71 dB	69 dB	51 dB	27 dB
C	66 dB	65 dB	50 dB	27 dB
D	64 dB	61 dB	52 dB	26 dB
E	63 dB	59 dB	51 dB	28 dB
F	59 dB	53 dB	49 dB	26 dB
G	58 dB	51 dB	48 dB	28 dB
Adanac Novelties				
H	86 dB	–	–	25 dB
I	83 dB	–	–	26 dB

*Mostly air-conditioning noise.
Note: Measurements made with GR 1551-C Sound-Level Meter set to "A" scale.

On returning to your office you write a progress report for Peter Bell. (To do this, you may want to use some of the information in Part 3).

Part 3. You consider possible ways to reduce the sound levels in Mansask Insurance Corporation's offices:

1. Erect a false wall, insulated internally with Corrugon, from floor to ceiling on Mansask's side of the wall between the two companies.
2. Glue ¾ in. (19 mm) black cork panels on the Mansask side of the wall.
3. Install carpet throughout the Mansask office.

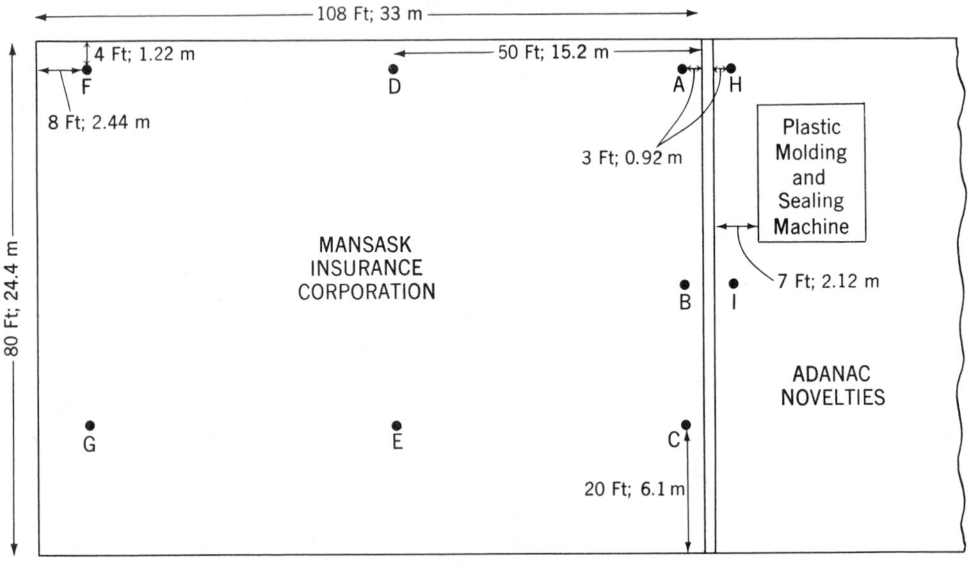

4. Mount Adanac's molding machine on Vib-o-Rub (insulating rubber that eliminates transmission of vibration from machine to building structure).

You recognize that remedies (1) and (2) are alternatives (they both deal with sound transmitted through the air). Remedies (3) and (4) also are alternatives (they both dampen vibrations and sound carried through the structure). Remedy (3) also quite effectively dampens internal office noise.

You consider the approximate costs:

Remedy (1)—$6,800
Remedy (2)—$1,400
Remedy (3)—$13,800
Remedy (4)—$800

You consider possible problems each remedy may offer:

1. Corrugon is in short supply; delivery time would be a minimum of 3 months.
2. To some people, cork has an offensive smell; this can be partly corrected by treating the cork with polymethynol.
3. The carpet must be dense and have a good quality rubber underlay (included in the approximate cost above).
4. Depends on cooperation of Adanac Novelties' manager.

You calculate probable noise reductions for each method:

Remedy 1: 6 to 10 dB
Remedy 2: 4 to 7 dB
Remedy 3: 8 to 12 dB
Remedy 4: 3 to 5 dB

(These anticipated reductions apply to the Mansask office only, when both businesses are working.)

You consider which alternatives to recommend to Mansask Insurance Corporation and then write an investigation report, either in letter form or as a semiformal report accompanied by a brief letter summarizing your findings. Prepare the letter for Peter Bell's signature.

Note: Because David Wizowaty knows little about noise and its effects, you may want to include some explanatory information for him. You would also be wise to research and document such information at a learning resources center, to establish positive evidence for the statements you make in your report.

The following projects in Chapters 5 and 6 can also be used as investigation reports:

Chapter 5: Projects 1, 2, and 3.
Chapter 6: Projects 2, 3, and 4.

5

INFORMAL REPORTS DESCRIBING IDEAS AND CONCEPTS

The previous chapter discussed reports that deal almost entirely with facts, known data, or tangible evidence. Only at the end of any of these reports does the writer have the opportunity to introduce new ideas or concepts. The reports in this chapter have a reverse flow: they start with an idea or a concept and then develop it, using both tangible and intangible evidence to reach a conclusion based on facts and probabilities. They are written in a fluent narrative style that is both persuasive and convincing; their writers have new concepts and techniques to present, and they want their readers to understand their line of reasoning.

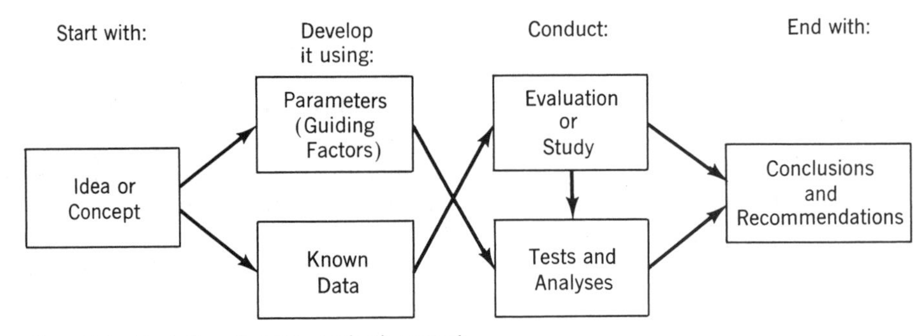

Figure 5-1. Basic flow diagram—evaluation report.

Evaluation Report

An evaluation report starts with an idea or concept its author wants to develop (see Figure 5-1). The writer establishes guidelines that will keep its development within prescribed bounds, and researches known data to analyze the idea. He or she then evaluates this concept in the light of the parameters that have been set and the data that has been amassed, conducting tests and analyses to prove or disprove the theories. At the end of the study the writer is able to draw the conclusion that the concept either is or is not feasible, or perhaps is feasible in a modified form.

The Robertson Engineering Company report on training methods in Figure 5-2 is an evaluation report because it deals with ideas and concepts rather than with facts and events. The *idea* is management's belief that new training methods will increase production; the *known data* are the four training methods discussed by Fred Stokes; the *evaluation* is his assessment of each method, or combination of methods, in the light of the company's training needs. The standard parts that shape the narrative of this report are, however, the same as those for the investigation report (see Chapter 4).

The comments below identify the standard parts that Fred Stokes has used and discuss briefly how they have helped him fashion an effective report.

Paragraph 1 is a brief *summary* of the evaluation and main conclusions.

Rather than write a summary of the main points of the whole report, Fred homes in on the one point of immediate concern to management: How much is all this training going to cost? He has followed the first rule of summary writing: Identify what is most important to your readers *before you write.*

Paragraphs 2 and 3 provide *background information* and details of events that led up to the evaluation.

Fred knows that the person to whom the report is directed will already be aware of much of this information, but he also recognizes that there may be other readers to whom it may be new.

Paragraph 4 is a *guiding factor* that limits the extent of the evaluation.

Fred probably discussed these limitations with Wayne Robertson when he first established his terms of reference. No engineer or technician should attempt to write a report without first clearly establishing the objectives.

Paragraphs 5 through 9 comprise *investigation details.* Paragraph 5 outlines the *investigation method,* while paragraphs 6, 7, 8, and 9 each evaluate one of the training aids.

Fred's analysis is interesting and logical. He uses a three-step approach that makes the evaluation of each training method a small report in itself: (1) a brief *introduction* of the training method and how it is used; (2) a *discussion* of its application to industry and the implications of using it; and (3) a *conclusion* that sums up the value of the training method to Robertson Engineering Company.

Paragraphs 10 through 15 *discuss evaluation results.*

By first introducing the guiding factor that will most influence his selection, Fred is able to eliminate one training method quickly, identify another that has to be used, and establish that either of the two remaining methods can be used with it. His discussion is unquestionably logical, and follows naturally from the conclusions drawn at the end of the four analyses.

In paragraph 14 Fred outlines his preference and gives reasons for it, to prepare readers for the recommendation he will make. Although he knows management wants him to indicate his preference (see paragraph 3), he also recognizes that management may choose not to accept his recommendation (see paragraph 15).

robertson
engineering
company

MEMORANDUM

From: F Stokes, Chief Engineer

Date: 20 November 19__

To: W D Robertson, President

Subject: Updating Our Training Methods

Summary

1. If we are to meet our production schedules with the semiskilled labor currently available, we will have to augment the informal on-the-job training method we currently use by introducing more comprehensive instruction, either in a classroom or by closed circuit television. A classroom would be less costly than a television system ($11,000 vs $36,000 to set up; $28,000 vs $31,000 annually to operate), but CCTV offers many peripheral advantages which make it the better choice.

Introduction

2. Over the past five years the scientific instruments we manufacture have become increasingly complex and have demanded highly sophisticated manufacturing techniques. At the same time our source of skilled labor has decreased, so that we have had to use semiskilled and unskilled labor for many positions on the assembly lines. This has resulted in a gradual slowdown of production, an increase in the number of manufactured units rejected by Quality Control, and a consequent increase in manufacturing costs.

3. At the management meeting held on 20 September 19__ it was recognized that a better training program than we now have would do much to improve the manufacturing skills of our employees. I was requested to evaluate alternative training methods and to recommend a training technique that might increase productivity.

4. I have confined my report to an analysis of training methods that could be used in our plant. Development of a specific training program I will discuss in a separate report to be prepared after the training method has been selected.

Analysis of Industrial Training Methods

5. There are four basic training methods, ranging from inexpensive established methods to costly new techniques, that we could use to train our staff. The suitability of each is discussed generally here and is supported by a comparison chart attached as an appendix.

6. On-the-Job Training (OJT)

6.1 This time-honored training method is used extensively in industry and is the method currently used on our production lines. It calls

Figure 5-2. An evaluation report.

for trainees to learn manipulative tasks in which a sequence of movements has to be mastered. The trainees "learn by doing" under the watchful eye of an instructor, who often is another fully trained employee. When they have learned the task, they are given a proficiency test to assess whether they can work on their own.

6.2 Every time we hire new employees, or transfer existing employees from one production line to another, they are given OJT which continues until they are fully able to perform the new task. It is an effective training method, but is slow and has hidden costs: trainees are poor producers for a long time, and a higher-than-normal proportion of their work is below standard and has to be rejected.

6.3 To continue relying solely on OJT would be to perpetuate a training method that has already been proved inadequate for our needs.

7. Classroom Instruction (C.I.)

7.1 The classroom situation is the form of training most of us know best. Employees recognize this type of teaching and, even though some may resent the schoolroom atmosphere it evokes, they usually respond quite readily to it. Courses follow a basic pattern, with the instructor setting objectives, preparing instructional material to meet them, then presenting lectures. An end-of-course test demonstrates whether the trainees have learned enough to be able to perform their tasks under supervision.

7.2 To be effective, C.I. demands the proper atmosphere, a good instructor, and receptive trainees. This means providing space for a properly equipped training room rather than just a corner of the workshop; it means keeping an instructor who can teach specialist subjects on the payroll, or drawing instructors from the supervisory line staff (this must be done carefully because not all supervisors make good instructors); and it means making sufficient employees available for training at the same time.

7.3 The cost of classroom instruction is high in relation to the degree of training it provides. It is particularly suitable for teaching theory, but of less value when purely manipulative skills must be learned. Hence it must be supplemented by practical training such as OJT.

8. Programmed Instruction (P.I.)

8.1 Programmed instruction is a relatively new technique that only recently has been generally accepted as a useful training aid. Because it permits students to work alone and at their own speed, it is invaluable for teaching routine tasks to very small groups or to individuals.

8.2 The type of programmed instruction most of us recognize is the programmed textbook that teaches a small piece of information at a time, tests the trainee on what he has just learned, provides correction if he still has not understood, and then proceeds to the next piece of information. Trainees progress quickly or slowly through the program, depending on their previous knowledge and ability to learn rapidly. They are tested periodically, and at the end they are given a proficiency test. Except for monitoring the trainees' progress and marking their final tests, an instructor is not needed.

8.3 P.I. textbooks are ideal for teaching specialist subjects and new techniques, but seldom can be used to teach manipulative skills. Therefore, any P.I. manuals we produce would have to be supplemented by some verbal instruction and practical training. Preparation of P.I. materials is also expensive and very slow -- too slow to meet the rapid training needs of many of our production lines.

9. Closed Circuit Television (CCTV)

9.1 As an educational training aid, CCTV is rapidly gaining popularity with departments of education across the country. Because its initial cost is high, until recently it has been limited primarily to colleges and schools. But there is no reason why the new, less expensive systems should not be used in industry.

9.2 Training lessons showing manufacturing processes, assembly techniques, packaging methods, etc, could be prepared as TV "programs" and stored on videotape ready for playback to trainees whenever the need arises. We would be able to show programs repeatedly, either to groups or individuals, at virtually no cost. We could demonstrate the right and wrong ways of doing things, and enlarge tiny objects for all to see simultaneously. And we could even install monitors right on the production line for employees to follow step-by-step as they assemble components. CCTV would give us the ability to prepare and record instructional programs quickly, so that teaching of new assembly methods would be able to start very early in future projects.

9.3 The initial cost of setting up CCTV would be much greater than for the other three methods, and operating costs would also be higher. A staff member would have to be trained in TV production techniques, and a small room would be required for equipment storage and playback. The advantages of CCTV should, however, outweigh many of the objections to its high cost. The familiarity of TV, combined with its uniqueness as an inplant teaching method, would promote learning and employee acceptance of training courses. And the image of a progressive company that it would convey would be invaluable for building employee morale.

9.4 CCTV would not, however, be able to stand alone as a training medium. We could use it to teach theory and to demonstrate manipulative skills, but it would not entirely replace the final step: practice. This would still have to be provided by OJT.

Selection of a Training Method

10. The training method we select must teach both theory and manipulative skills, and also provide opportunity for practice. Since none of the training methods I have discussed meet all of these requirements, we will have to use a combination of methods that provides the best training for the least cost.

11. OJT must be retained because it is the only method which provides the practice that is essential before a trainee can become fully productive. But because OJT has so many hidden costs, it must be reduced to the minimum time possible by combining it with the most effective teaching method.

12. P.I. can be eliminated because of its high costs, slowness of preparation, and inability to teach complicated manipulative skills.

13. Either C.I. or CCTV could be combined successfully with OJT to provide a comprehensive training program. Both would significantly reduce the time required for OJT. The C.I./OJT combination would cost $45,000 the first year, and $34,000 per year thereafter. The CCTV/OJT combination would cost $73,000 the first year, and $37,000 per year thereafter. CCTV, however, offers many peripheral advantages that C.I. does not.

14. I consider that the advantages of the CCTV/OJT combination outweigh those of the C.I./OJT combination. The image that CCTV would convey and the enthusiasm it would spark are intangible factors that cannot be measured in dollars. CCTV equipment could also be used for activities such as sales seminars, management training, time and motion studies, and advertising, so that maximum value would be obtained from our investment.

15. Whichever plan is adopted, I will prepare specifications for purchasing the equipment, and will develop a detailed training implementation program.

Conclusions

16. Training methods can be improved at Robertson Engineering Company by combining the existing on-the-job training with either:

    16.1 classroom instruction, at moderate installation and operating cost, or

    16.2 a closed circuit television system, at high installation but moderate operating cost.

Recommendation

17. I recommend that we update our training methods by installing a CCTV system costing $36,000, and budget for an annual operating cost of $37,000.

F Stokes

APPENDIX: COMPARISON OF FOUR TRAINING METHODS

	OJT	C.I.	P.I.	CCTV
What is initial purchase and set up cost?	Nil	$11,000	$2,000	$36,000
What are annual operating costs? (including instructors salaries)	$6,000	$28,000	$64,000	$31,000
Can training method teach both theory and practice?	No; mostly practical skills	No; mostly theory	No; almost entirely theory	No; mostly theory, some practice
How quickly do trainees learn?	At moderate speed	At moderate speed	Depends on student & program	At moderate speed
How readily do trainees accept training method?	Readily	Usually readily	Varies; resisted by some	Usually readily
Can training be done on production line?	Yes	No; generally not	No; rarely	In part (direct assembly training)
How quickly can a specific training program be set up?	Rapidly	Fairly quickly	Slowly	Fairly quickly
Can students be taught independently or must they work in groups?	Mostly independently	Generally in groups	Only independently	Either
Can training method be used without an instructor?	Some of the time	No	Yes	Yes
Can equipment/facilities be used for other purposes?	No	Yes	No	Yes
Has training method any special requirements or limitations?	Interferes with production	Requires skilled instructor	Needs specialist writer; students must read well	Needs TV producer/ instructor

Data source: File ENG/26/3107

Paragraphs 16 and 17 state the main *conclusions* and make a *recommendation.*

Fred ends his report by briefly outlining the two alternatives, and then stating clearly which he recommends. This section of a report is sometimes referred to as the *terminal summary.*

The *attachment* compares the methods in an easy-to-digest form.

Rather than support his short report with numerous pages of data, Fred has chosen to compare the training methods very generally in a single table and to refer readers who want more substantive data to the source of his information. (This can be done only when the source is readily available to readers.)

Note that Fred Stokes does more than simply describe each training method in the investigation details section. By discussing the advantages and disadvantages of each method, and then skillfully inserting a persuasive concluding statement at the end of each description, he conditions you to agree with his conclusions *before* you read them. This does not mean he can allow personal bias to creep into his work: his evaluation must be objective (for the good of the company), even though it may seem to be persuasive.

The first draft of Fred's report contained no headings: it was continuous narrative, just like an ordinary memorandum. He inserted them on the advice of Anna King, who told him: "In a multipage letter or memorandum, headings can help readers *see* how you have organized your thinking." The headings Fred uses are those in Figure 4-8 in the previous chapter. Paragraphs 5 through 15 of his report are his *discussion.*

The difference between an evaluation report and an investigation report can be seen more readily by comparing two similar situations that call for different types of reports. The previous chapter described a situation in which Harvey Winman *has decided* to purchase an executive jet aircraft, and assigns two engineers to *investigate* which type of aircraft would be most suitable. Now suppose that Harvey only *has an idea* that an executive jet might be useful to him and his company. He assigns an engineer to *evaluate* whether purchase of a private aircraft would be practicable and possibly to recommend what action should be taken.

This engineer's first step is to establish parameters: How often would an aircraft be used? How many persons would use it? What sort of budget allocation can be made available for its purchase and operation? What is the likely route length? The engineer then identifies the types of aircraft available, collects data on each (such as purchase price, operating costs, and performance), and looks at leasing as an alternative to purchase. Armed with all this data, he or she evaluates the advantages of having an executive jet under varying conditions, developing cost analyses for each situation. (In practice this would be done by computer, with the results possibly expressed as a series of benefit-cost ratios.) From this evaluation the engineer establishes whether the idea is economically feasible; if it is, whether it would be better to purchase or rent an aircraft; and which type of executive jet would be most suitable for the company's operations. The report submitted to Harvey Winman would be an evaluation report because it starts by outlining an idea, develops it by using known information, and ends by drawing definite conclusions based on an objective analysis of concepts, data, and parameters.

Feasibility Study

A feasibility study is very similar to an evaluation report. It also starts with an idea or concept, and then develops and analyzes it to assess whether it is technically or economically feasible. The chief difference lies in the name and application of each document. An evaluation report is generally based on an idea that is originated and evaluated within the same company; hence it is nearly always informal. A feasibility study is normally prepared at a slightly higher level: the management of company A requests company B to conduct a feasibility study for it, because company A does not have staff experienced in a specific technical field. It is unlikely, for instance, that a company engaged solely in wholesale distribution of dry goods would have the capability to assess the feasibility of purchasing an executive jet. The company would seek advice from a firm of management consultants, who would submit their findings in a report called a feasibility study.

When a feasibility study is prepared for a client, the document that results may be either a letter or a formal report. A letter is used when the project is relatively small and there is no need to present the information formally. If Fred Stokes's memorandum report evaluating training methods (Figure 5-2) had been prepared for a client, it could have been called a feasibility study and written as a letter report. There would have been a few minor changes in narrative and format, but the basic information would have been the same. Alternatively, it could have been made more impressive if it had been prepared as a formal report, like the "Evaluation of Sites for Montrose Residential Teachers' College" in Chapter 6. The margins of difference among the feasibility study, the evaluation report, and the technical brief (described in the next section) are so narrow that it is frequently personal preference that dictates which label is applied.

Technical Brief

The purpose of a technical brief is to introduce a new idea and to develop it in sufficient depth to enable the reader to assess its practicability. The idea should be innovative, logical, and practical, and the brief should indicate that it is worth the reader's time to pursue the subject further. It can be originated by almost anyone and cover almost any topic or situation: a supervisor with a novel scheme for calibrating test instruments, a salesperson with ideas for strengthening a product line, a technician with a new method for testing mechanical components, and an engineer with plans for a unique product can all present their ideas to management in a technical brief.

Scientists, supervisors, technicians, engineers, and technologists write memorandums every day to their managers suggesting new ways of doing things. These memorandums become technical briefs if their writers have developed their ideas sufficiently to convince management that their proposals are sound. A memorandum that introduces a new idea without demonstrating fully how it can be applied is no more than a suggestion. A memorandum that introduces a new idea, discusses its advantages and disadvantages, demonstrates the effect it will have on present methods, calculates cost and time savings, and finally suggests what further developments might accrue, proves that its originator has done a thorough research job before attempting to put pen to paper. Only then does a suggestion become a technical brief.

The technical brief in Figure 5-3 promotes development of an audiovisual training aid called the APL System. Although its originator, Ron Brophy, could have written his brief entirely as an interoffice memorandum, he has chosen to prepare it as a separate document and to preface it with a short covering memo to Jim Perchanski, his department head:

From: R. S. Brophy　　*Date:* January 13, 19____
To: J. Perchanski　　*Subject:* New product design for the education field

The attached technical brief describes "The APL System," which is a method of Audiovisual Programmed Learning I have devised. I believe it could be developed into a marketable product for the education field.

Its development grew out of a conversation I had with Les Walters, head of the Training Aids Department of Montrose Community College, when we were designing modifications for their remote control room. He would be interested in evaluating a prototype system in a small classroom, which would give us a chance to assess its potential without excessive development costs. An initial system for up to 16 students could be built for about $3200.00.

If we do install a small system, Les Walters has promised to develop a short program to test it.

Ron Brophy

This arrangement allows additional copies of the brief to be printed without the memo, for distribution to readers within and outside H. L. Winman and Associates. The memorandum opens the door for further development by suggesting where the APL System can be tested and how much money would be involved in initial development.

Within the brief, Ron describes why there is a need for such a system and establishes parameters for its development. Since he is dealing with a concept, he discusses how the system can be applied, describes clearly how it works, and estimates what it would cost. Then, when he has presented enough evidence to convince Jim of the system's capabilities and practicability, he draws a brief conclusion encouraging further development. His approach is generally similar to that used for an investigation report (Chapter 4).

Not all technical briefs are generated through the writer's own initiative. An idea may spring unexpectedly out of a conversation between engineers and

H L Winman and Associates

Technical Brief

DESIGN FOR AN INEXPENSIVE AUTOMATED TEACHING AID:
"THE APL SYSTEM"

Prepared by

Ron S. Brophy
Development Engineer

Audiovisual Programmed Learning (APL)is a system I have devised which would combine the simplicity of audiovisual training aids with existing programmed learning techniques to provide an effective method for automatic teaching of groups of students. Cost for a 40-student system would be about $5500.

Introduction

Before developing the APL System I briefly studied the automated teaching machines currently available in the educational field, or under development, and found they are divided into two general groups. The first comprises a fairly wide variety of simple, fairly cheap machines which can be used by one person at a time and normally require a written response. The second is a smaller range of very sophisticated, highly expensive systems that simulate complex conditions and situations, often presenting information on a television screen; these vary from one-operator units to complete classroom installations.

I considered that between these groups there is a need for an inexpensive, basically simple system that can be used by a whole class simultaneously. Working on this principle, I established the following parameters as representative requirements for an "ideal" system:

Equipment: The system must handle up to 40 students simultaneously and use off-the-shelf hardware to keep development costs low. It should also be:

* Mobile, or a fixed installation that does not interfere with the classroom's normal role.

* Robust and tamperproof.

* Simple to maintain (i.e. it should not need costly test equipment or specialist repair skills).

1

Figure 5-3. A technical brief.

Operation: The system should be simple to set up and operate, for both teacher and students, and preferably be fully automatic. It should also:

* Demand continual student participation.

* Monitor student participation individually.

* Test student understanding at regular intervals.

* Detect student misunderstanding and provide immediate remedial instruction.

* Eliminate the need for written answers.

* Provide test scores.

Price: The system, and the programs to be used with it, should be cheap enough to make APL a desirable A/V aid for a school to acquire. As a rough guide: no more than $6000 for an installed system, and $50 for a 30-minute program (lesson).

General System Description

The system I have planned consists of a series of 35-mm slides projected onto a screen, accompanied by a spoken commentary heard by each student through a pair of headphones. Periodically the slides and commentary contain multiple-choice questions which the students have to answer by depressing one of three pushbuttons, corresponding to three possible answers. A multiple-track commentary then informs each student separately whether or not his answer is correct and, if incorrect, provides further instruction to correct his knowledge. Simultaneously, a counter registers his total number of correct responses.

The complete APL System comprises:

*** At the teacher's position: a tape recorder reproducer (wired for playback only and modified for simultaneous reproduction of all four channels of 4-track quarter-inch tape), an automatic slide projector, and a screen.

*** At each student position: a pair of headphones with an individual volume control, three pushbuttons for selecting answers to questions, and a small three-digit counter for recording the number of questions answered correctly. (There is no limit to the number of student positions which could be established; for practical purposes, 50 would seem a likely maximum.)

These components are itemized in attachment 1 and shown as part of an overall system in the diagram on page 3.

APL System Operation

The slide tray and the 4-track tape are loaded into the projector and the tape deck. At each student position the counter is reset to "000" (counters are the pushbutton "total reset" type, therefore cannot be advanced by the students). The ouput from each of the four tape recorder tracks is applied as follows:

1. The output from track 4 is fed only to the slide projector. This track carries "beeps" (low frequency tone signals, 100 to 200 Hz) which cause the projector to change slides at the correct points in the spoken commentary.

2. The audio outputs from tracks 1, 2, and 3 are fed to pushbuttons "A", "B", and "C" at each student position. Individual students hear only the commentary carried by the track they have selected by depressing one of the pushbuttons.

3. The counters at each student position are triggered by inaudible low frequency tone signals inserted immediately after each question, but placed on only one of the three audio tracks (always the track carrying the commentary signifying that a "correct" answer has been selected). Thus counters advance one digit only at those student positions where the depressed pushbutton corresponds to a correct answer.

Initially, the same commentary is carried on all three tracks. Slides projected onto the screen change as the commentary progresses, until a point is reached when the program needs to test student assimilation of the instruction.

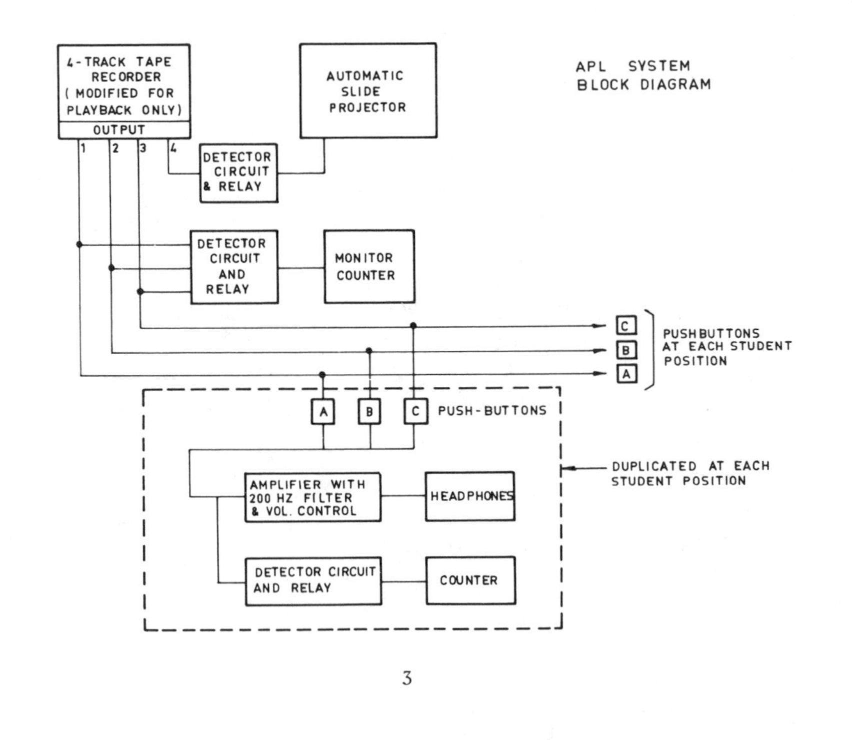

3

The next slide projected onto the screen poses a question and provides three possible answers, coded "A", "B", and "C". The commentary asks students to read the question and depress the pushbutton corresponding to the answer they consider correct. (For the example below, assume answer "B" is correct.)

Just before the answer commentaries start, a tone signal inserted on track 2 (answer "B") causes the counters of all students who have selected the correct answer to advance one digit. Different commentaries are then carried simultaneously on the three tracks:

Tracks 1 and 3 (Answers "A" & "C")	The commentary informs students that the answer they have selected is not correct. It then describes why it is incorrect and why they should have selected answer "B".
Track 2 (Answer "B")	The commentary informs students they have selected the correct answer. It then explains briefly why the answer is correct in case a student was unsure of the correct answer and hopefully selected answer "B".

Further instructional slides and a general commentary common to all three tracks continue until subject assimilation again needs to be checked. This procedure continues throughout the lesson, instructional material alternating with test questions to provide positive feedback of student response. At the end of the lesson individual results are read from the counter at each student position.

APL System Costs

Because the need for extensive wiring would make a mobile system impractical, I have priced only a permanent installation.

I estimate equipment costs, based on a permanently installed 40-student system, would be $5500.00, made up of:

Instructor's equipment, including the wiring network:	$1700.00
Students' equipment (40 sets @ $45.00):	$1800.00
Installation costs:	$2000.00

Program costs are more difficult to estimate because they depend on intangible factors such as complexity of subject, cost of writer and artist, and number of copies to be made. I estimate each program, packaged in a slide tray and cassette, could be priced at $50.00, based on an average development cost of $7500.00 and reproduction costs of $2500.00 for 200 units.

A detailed cost breakdown is contained in attachment 2.

Conclusions

The foregoing represents a first assessment of a simple, practical method of whole-class automated teaching. Because it uses equipment and methods that are already accepted as valuable classroom training aids, it should not meet with the resistance sometimes accorded by the teaching profession to automated teaching systems. The combination of relatively low price and simplicity of operation should make it attractive to both teachers and school boards. Thus I believe it has the prerequisites of a marketable product.

R. S. Brophy

R. S. Brophy
April 1, 19__

NOTE

Because of their length, Ron Brophy's comprehensive supporting details have been omitted from this copy of his brief. They comprised:

Attachment 1 -- A detailed list of components (almost a specification sheet).

Attachment 2 -- A detailed cost analysis.

5

technologists, resulting in one person being asked to prepare a brief for management's perusal. Or an innovative technician may construct a test jig simply to speed up part of a job, and it may prove so effective that he or she is prompted by colleagues to "write it up" for the boss. The write-up, if it is developed properly, becomes a technical brief.

Technical Proposal

There are two kinds of technical proposals: those written on a personal level between individuals, and those prepared at company level and presented from one organization to another. The former are usually informal memorandums or letters; the latter are more often formal letters or reports. Both are similar to a brief, but both take the brief one step further: *they recommend what action should be taken.*

A proposal follows the basic **Summary-Background-Facts-Outcome** arrangement suggested for reports:

1. A brief **Summary** of the situation or problem and how it can be resolved.
2. An **Introduction** that defines the situation and establishes why it is undesirable or is a problem.
3. A **Discussion** of how the situation or problem can be resolved:
   - 3.1 **The Objective:** A definition of the parameters or criteria for an ideal (or good) solution.
   - 3.2 **The Proposed Solution:** What it is, the result it would achieve, how it would be implemented, its cost, and its advantages and disadvantages.
   - 3.3 **Alternative Solution(s):*** For each: what it is, the result it would achieve, how it would be implemented, its cost, its advantages and disadvantages, and why it is not as effective as the proposed solution.
   - 3.4 **Evaluation:** A brief tradeoff analysis, with reference to the **Objective,** comparing the effects of: (a) adopting the proposed solution, (b) adopting an alternative solution,* and (c) taking no action.
4. A **Recommendation** for action, which should be a strong, positive statement or a direct request.

**Omitted if there are no alternative solutions.*

Ron Brophy's technical brief (Figure 5-3) contains some of these parts. To change it into a proposal, Ron would have to state in the summary that he is requesting approval to build a prototype APL System, insert in the narrative that the system will be installed and evaluated at Montrose Community College, and (most important) follow the conclusion with a strong request for action (his Recommendation).

Formal technical proposals are prepared by companies seeking to impress or convince another company, or the government, of their technical capability to perform a specific task. They are normally impressive documents prepared under extreme pressure, and call for techniques beyond the scope of most courses you are likely to encounter.

Assignments

PROJECT NO. 1: EVALUATING MICROFILM READERS

At a meeting held on the 26th of last month, the H. L. Winman and Associates' department heads decided to purchase 20 microfilm readers. All company drawings and most major documents are to be converted to 16 mm microfilm, and June Toska, the company librarian, wants to order future reference information in microfilm form rather than purchase the information in conventional paper form and then convert it to microfilm. The Special Projects Department was asked to investigate available microfilm readers and then place the order. Department Head Andy Rittman assigned the project to you.

Andy tells you that:

1. Of the 20 microfilm readers:
   - 5 are for head office.
   - 3 are for Robertson Engineering Company.
   - 12 are for branch offices.
2. He hopes a bulk order will achieve a lower price per unit.
3. The company has budgeted $35,000 to purchase the 20 units.
4. He wants you to write him an informal evaluation report describing your findings.

You know little about microfilm readers, so you talk first with June Toska. She tells you she needs readers which:

Can accept both 16 mm and 35 mm roll film.

Can be adapted to accept 16 mm cartridges and microfiche cards.

Have clear instructions on the front of the reader (she doesn't want to be continually interrupted to show users how to use machines).

Have bulbs which can be changed easily. ("Some instruments seem to need a genius simply to change a bulb," she quips.)

Have a local repair center—with parts in stock—so unserviceable readers won't be out of commission for long periods.

You question June about cartridges and microfiche: "I guess you can classify cartridges as nice to have," she says. "We probably wouldn't use them much. But microfiche—that's a different story. A lot of technical journals and books are on microfiche—it's the easiest way to store them."

She agrees that not all readers will need microfiche adapters. "At least, not at first," she continues. "Maybe two would be enough: one for my library, and one for Robertson Engineering Company in Toronto."

June agrees to accompany you to the public library, where several readers are in use, and to local suppliers of microfilm readers. At the public library you talk to Ms. Francine Wong, who is head of Information Retrieval.

"The main things we look for in a reader," she says, "are robustness and easy accessibility of controls. We get all sorts of people using it. It has got to be tough, and easy to use."

From a technical point of view, Ms. Wong says a large viewing area is important. "Users get annoyed if they can't see the whole frame at once and have to keep moving the controls to reposition the image." She adds that ability to rotate the image through 180 degrees is essential; through 270 degrees is useful, but not as essential.

"A fixed take-up reel is an important advantage from your point of view," she says to June. "It forces the viewer to rewind the film before removing it. You will always know your films have been properly rewound."

You then visit three suppliers and examine four different microfilm readers, which are described below. The major difference between the readers is the position of the viewing image: some are vertical, immediately in front of the viewer; others are horizontal, so that viewer looks down onto the image.

You suggest to June that you should ask around to see if company engineers have a preference for vertical or horizontal viewing, before deciding which to choose. You also notice that some machines have a manually operated film advancement mechanism whereas others have a motorized mechanism to speed up film movement and frame selection.

The four models you examine are listed below.

Vancourt Manufacturing "25 SQUARE"

Price: $1650.00 (5% less for orders of 10 or more)

Screen: 25 × 25 in. (64 cm sq); vertical; whole image can be viewed on screen.

Mechanism: Manual operation; controls readily accessible.

Accepts: 16 mm and 35 mm roll film; will accept microfiche (adapter cost: $200.00 extra); will not accept cartridges.

Image rotation: 180°.

Magnification factor: 18×.

Lamp replacement: awkward.

Fixed take-up reel: yes.

Instruction panel on front: yes

Delivery: 6 weeks from receipt of order.

Warranty: 3 years parts; 90 days labor.

Supplier: City Business Supplies, 206 Mountain Avenue; only minor repairs done locally; major repairs by manufacturer (Detroit).

Multiple Industries "Midget"

Price: $1295.00 (no discount available).

Screen: 12 × 15 in. (30 × 38 cm); horizontal; cannot display whole image.

Mechanism: Manual operation; some controls rather inaccessible.

Accepts: 16 mm and 35 mm roll film only.

Image rotation: 180°.

Magnification factor: 15×.

Lamp replacement: simple.

Fixed take-up reel: yes.

Instruction panel on front: no

Delivery: 4 weeks from receipt of order.

Warranty: 3 years parts; 90 days labor.

Supplier: Montrose Office Systems, 1600 Western Avenue; all repairs done locally; full range of parts available.

Multiple Industries "Major"

This is a beefed-up version of the "Midget." Details are the same, except for:

Price: $1795.00 (no discount available).

Screen: 24 × 24 in. (61 cm sq); horizontal; displays whole image.

Mechanism: Motorized; controls nicely grouped, readily accessible.

Accepts: 16 mm and 35 roll film, and microfiche.

Magnification factor: dual lens: 18× and 24 ×.

Fixed take-up reel: yes.

Instruction panel on front: yes.

Technetron Inc. "View-It"

Price: $2200.00 (10% discount for orders of 5 or more).

Screen: 24 × 30 in. (61 × 76 cm); vertical; displays whole image.

Mechanism: Motorized; all controls readily accessible.

Accepts: 16 mm and 35 mm roll film, cartridges and microfiche.

Image rotation: 270°.

Magnification factor: triple lens: 12×, 18×, 24×.

Lamp replacement: moderately easy.

Fixed take-up reel: yes.

Instruction panel on front: yes, clear.

Delivery: 10 weeks from receipt of order.

Warranty: 3 years parts; 1 year labor.

Supplier: Technetron Inc. Sales and Service, 1835 Villa Road.

For ease of comparison, you combine information on the four machines into a comparison table, which you can attach to your report. You also survey 100 company engineers and technicians, both at head office and at Robertson Engineering Company, to determine their preference for screen position. Results of your mail survey are:

Vertical screen preferred: 17
Horizontal screen preferred: 23
No preference: 28
No answer: 32.

Finally, you assemble your information into an informal report addressed to Andy Rittman. In it, you discuss the factors affecting selection, establish criteria for the best choice, compare the available models against the criteria, draw conclusions, and make a recommendation.

PROJECT NO. 2: CHOOSING A SITE FOR AN AUTO-MART

Auto-Marts Inc has commissioned the local branch of H. L. Winman and Associates to recommend a site for an Auto-Mart to be built in the southern area of your city. (Auto-Marts are drive-in convenience food stores.) The requirements are spelled out by Sharon M. Wilkinson, vice-president of Market Development for Auto-Marts Inc, who in a letter to H. L. Winman and Associates' Branch Manager Vern Rogers, says:

> Aside from exterior facade and foundation details, Auto-Marts are built to a standard 80- by 30-foot pattern (24.5 by 9.2 m) with an order window at one end of the 80-foot (24.5 m) wall, and a delivery window at the other. Auto-Marts carry only a limited selection of grocery items but boast 90-second service from the time of placing an order to its delivery at the other end of the building. For this reason, Auto-Marts attract the householder hurrying homeward from work, the impulse buyer, and the late-night shopper, rather than the selective shopper. The store depends more on volume of customers than on the volume of goods sold to each individual.
>
> The chief consideration in selecting a site must therefore be a location on the homeward-bound side of a main trunk road into a large residential area. The site must have quick and easy entry onto and exit from this road, even during peak rush hour traffic. The residential area should be occupied mainly by upper and middle class people who own cars and have money to spend. There should be little competition from walk-in grocery stores.
>
> Our market studies show that a location on Evergreen Drive, between the downtown business area and Grover Green residential area, has high sales potential. We want H. L. Winman and Associates to identify possible sites, to analyze them for accessibility, visibility, and ease of entry and exit, and to establish a firm purchase price for each.

Vern Rogers assigns the project to you, with instructions to:

Find suitable sites.

Draw up a plot plan for each, showing the Auto-Mart and entry and exit lanes.

Consider their accessibility both to the homeward-bound shopper and the late-evening shopper driving in from the residential areas.

Consider the positions of advertising signs.

Consider availability of services (water, gas, electricity, and sewers).

Compare land costs.

Write a report of your findings, which both you and he will sign.

You tour Evergreen Drive and identify two possible sites of comparable size and probable equal suitability. They are (see sketches):

Site A–on the west side of Evergreen Drive, at the junction of Alexander Street.

Site B—on the east side of Evergreen Drive, at the junction of Malcolm Street.

Some factors you examine are:

1. Distances from downtown:
   Site A: 1.6 miles (2.6 km)
   Site B: 2.4 miles (3.9 km)
   (Grover Green is 3.8 miles [6.1 km] from downtown.)
2. Zoning: both are zoned for commercial use.
3. Services: both are fully serviced.
4. Condition:
   Site A: site of ice-cream parlor (closed); cost to clear: $8600.
   Site B: clear; no buildings on it.

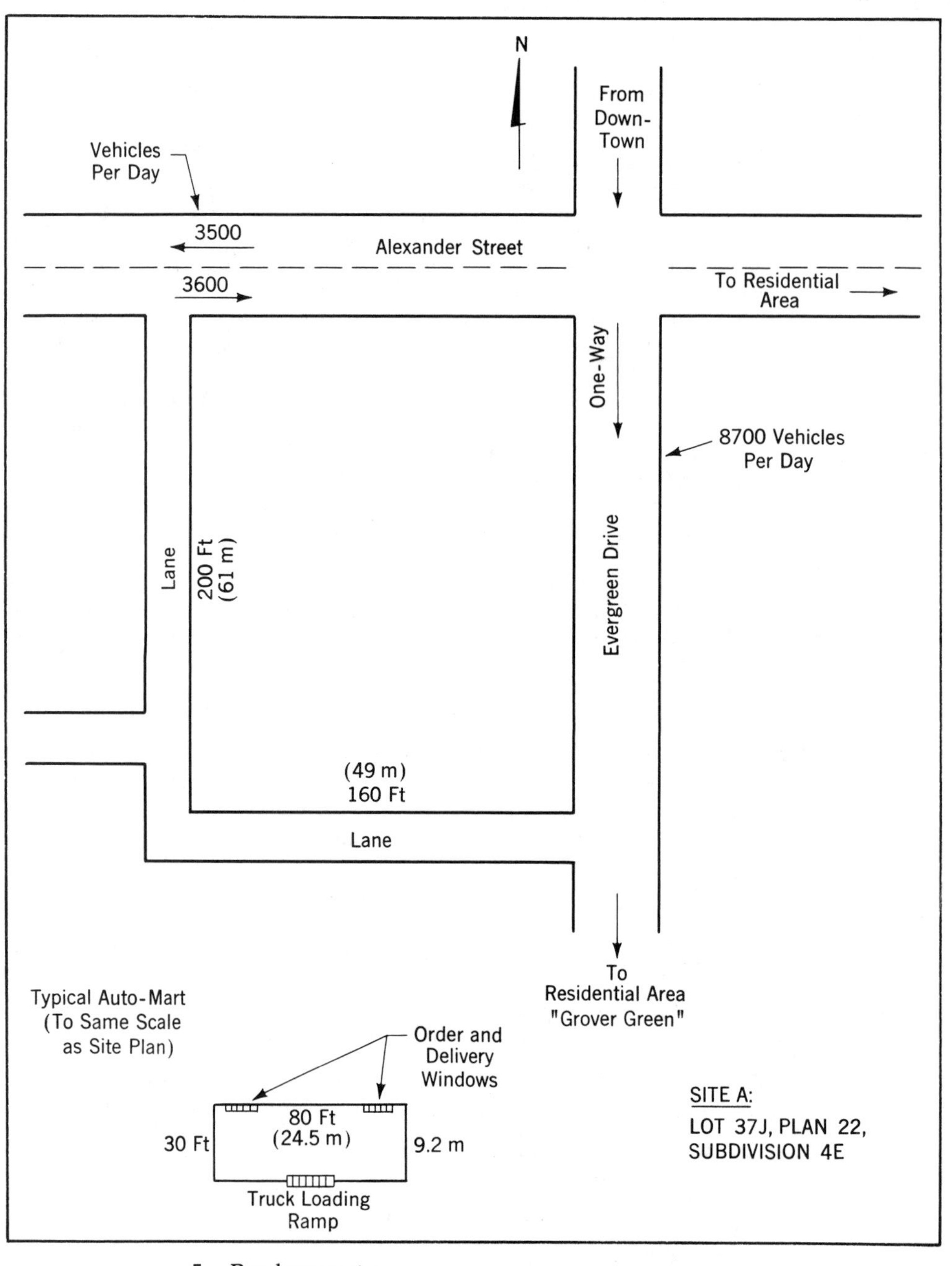

5. Purchase costs:
   Site A: $140,000
   Site B: $168,000
   Site B costs more because it has much greater frontage on Evergreen Drive.
6. Nearest competition from walk-in food convenience stores:
   Site A: 0.7 mile (1.1 km) east
   Site B: 0.5 mile (0.8 km) south

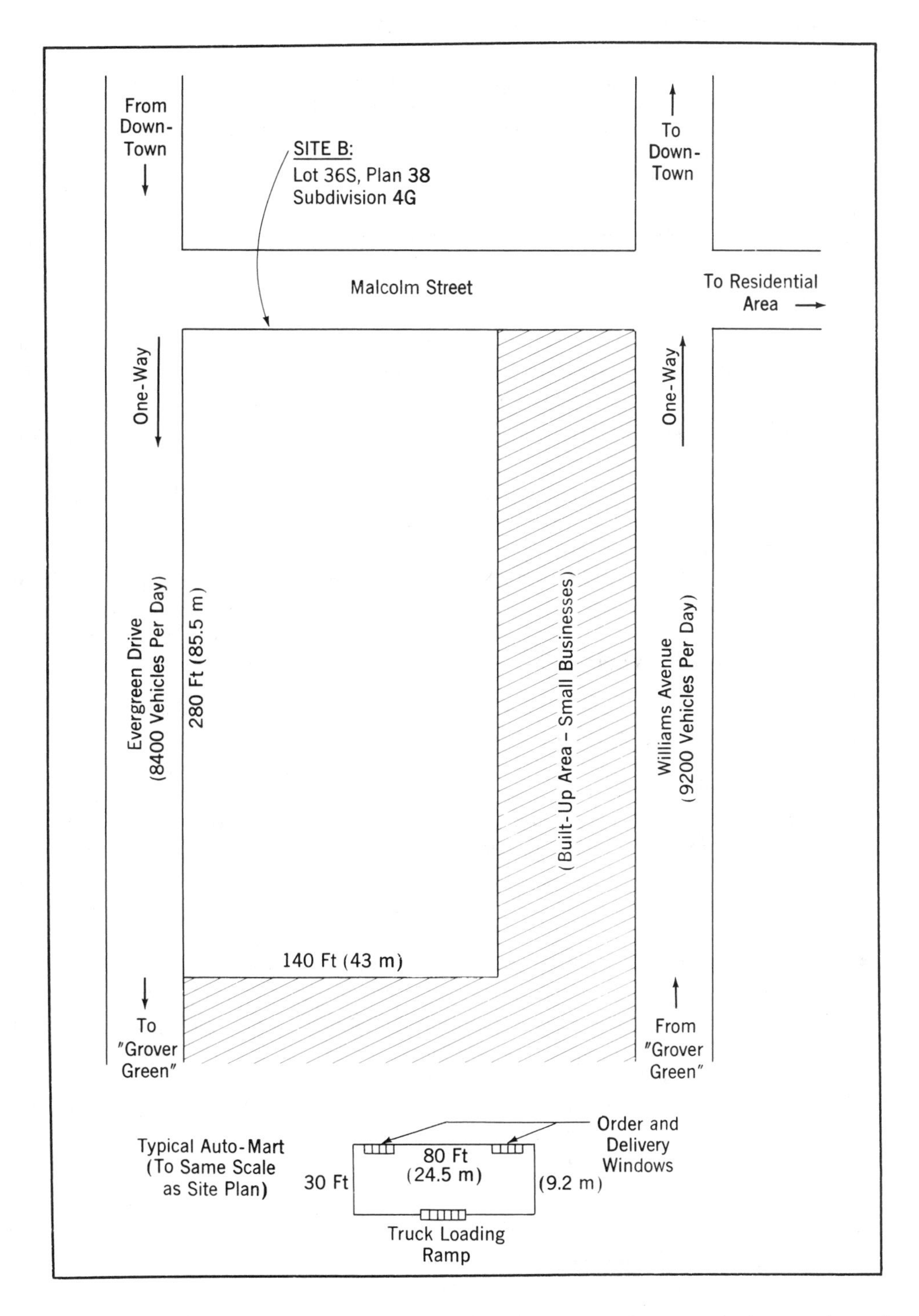

You plot the best position for the Auto-Mart on each site (use the sketches), bearing in mind that:

City by-law 261 stipulates that no exits may be placed within 100 feet of a major intersection, that separate cuts must be used for both entry and exit, that entries must turn off the road at a 30° angle, and that exits must approach the road at a 60° angle, as shown below:

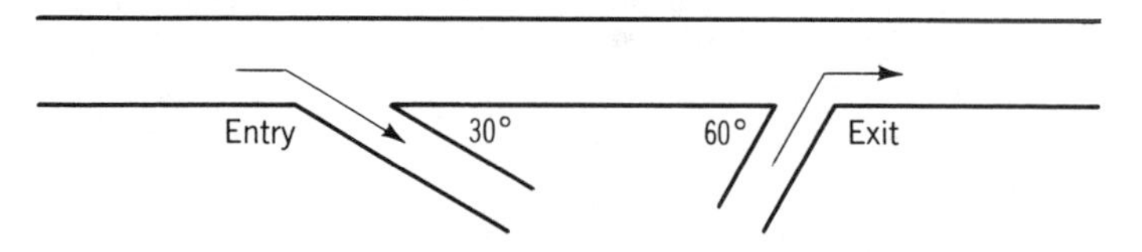

The driver's side of each car must be adjacent to the order and delivery windows.

There must be space for a truck or semitrailer to maneuver up to the truck loading ramp.

You are concerned about the wide price variance and discuss it with Vern Rogers.

"Ms. Wilkinson quoted an approximate maximum of $150,000," he comments. "But she said they might be willing to pay more for a site that has superior potential."

You then compare the suitability of each site from all points of view and select the one you will recommend to Auto-Marts Inc. Your report should, however, discuss both sites in depth, so that your client has a clear view of what is available and why you have a preference for a particular location.

PROJECT NO. 3: TESTING HIGHWAY MARKING PAINT

For the past seven years the Highways Department in your area has been using a paint known as "Centrex CL" for marking highway pavement center lines and lanes. Recent advances in paint technology, however, have brought several new products onto the market, which their manufacturers claim are better than Centrex CL. To meet this challenge, Centrex, Inc has also developed a new paint ("TL") and has recommended that the Highways Department use it in place of CL.

Chief Highways Engineer Claude Auger wants to test Centrex TL and the other paints before considering any change, but has insufficient staff or facilities to carry out the tests. So he commissions the local branch of H. L. Winman and Associates to carry out the tests for him under contract H64506. Branch Manager Vern Rogers assigns the project to you.

You start your project by obtaining samples of white and yellow highway paint from six manufacturers, transferring the samples into unmarked cans and then coding the cans like this:

Manufacturer	*Paint Coding* White	*Paint Coding* Yellow
1. Centrex Inc, Hartford, Connecticut Paint type: CL (the "old" paint)	WA	YL
2. Marvin Labs, Portland, Oregon Paint type: 445	WB	YM

3.	Vancourt Manufacturing, Reece, Minnesota Paint type: HILITE	WC	YN
4.	Multiple Industries Corporation, Montrose, Ohio Paint type: 601	WD	YO
5.	Wishart Incorporated, Windsor, Ontario Paint type: ROADMARK	WE	YP
6.	Provincial Paint Company, Pittsburgh, Pennsylvania Paint type: 81-324	WF	YQ
7.	Centrex Inc, Hartford, Connecticut Paint type: TL ("new" paint)	WG	YR

You then place the coding list into a sealed envelope, and lock it away in a safety deposit box of a local bank.

You decide to paint sample stripes on two regularly traveled stretches of highway and to assess the samples in four ways:

Spraying characteristics.
Drying time.
Visibility after three months.
Visibility after six months.

You assess spraying characteristics as Excellent, Very Good, Good, Fair, and Poor. The ratings are:

Very good: WB, WD, WG, YO, YR
Good: WA, WC, WE, WF, YL, YM, YN, YQ
Fair: YP

You assess drying time in minutes:

WA –16	WE –14	YL –13	YP –10
WB –33	WF –13	YM –26	YQ –12
WC –18	WG –11	YN –17	YR –15
WD –11		YO –14	

After three months you assess visibility by day and by night. You use five persons (one is yourself) to rate the stripes independently and to place the stripes' visibility on a scale of 1 to 10. You then average the five assessments:

Paint Code	*Concrete Pavement*		*Asphalt Pavement*		*Paint Code*	*Concrete Pavement*		*Asphalt Pavement*	
	Day	*Night*	*Day*	*Night*		*Day*	*Night*	*Day*	*Night*
WA	8	8	8	9	YL	8	9	9	9
WB	7	8	7	7	YM	7	7	7	9
WC	8	9	8	8	YN	7	9	7	8
WD	9	9	8	9	YO	8	8	7	8
WE	7	6	7	8	YP	7	8	6	8
WF	8	9	9	9	YQ	5	6	4	6
WG	7	7	7	8	YR	9	10	9	9

After another three months the same five persons again assess stripe visibility, with these results:

Paint Code	Concrete Pavement Day	Concrete Pavement Night	Asphalt Pavement Day	Asphalt Pavement Night
WA	6	6	6	7
WB	4	5	5	5
WC	7	8	7	8
WD	8	9	8	8
WE	6	6	6	7
WF	6	7	5	6
WG	6	7	7	7

Paint Code	Concrete Pavement Day	Concrete Pavement Night	Asphalt Pavement Day	Asphalt Pavement Night
YL	7	9	8	8
YM	6	7	6	8
YN	7	8	6	7
YO	7	7	6	8
YP	3	4	4	5
YQ	2	3	3	4
YR	8	9	8	9

You consolidate all your results into two tables, one for white paint, one for yellow paint, and then:

Reject any unacceptable paints (see guidelines below).
Rank acceptable paints in order of suitability.
Identify the best paint(s) to use for highway marking.
Retrieve the paint coding list from the bank deposit box.
Write your report.

Some factors you use to conduct your study and to write your report are:

1. Chief Highways Engineer Claude Auger's address is 416 Inkster Building, 2035 Perimeter Road.
2. The paint stripes were painted on two stretches of highway:
   2.1 Highway 101 (concrete surface), 1½ miles (2.4 km) north of the intersection with highway 216.
   2.2 Highway 216 (asphalt surface), ½ mile (0.8 km) west of the intersection with highway 101.
3. You are unable to borrow the regular highway paint stripe applicator from the Highways Department. Instead, you mount a regular applicator on a small garden tractor. The paint stripes are applied at night, between midnight and 6:00 A.M.
4. Paint Manufacturers' Association specification PMA-28H states that spraying characteristics for fast-drying highway paint should be at least "Good," and preferably "Very Good."
5. You refer to specification ASTM D-711 to establish the maximum acceptable paint drying time, which is 28 minutes.
6. Guidelines you give to the persons assessing paint visibility are:
   10: excellent visibility
   8: very good visibility
   6: good visibility
   4: fair visibility
   2: poor visibility
7. You establish minimum acceptable visibility levels for the paints to be:
   After three months' traffic wear: 7
   After six months' traffic wear: 6

Note: Calculate real dates for each stage of the study and quote them in your report. Include an imaginary date for Claude Auger's letter of project assignment to H. L.Winman and Associates.

Your report should not only present the results of your tests, but also analyze them, draw conclusions, and make a recommendation.

PROJECT NO. 4: A VISIT TO WAKELING PROCESSORS INC.

Wakeling Processors Inc has its head office in Cleveland, Ohio, and numerous manufacturing plants throughout the surrounding area. Oliver R. Wakeling, president and general manager, calls in H. L. Winman and Associates to help him resolve a management problem with technical implications. The problem concerns the Wakeling Processors Inc plant at Weston, 95 miles (153 km) from your city, and centers around the operation of the power house. This subsidiary is engaged in vegetable oil processing, for which the power house has to produce large supplies of hot water and a moderate supply of compressed air.

Mr. Wakeling is concerned because there has been a continuous upward increase in power house costs at the Weston plant over the past two years. Fuel costs have risen by 18% and numerous breakdowns have occurred which have interfered with production, with a consequent increase in overhead costs. Yet management visits have revealed little that could be attributed to poor operation; in fact, the power house has always been immaculate.

Mr. Wakeling believes that for him to make another inspection will not produce the answers he requires. He calls on Martin Dawes, at the head office in Cleveland, and outlines his problem. Martin suggests sending someone at the technician or technologist level who will be able to talk on technical subjects to all the people in the power house and at the same time evaluate production problems.

He telephones Vern Rogers, branch manager of the H. L. Winman and Associates office in your city, who in turn suggests assigning you to the project. They arrange a "conference call" between yourself, Martin, Vern, and Mr. Wakeling, during which Mr. Wakeling gives you a brief outline of the situation:

> The present chief engineer at the Weston Plant is David Skyla, and he is to retire in three months. Management has to decide whether to promote Barry Bishop, the existing senior shift engineer, or to bring in a new chief engineer from outside the plant. On paper, Bishop is ideal for the job: He has worked in the plant for 15 years (he is now 36) and always under Skyla, so his knowledge of the power house and its operating conditions cannot be challenged. Yet the rising costs indicate that all is not correct in the plant, and management wants to be sure that the new chief engineer does not perpetuate the present conditions.

Mr. Wakeling says he will advise Skyla and Bishop that he has engaged H. L. Winman and Associates to study the hot water and power generating system at their plant and that they should expect you. You visit the plant one week later.

During your talks to plant staff and tours of the powerhouse you make the following notes:

1. Housekeeping excellent—whole place shines (but is this only surface polish for impression of visitors?).
2. Maintenance logs are inadequately kept—need to be done more often. Need more detail. Equipment files not up to date and not properly filed.
3. Boiler cleaning badly neglected. Firm instructions re boiler cleaning need to be issued by head office.
4. Flow meters are of doubtful accuracy. May be overreading. Not serviced for three years. Manufacturer's service department should be contacted (these are Vancourt meters). Manufacturer needs to be called in to do a complete check and then recalibrate meters.
5. Overreading of meters could give false flow figures—make plant seem to produce more steam than is actually produced.
6. Good housekeeping obviously achieved by neglecting maintenance. Incorrectly placed emphasis probably caused by frequent visits from company president, who likes to bring in important visitors and impress them. Skyla likes reflected glory (so does Bishop).
7. Shift engineers are responsible for maintenance of pumps and vacuum equipment. Not enough time given over to this. They seem to prefer straight replacement of whole units on failure rather than preventive maintenance. Costly method! Obviously more breakdowns: they wait for a failure before taking action. A preventive maintenance plan is needed.
8. Bishop seems O.K. Genial type; obviously knows his power house. Proud of it! But seems to resist change. Definitely resents suggestions. Does he lack all-round knowledge? Is he limited only to what goes on in his plant? Is he afraid of new ideas because he doesn't understand them? Young staff hinted at this: too loyal to say it outright, but I felt they were restive, hampered by his insistence that they use old techniques that are known to work but are slow. Nothing concrete was said—I just "felt" it.
9. Skyla's done a good job training Bishop. Made him a carbon copy. Skyla doesn't do much now. Bishop runs the show, and has for over a year. He *expects* to get the job when Skyla retires. It'll be a real blow to him if he doesn't! Wakeling might even lose a good company man.
10. Discussed RAMSORT 2300 power panel with staff. Young engineers had read about it in "Plant Maintenance"—eager to have one installed (I described the one I'd seen at Abotinam Pulp and Paper). But Skyla and Bishop knew nothing about it—weren't interested. Obviously not keeping up to date with technical magazines.

When you return to head office you inform Vern Rogers verbally of your findings. He asks you to write an evaluation report for Mr. Wakeling.

PROJECT NO. 5: TO BUY OR TO LEASE?

You are employed by the local branch of H. L. Winman and Associates. Branch Manager Vern Rogers informs you that the company is considering buying or leasing three automobiles for use by staff members. Currently, staff members use their own cars when they drive locally, and sometimes for considerable distances, on consulting projects. They submit expense claims and are reimbursed for the mileage they cover. Recently, insurance regulations have been introduced which require car owners who use their automobiles for part-time business purposes to buy *business* automobile insurance. Twenty-three staff members fall into this category, and this year their insurance premiums have risen by 40%.

Vern asks you to carry out a cost analysis to assess the feasibility of purchasing or leasing three vehicles. He suggests you base your analysis on buying an eight-cylinder station wagon for major projects and long-distance projects, a six-cylinder "intermediate" car for normal assignments, and a four-cylinder runabout for local use. All should have four doors and automatic transmission, and the two larger vehicles should have power steering.

"Radios?" you ask.

"Yes, but AM only," Vern agrees. "And steel-belted radial tires for all three—they last twice as long."

You start the project by establishing how much it currently costs the company to reimburse employees who drive their own cars on company assignments. Walter Chan, the cost accountant, tells you:

On average, employees' claims total 3200 miles (5150 km) per month.

The reimbursement rate is 20.5 cents per mile (12.75 cents per km).

For each of the three vehicles you require, you survey available automobiles and select the one you consider most suitable. From an automobile dealer, you then obtain:

A price for outright purchase of the vehicle.

A price (per month) for a three-year lease, in which servicing, maintenance, licensing, oil, and gasoline are *not* included.

You then use all these figures to carry out a cost analysis, so that you can compare the average cost per year of:

1. Continuing the present system (mileage reimbursement).
2. Purchasing three automobiles.
3. Leasing three automobiles.

To obtain a realistic comparison, you project costs forward for six years. For the purpose of your study you also assume that:

Purchased automobiles will be kept for six years, after which they will be "written off" (considered to have no trade-in value).

Leased automobiles will be replaced every three years.

Automobile prices, lease prices, and interest rates will remain constant for the six years.

The value of each purchased automobile will depreciate 30% per year— e.g. an $8000 automobile will depreciate by $2400 (to $5600) after one year, by a further $1680 (30% of $5600) after two years, and so on.

Because purchased automobiles tie up capital, interest should be added to the cost of company-owned vehicles. You calculate this interest at 10% of the annual (depreciated) value of each automobile the company would buy.

The cost of maintenance, servicing, licensing, insurance, and consumables (gasoline, oil, tires) must be added to the cost of both buying and leasing.

When you have finished your analysis, you write a feasibility study which shows which is the most economical method to use, and submit it to Vern Rogers.

Additional notes:

1. You will have to obtain some of the information you need for this project from outside sources (car dealers, for example).
2. This assignment may be broadened to include additional data which you consider important. For example, it can include a comparative analysis of different makes of vehicles.

PROJECT NO. 6: DESIGNING AN ELECTRONIC MEETING TIMER

Recently you came across an old H. L. Winman and Associates' file containing a proposal for an electromechanical meeting timer. It was a June 18, 1970, memorandum addressed to Andy Rittman, who was then a senior project engineer. Its author, Michael Henning, described a meeting timer he had designed (see sketch), which he felt could be a marketable product. He proposed that a prototype be built and tested, and that Robertson Engineering Company in Toronto be encouraged to build production units.

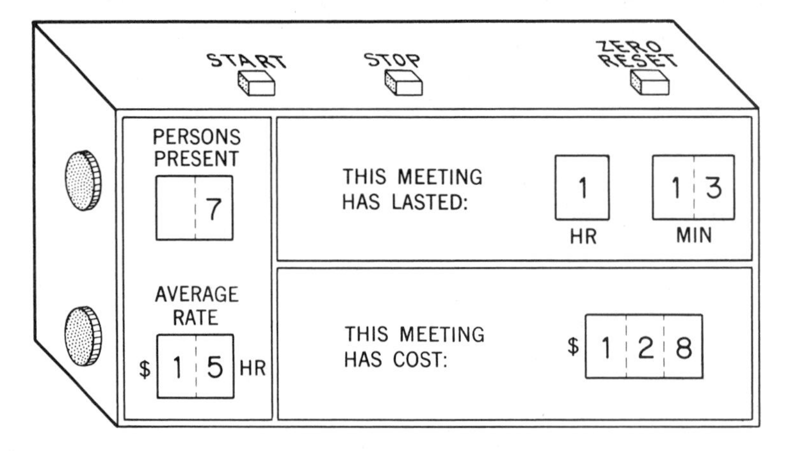

Michael had designed the timer in response to a memo from Harvey Winman to all department heads, in which Harvey criticized the length of meetings they held, suggested ways they could shorten them, and circulated a bulletin describing how to run and take part in meetings (similar to the ideas presented in Chapter 8). Michael suggested that the timer be prominently displayed in the conference room, where it would continually remind meeting participants how much each meeting has cost at any particular moment. It would multiply time elapsed by the average hourly salary (summated) of every person present.

There were several pieces of correspondence in the file, indicating some enthusiasm (Harvey Winman, Andy Rittman), but more doubt (Martin Dawes, John Wood, Peter Bell, Vic Braun). Apparently the doubters felt that the timer would intrude too much and possibly dampen contributions from meeting participants. They were also concerned that it would carry too high a price tag to make it a viable product for the business market.

You are intrigued by Michael Henning's ideas and recognize that by using microprocessor technology you could readily convert his design for an electromechanical device into a design for a fully electronic device. It would be smaller (even pocket sized), could be AC or DC (battery) powered, and would have either an LED (light emitting diode) display, or even an LCD (liquid crystal display). And, built in quantities, it would be much cheaper than the production model of an electromechanical timer.

You devise a preliminary design and calculate an approximate cost for a prototype. Then you write a proposal to your department head (Jim Perchanski), in which you:

Describe your idea.
Demonstrate that it is technically feasible.
Ask for approval to design and build a prototype.
Discuss the anticipated costs in labor (hours) and materials (dollars).
Describe how you would test the timer, both technically and in use.
Comment (if possible) on a likely price for a production model.

Some notes contained in Michael Henning's original proposal are:

Approximate size: 12 in. W × 8 in. H × 4 in. D (307 × 205 × 102 mm).
Mounting Details: For wall, table, or shelf.
Knobs at left: For setting number of persons present, and an average salary for the group.
Cost of prototype: $300 in 1970 dollars (current equivalent $650).

6

FORMAL REPORTS

A formal report requires more careful peparation than the informal reports described in the previous chapters. Because it will be distributed outside the originating company, its writer must consider the impression it will convey of the entire company. Harvey Winman recognized long ago that a well-written, esthetically pleasing report can do much to convince prospective clients that H. L. Winman and Associates should handle their business, whereas a poorly written, badly-presented report can cause clients to question the company's capability. Harvey also knows that the initial impression conveyed by a report can influence a reader's readiness to plough through its heavy technical details.

The presentation aspect must convey the originating company's "image," suit the purpose of the report, and fit the subject it describes. For instance, a report by a chemical engineer evaluating the effects of diesel fumes on the interior of bus garages would most likely be typed on standard bond paper, and its cover, if it had one, would be simple and functional. At the other end of the scale, a report by a firm of consulting engineers selecting a college site for a major metropolis might be printed professionally and bound in an artistically designed book-type folder. But regardless of the appearance of a report, its internal arrangement will be basically the same.

Formal reports are made up of several standard parts, not all of which appear in every report. Each writer uses the parts that best suit the particular subject and the intended method of presentation. There are six major and several subsidiary parts on which to draw, as shown in the table on page 167. Opinions differ throughout industry as to which is the best arrangement of these parts. The two arrangements suggested in the section on The Full Report later in this chapter, and illustrated in the mini-reports in Figures 6-4 and 6-5, are those most frequently encountered. Knowledge of these parts and the two basic arrangements will help engineers and technicians adapt quickly to the variations in format preferred by their employers. Listed below are the parts as they appear in the most widely used format.

FORMAL REPORT: (**Major Parts** and *Subsidiary Parts*)

Cover or *Jacket*
Title Page
Summary
Table of Contents
Introduction
Discussion
Conclusions
Recommendations
References or *Bibliography*
Appendix (es)

A formal report is often accompanied by a *Cover Letter* or *Letter of Transmittal,* and may also contain other small parts such as a *List of Illustrations* (which follows the *Table of Contents*), a *Distribution List,* and an *Acknowledgments* paragraph.

Major Parts

Because the six major parts form the central structure of every formal report, technical writers must understand their purpose and function if they are to use them effectively.

SUMMARY

The summary is a brief synopsis which tells readers quickly what a report is all about. Normally it appears immediately after the title page, where it can be found easily. It identifies the purpose and most important features of the report, states the main conclusion, and sometimes makes a recommendation. It does this in as few words as possible, condensing the narrative of the report to a handful of succinct sentences. It also has to be written so interestingly—so enthusiastically—that it encourages readers to read further.

The summary is considered by many to be the most important part of a report—and the most difficult to write. It has to be informative, yet brief. It has to attract the reader's attention, but must be written in simple, nontechnical terms. It has to be directed to the executive reader, yet it must be readily understood by almost any reader.

Generally, the first person in an organization who sees a report is a senior executive, who may have time to read only the summary. If the executive's interest is aroused, he or she will pass the report down to the technical staff to read in detail. But if the summary is unconvincing, the executive is more likely to put the report aside to read "later"; if this happens, the report may never be read.

Because a summary is so important, always write it last, after the remainder of your report has been written. Only then will you be fully aware of the report's highlights, main conclusions, and recommendations. To write the summary first can prove difficult and frustrating, because you will not yet have hammered out many of the finer points. You need the knowledge of a soundly developed report firmly fixed in your mind if you are to fashion an effective summary.

For a summary to be interesting, it must be informative; if it is to be informative, it must tell a story. It should have a beginning, in which it states why the project was carried out and the report written; a middle, in which it highlights the most important features of the whole report; and an end, in which it reaches a conclusion and possibly makes a recommendation. The example below illustrates how the interest is maintained in an informative summary:

Informative Summary

> A specimen of steel was tested to determine whether a job lot owned by Northern Railways could be used as structural members for a short-span bridge to be built at Peele Bay in northern Alaska. The sample proved to be G40.12 structural steel, which is a good steel for general construction but subject to brittle failure at very low temperatures.
>
> Although the steel could be used for the bridge, we consider that there is too narrow a safety margin between the −51°C temperature at which failure can occur, and the −47°C minimum temperature occasionally recorded at Peele Bay. A safer choice would be G40.8C structural steel, which has a minimum failure temperature of −62°C.

Other informative summaries preface the two formal reports at the end of this chapter, and the semiformal report in Figure 4-11, Chapter 4.

Some writers prefer to write a topical summary for reports that describe history or events, or that do not draw conclusions or make recommendations. As its name implies, a topical summary simply describes the topics covered in the report without attempting to draw inferences or captivate the reader's interest:

Topical Summary

> Construction of Alaska's Minnowin Point Generating Station was initiated in 1978, and first power from the 1340 MW plant is scheduled for 1984. A general description of the structures and problems peculiar to construction of this large development in an arctic climate is presented. The river diversion program, permafrost foundation conditions, and major equipments are described. The latter include the 16 propeller turbines, among the largest yet installed, each rated at 160,000 horsepower.

Because they are less results-oriented, topical summaries are *not* recommended for most formal reports.

In a formal report the summary should have a page to itself, be centered on the page, and be prefaced by the word "Summary" (or, sometimes, "Abstract"). If it is very short, it may be indented equally on both sides to form a roughly square block of information.

INTRODUCTION

Some report writers confuse the summary with the introduction. There is no need for this if they remember that the summary provides a *brief synopsis of the whole report,* whereas the introduction *introduces the subject* to readers so that they may read the remainder of the report more intelligently.

The introduction begins the major narrative of the report by preparing readers for the discussion that follows. It orients them to the purpose and scope of the report and provides sufficient background information to place them mentally in the picture before they tangle with technical data. A well-written introduction contains exactly the correct amount of detail to lead readers quickly into the major narrative.

The length of an introduction and its depth of detail depend mostly on the reader's existing knowledge of the topic. A writer who knows that the ultimate reader is technically knowledgeable will want to use technical terminology, but at the same time the writer has to cater to the executive reader who is probably only partly technical. To overcome this disparity in reading levels, skillful report writers often write the summary in a lay person's language and the introduction, conclusions, and recommendations in semitechnical language. They thus permit the semitechnical executives to gain a reasonably comprehensive understanding of the report without devoting time and attention to the technical details contained in the discussion.

Most introductions contain three parts: *purpose, scope,* and *background information.* Frequently, the parts overlap, and occasionally one of them may be omitted simply because there is no reason for its inclusion. The introduction normally is a straightforward narrative of one or more consecutive paragraphs; only rarely is it divided into distinct sections preceded by headings. It always starts on a new page (normally identified as page 1 of the report) and is preceded by the report's full title. This is followed by the single word "Introduction," which can be either a center heading or a side heading, as shown in the illustration.

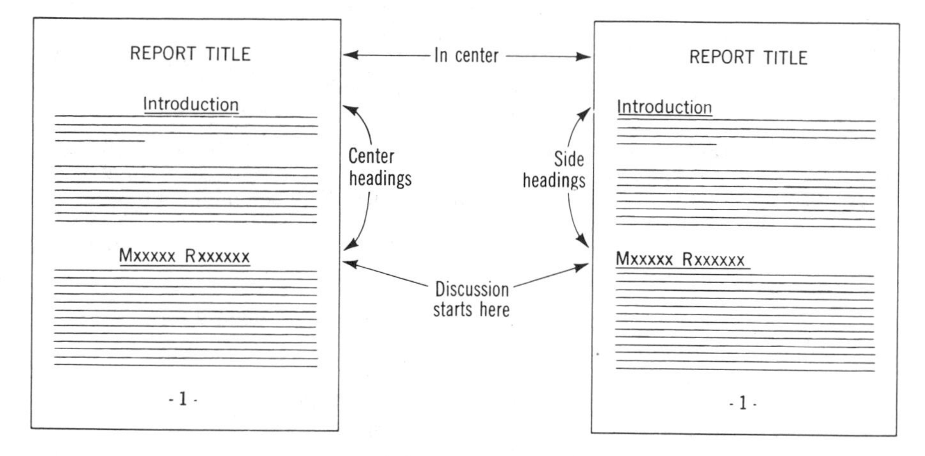

The *purpose* explains why the project was carried out and the report is being written. It may indicate that the project has been authorized to investigate a problem and recommend a solution, or it may describe a new concept or method of work improvement that the report writer believes should be brought to the reader's attention.

The *scope* defines the parameters of the report. It describes the ground covered by the report and outlines the method of investigation used in the project. If there are limiting factors, it identifies them. For example, if 18 methods for improving packaging are investigated in a project but only 4 are discussed in the report, the scope indicates what factors (such as cost, delivery time, and availability of space) limited the selection. Sometimes the scope may include a short glossary of terms that need to be defined before the reader starts to read the discussion.

Background information comprises facts readers must know if they are to fully understand the discussion that follows. Conditions or events that caused the project to be authorized, plus details of previous investigations and studies on the same or a closely related subject, fall under this heading. In a very technical report, or when a significant time lapse has occurred between it and previous reports, background information may include a brief theory review and references to other documents. If the theory review is lengthy, it is placed in the appendix and only a brief summary of the theory is inserted in the introduction; similarly, detailed references are listed at the end of the report rather than in the introduction.

The introductions shown here, plus those forming part of the two sample reports at the end of this chapter, are representative of the many ways a writer can introduce a topic.

Introduction No. 1

Northern Railways plans to build a short-span bridge one-half mile north of Lake Peele in northern Alaska and has a job lot of steel they want to use in its construction.	*Background*
In letter NR-70/LM dated March 20, 19__, Mr. David L. Harkness, Northern Area Manager, requested H. L. Winman and Associates to test a sample of this steel to determine its properties and to assess its suitability for use as structural members for a bridge in a very low temperature environment.	*Purpose*
The test performed was the Charpy Impact Test. Two series of tests were run, one parallel to the grain of the test specimen, and the other transverse to the grain, at 10° increments from +22°C down to −50°C.	*Scope*

Introduction No. 2

H. L. Winman and Associates was commissioned by Mr. D. C. Scorobin, President and General Manager of Auto-Marts Inc. of Dallas, Texas, to select a Cleveland site for the first of a proposed chain of drive-in groceries to be built in the north-eastern United States. This area was chosen since it represents an average community in which to assess customer acceptance of this revolutionary form of service.	*Purpose*

Aside from exterior facade and foundation details, Auto-Marts are built to a standard 22- by 7.3-meter pattern with an order window at one end of the 22-meter wall, and a delivery window at the other. Auto-Marts carry only a limited selection of household goods, but boast 90-second service from the time of placing an order to its delivery at the other end of the building. For this reason, Auto-Marts attract the householder hurrying homeward from work, the impulse buyer, and the late-night shopper, rather than the selective shopper. The store depends more on the volume of customers than on the volume of goods sold to each individual. — *Background*

The chief consideration in selecting a site must therefore be a location on the homeward-bound side of a main trunk road into a large residential area. The site must have quick and easy entry onto and exit from this road, even during peak rush hour traffic. The residential area should be occupied mainly by upper and middle class people who own cars and have money to spend. And there should be little competition from walk-in grocery stores. — *Scope*

DISCUSSION

The discussion, which normally is the longest part of a report, presents as much evidence (facts, arguments, details, data, and results of tests) as readers will need to understand the subject. The writer must develop this evidence in an organized, logical manner to avoid confusing readers, and must present it imaginatively to retain the readers' interest.

In the discussion the writer describes fully what he set out to do, how he went about it, what he actually did, and what he found out as the result of his efforts. He may consider the writing of these facts a chore, and simply record them dutifully but unenthusiastically. Or he can respond to the challenge by organizing the information so interestingly that the readers feel almost as though they are reading a short story.

There is no reason why the discussion should not have a plot. It can have a beginning, in which the writer describes the problem and outlines some background events; a middle, which tells how the writer tackled the problem; dramatic effect, which can be used to hold the readers' interest by letting them anticipate the writer's successes, as well as share the disappointments when a path of investigation results in failure; a climax, in which the readers are permitted to share the writer's pleasure in a description of how the problem was eventually resolved; and a denouement (a literary word for final outcome), in which the writer ties up the loose ends of information and evaluates the final results. Storytelling and dramatization can be such important factors in holding reader interest that no report writer can afford to overlook them.

There are three ways you can build the discussion section of a report:

By chronological development—in which you present information in chronological sequence (in the order that events occurred).

By subject development—in which you arrange information by subjects, grouped in a predetermined order.

By concept development—in which you organize information by concept, presenting it as a series of ideas which reveal imaginatively and coherently how you reasoned your way to a logical conclusion.

Reports using the chronological or subject method offer less room for imaginative development than those using the concept method, mainly because they depend on a straightforward presentation of information. The concept method can be very persuasive. Identifying and describing your ideas and thought processes helps your readers organize their thoughts along the same lines.

As a report writer, you must decide early in the planning stages which method you intend to use, basing your choice on which is most suitable for the evidence you have to present. Use the following notes as a guide.

1. Chronological Development. A discussion which uses the chronological method of development is simple to organize and write. It requires minimal planning. You simply arrange the major topics in the order they occurred, eliminating irrelevant topics as you go along. It is useful for very short reports, for laboratory reports showing changes in a specimen, for progress reports showing cumulative effects or describing advances made by a project group, and for reports of investigations that cover a long time and require visits to many locations to collect evidence.

But the simplicity of the chronological method is offset by some major disadvantages. Because it reports events sequentially, it tends to give emphasis to each event regardless of importance, and this may cause readers to lose interest. If you read a report of an astronaut's third day in orbit, you do not want to read about every event in exact order. It may be chronologically true to report that he rose at 7:15, ate breakfast at 7:55, sighted the second stage of his rocket at 9:23, carried out metabolism tests from 9:40 to 10:50, extinguished a cabin fire at 11:02, passed directly over Cape Kennedy at 11:43, and so on, until he retired for the night. But it can make deadly dull reading. Even the exciting moments of a cabin fire lose impact when they are sandwiched between routine occurrences.

When using chronological development, you must still manipulate events if you are to hold your readers' attention. You must emphasize the most interesting items by positioning them where they will be noticed, and deemphasize less important details. Note how this has been done in the following passage, which groups the previous events in descending order of interest and importance:

> The highlight of Sam Smith's third day in orbit was a cabin fire at 11:02. Rapid action on his part brought the fire, which was caused by a short circuit behind panel C, under control in 38 seconds. His work for the day consisted mainly of metabolism tests and. . . .
>
> He sighted the first stage of his rocket on three separate occasions, first at 9:23, then at. . . , and passed directly over Cape Kennedy at 11:43. His meals were similar to those of the previous day.

This narrative uses *chronological* order to describe the events on a day-to-day basis, and a modified *subject* order to describe the events for each day.

Over a five-year period H. L. Winman and Associates has been investigating the effects of salt on concrete pavement. Technical editor Anna King has suggested that the final report should have chronological development, because the investigation recorded the extent of concrete erosion at specific intervals. She wanted the engineering technician writing the report to describe how the erosion increased annually in direct relation to the amount of salt used to melt snow each year, and for the final conclusion to demonstrate the cumulative effect that salt had on the concrete.

2. Subject Development. If the previous investigation had been broadened to include tests on different types of concrete pavements, or if both pure salt and various mixtures of salt and sand had been used, then the emphasis would have shifted to an analysis of erosion on different surfaces or caused by various salt/sand mixtures, rather than a direct description of the cumulative effects of pure salt. For this type of report Anna King would have suggested arranging the topics in *subject* order.

The subject order could be based on different concentrations of salt and sand. The technician would first analyze the effects of a 100% concentration of salt, then continue with salt/sand ratios of 90/10, 80/20, 70/30, and so on, describing the results obtained with each mixture. Alternatively, he could select the different types of pavement as the subjects, arranging them in a specific order and describing the effects of different salt/sand concentrations on each surface.

Chief draftsman Barry Brewster has been investigating blueprint machines and he plans to recommend the most suitable model for installation in his department. He writes his report using the subject method. First, he tells readers that he intends to evaluate the speed, economy of operation, and usefulness of each blueprinter, and then he emphasizes the most suitable machine by discussing it either first or last. If he chooses to describe it first, he can state immediately that it is the best blueprinter, and say why. He can then discuss the remaining models in descending order of suitability, comparing each to the selected machine to show why it is less suitable. If he prefers to describe the best machine last, he can discuss the others in ascending order of suitability, stating why he has rejected each one before he starts describing the next.

Also suitable for subject development is an analysis of insulating materials being conducted by laboratory technician Rhonda Moore. In her report, she arranges the materials into groups having similar basic properties or falling within a specific price bracket. Within the groups she describes the test results for each material and draws a conclusion as to its relative suitability for a specified purpose. Like Barry, she describes the materials in ascending or descending order of suitability.

The subject method of development permits Barry and Rhonda either to hide their personal preferences until almost the end of their reports or, if they prefer, to let their preferences show all the way through. They have the choice of using an impersonal, objective approach, or a personal, subjective approach similar to that for the concept method. These alternative approaches are illustrated in Figure 6-1.

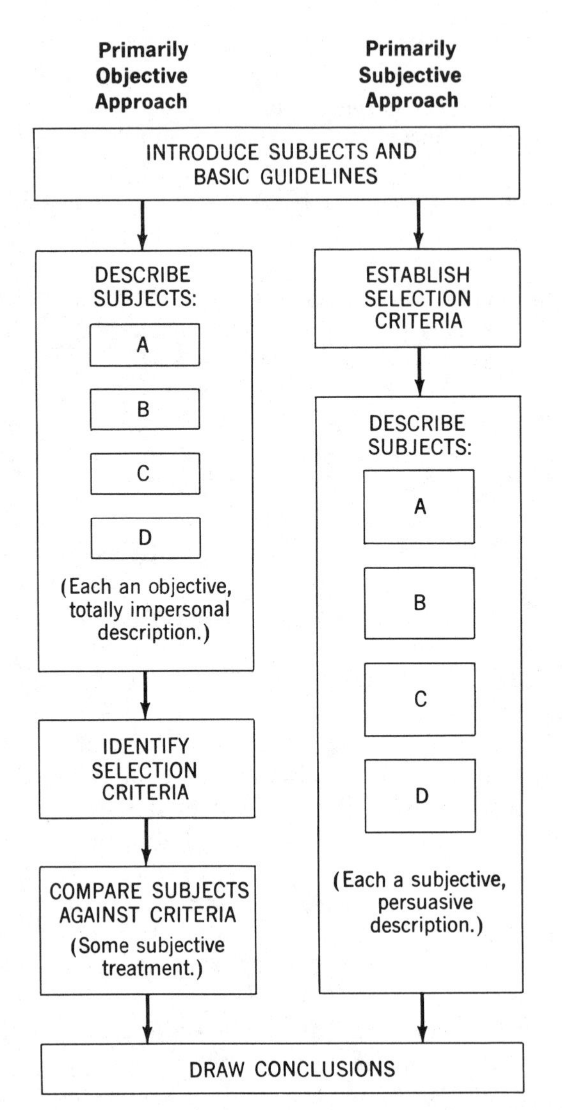

Figure 6-1. Alternative methods for a comparative analysis.

Rhonda prefers to use the primarily objective approach. She discusses each insulating material (i.e. subject) without using any words that might make her readers think she believes the material to be a good or bad choice:

> Material A has an R-17 rating. It is a dense, brown, tightly packed fibrous substance enclosed in a continuous, one-meter wide envelope, which is supplied in 10-, 20-, and 50-meter rolls. To cover a standard 4 × 3 meter area would cost $108.00. The fire resistance of Material A is. . . .

Barry prefers the more colorful subjective approach: he wants to persuade his readers to accept or reject each blueprinter (subject) as they read about it:

> The Image-copy is the only machine that meets all our technical requirements. It

> has a high speed of 25 feet per minute, an extended continuous-roll feature with automatic cutter, and both manual and automatic stacking capabilities. Although its purchase price of $5700.00 is $700.00 above budget, its high reliability (reported by users such as Multiple Industries and Apex Architectural Services) would probably cause less equipment downtime and consequent production frustrations than other models. Specific features are. . . .

Barry must establish the criteria or parameters which most affect his choice, and must convince his readers why they are important, *before* he describes each blueprinter; if he does not, readers will not readily accept his subjective statements. Rhonda has only to introduce the insulating materials, and possibly outline how they were selected, before she describes them. She need not identify her selection criteria until almost the end of the discussion, after she has objectively described each material but before she subjectively compares their main features.

Whichever approach you use for a comparative analysis, your readers should find that the discussion leads comfortably and naturally into the conclusions you eventually draw. The conclusions should never contain any surprises.

The report selecting a Residential Teachers' College for the City of Montrose (Report No. 1 later in this chapter) uses the subject method of development with a primarily objective approach for the comparative analysis. Its table of contents provides an immediate clue that the subject method has been used.

3. Concept Development. By far the most interesting reports are those using the concept method of development. True, they need to be organized more carefully than reports using either of the previous methods, but they give the writer a tremendous opportunity to devise an imaginative arrangement of the topic—which is the best way to hold the reader's attention.

They can also be very persuasive. Because the report is organized in the order in which the writer reasoned his or her way through the investigation, readers are able to understand the subject much more quickly. They will readily appreciate the difficulties encountered by the writer, and frequently draw the correct conclusions even before they read them. This helps readers to feel they are personally involved in the project.

In the elevator selection report (Report No. 2 in this chapter) author Barry Kingsley has gradually developed his topic one step at a time, so that readers can see the logic of his argument. There are many combinations of elevators that could fit the Merrywell Building, but we are ready to accept Barry's selection because he has effectively persuaded us that his choice is the only sound one. You can apply his approach to your reports by thinking of each project as a logical but forceful procession of ideas that form a total concept. If you are personally convinced that the results of your investigation are valid, and remember to explain in your report *how* you reached the results and *why* they are valid, then you will probably be using the concept method properly.

Always try to anticipate reader reaction. If you are presenting a concept (an idea, plan, method, or proposal) which readers are likely to accept, then use a straightforward four-step approach:

1. Describe your concept in a brief overview statement.
2. Discuss how and why your concept is valid; offer strong arguments in its favor, starting with the most important and working down to the least important.
3. Introduce negative aspects, and discuss how and why each can be overcome or is of limited importance.
4. Close with a restatement of your concept, its validity, and its usefulness.

But if you are presenting a controversial concept, or need to overcome reader bias, then modify your approach. Try to overcome objections by carefully establishing a strong case for your concept *before* you discuss it in detail.

Special Projects Department Head Andy Rittman used this approach in a report he prepared for Mark Dobrin, owner/manager of a company making extruded plastic and metal parts for the defense industry. Manufacturing costs had risen steeply over the past two years, and Mark's prices had rapidly become uncompetitive. Mark thought he should replace some of his older, less efficient equipment, so he asked H. L. Winman and Associates to evaluate his needs.

Andy Rittman quickly realized that if Mark was to avoid going completely out of business he would have to replace far more of his equipment than he expected, and would have to introduce electronic methods of production and inventory control. Because the cost would be high, he would have to lease rather than buy the equipment.

Here, Andy had a problem. Mark was as old-fashioned as some of his extruders and shapers, and throughout his life he had steadfastly refused to acquire anything that he did not own outright. He was unlikely to change now.

In his report, Andy used a carefully reasoned argument to prove to Mark that he needed a lot of new equipment and that the only feasible way he could acquire it would be to lease it. Throughout, Andy wrote objectively but sincerely of his findings, hoping that the logic of his argument would swing Mark around to accepting his recommendation. Very briefly, here is the step-by-step approach Andy used:

He opened with a summary which told Mark that to avoid bankruptcy he would have to invest in a lot of expensive equipment and make extensive changes in his operating methods.

Andy then produced financial projections to prove his opening statement, and discussed the productivity and profitability necessary for Mark to remain in business.

He discussed why Mark's equipment and methods were inefficient, introduced the changes Mark would have to make, established why each change was necessary, and demonstrated how each would improve productivity. (Andy referred Mark to an appendix for specific equipment descriptions, justifications, and costs.)

Andy introduced two sets of cost figures: one for making the minimum changes necessary for Mark's business to survive, and the second for more comprehensive changes that would ensure a sound operating basis for the future. He commented that both would require capital purchases likely to be beyond the financial resources of Mark's business.

He outlined alternative financing methods available to Mark, the implications and limitations of each, and the financial effect each would have on Mark's business.

(Although he introduced leasing, Andy made no attempt to persuade Mark that he would have to lease; he let the figures speak for themselves.)

Andy concluded his discussion by summarizing the main points he had made: new equipment *must* be acquired; to buy even the minimum equipment was beyond Mark's financial resources; and, of the various financing methods available, leasing was the most feasible. But not until his recommendation did Andy come right out and suggest that Mark should make comprehensive changes and lease the new equipment. By then, Andy had become so involved with Mark's predicament that he wrote strongly and sincerely in favor of his recommendation.

Even though the concept method challenges a writer to fashion interesting reports, it is not always the best reporting medium. For instance, the concept method could possibly have been used for the blueprinter report mentioned earlier, but it is doubtful whether the topic would have warranted full analysis of the author's ideas. When a topic is fairly clear-cut, there is no need to lead the reader through a lengthy "this is how I thought it out" discussion. Reserve the concept method for topics that are controversial, difficult to understand, or likely to meet reader resistance.

Whichever method you use, the discussion should not be cluttered with detailed supporting information. Unless tables, graphs, illustrations, photographs, statistics, and test results are essential for reader understanding while the report is being read, banish them to the appendix. But always refer to them in the discussion, like this:

> The test results attached at Appendix C show that aircraft on a bearing of 265°T experienced considerably weaker reception than aircraft on any other bearing. This was attributed to. . . .

Readers interested only in *results* would consider that this statement tells them enough and would continue reading the report. Readers interested in knowing *how the results were obtained,* who want to see the overall picture, will turn to Appendix C to find out how you went about performing the tests. They probably will not do so immediately, but will return to it when they have finished the report or reached a suitable stopping place. In neither case are readers slowed down by supporting information which, though relevant, is not immediately essential to an understanding of the report.

Unless the discussion is very short, it will consist of a series of sections, each containing information on a major topic. Insert an informative heading at the beginning of each section to indicate the section's contents. If the section is long or naturally subdivides into subtopics, also use subordinate headings. Major topics normally demand center or side headings, while minor topics need only side or paragraph headings, as shown in the illustration (also see Figure 10-4 in Chapter 10). These headings eventually form the table of contents for the report.

After each heading, start the section with an overview statement to describe what the section is about and suggest what conclusion will be drawn from it. Overview statements are miniature summaries which direct a reader's attention to the point you want to make. If the section is short, the overview statement

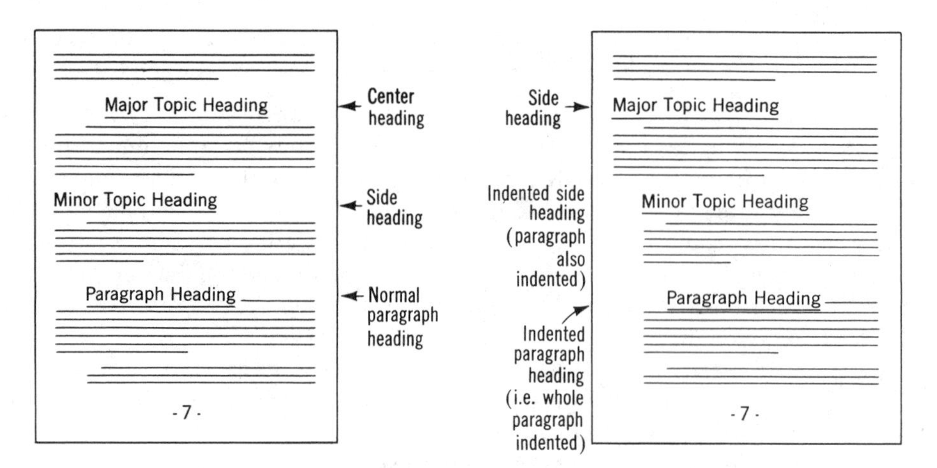

may be a single sentence: if the section is long, it will probably be a short paragraph.

At the end of each major section insert a concluding statement that summarizes the result of the discussion within that section. From these section conclusions you can later draw your main conclusions.

CONCLUSIONS

Conclusions state in brief terms the major inferences that can be drawn from the discussion. They must be based entirely on previously stated information, and never introduce new material or evidence to support your argument; they must offer no surprises. If there is more than one conclusion, the main conclusion is stated first, followed by the remaining conclusions in decreasing order of importance. This is shown in the two examples below, which present the same conclusions in both narrative form and tabular format.

Narrative Conclusion

The operational life of QK 3801 magnetrons can be increased 30% by installing modification XL-1 on our radar transmitters. This will save $8300 over the next 12 months, but will require us to increase magnetron run-in time from 4 to 8 hours.

Tabular Conclusion

Installation of modification XL-1 on our radar transmitters will result in:

1. A 30% increase in the operational life of QK 3801 magnetrons.
2. An $8300 saving over the next 12 months.
3. A change in operating procedures to increase magnetron run-in time from 4 to 8 hours.

Whether you use narrative or tabular conclusions depends on personal preference. Both are acceptable, although the advantage of the tabular arrangement is that its separate conclusions are easier to identify. Perhaps the best rule is to use narrative style when there are very few conclusions and to tabulate them when there are several.

Because conclusions are *opinions* (based on the evidence presented in the discussion), they must never suggest or recommend future action.

RECOMMENDATIONS

Recommendations appear in a report when the discussion and conclusions indicate that further work needs to be done, or if a specific solution to a problem has evolved from the writer's investigations. Write them in strong, definite terms to convince readers that the course of action you advocate is valid. Use the first person and active verbs, as has been done here:

> *Strong* I recommend that we build a five-station prototype of the Microvar system and test it operationally.

Compare this with the same recommendation written in the third person, using passive verbs:

> *Weak* It is recommended that a five-station prototype of the Microvar system be built and tested under operational conditions.

The first version is more convincing because the writer is personally involved in the recommendation. The second version is weak because there does not seem to be an author who really wants to build a prototype.

Unfortunately, in many industries, and often in government circles, personal involvement is severely frowned upon. (No one seems to know whether this stems from natural reticence supposedly displayed by technical writers, or from fear that a faulty recommendation might be traced to a specific individual.) As a result, many reports fail to contain really strong recommendations because their writers hesitate to say "I recommend. . . ," and so use the indefinite "It is recommended. . . ." Surely a technical person who has researched a subject well enough to advocate definite action can present recommendations more convincingly than this. It is not always necessary to use the personal "I" (most reports prepared for clients are expected to use the plural "we," to indicate that they represent the company's viewpoint). For example:

> *Strong* We recommend building a five-station prototype of the Microvar system. We also recommend that you:
>
> 1. Install the prototype in Railton High School.
> 2. Commission a physics teacher experienced in writing programed instruction manuals to write the first programs.
> 3. Test the system operationally for three months.

Because recommendations must be based solidly on the evidence presented in the discussion and conclusions, they can never introduce new evidence or new ideas.

APPENDIX

Related data not necessary to an immediate understanding of the discussion should be placed further back in the report, in the appendix. The data can vary from a complicated table of electrical test results to a simple photograph of a blown resistor. The appendix is a suitable place for manufacturer's specifications, graphs, comparative data, drawings, sketches, excerpts from other reports or books, and correspondence. There is no limit to what can be placed in the appendix, providing it is relevant and reference to it is made in the discussion.

The term "appendix" applies to only one set of data. In practice there can be many sets, each of which is called an appendix and assigned a distinguishing letter, such as A, B, or C. The sets of data are collectively referred to as either "appendixes" or "appendices," the former term being preferred in the United States, and the latter in Canada and Great Britain.

The importance of an appendix has no bearing on its position in the report. Whichever set of data is mentioned first in the discussion becomes Appendix A, the next set becomes Appendix B, and so on. Each appendix is considered a separate document complete in itself and is page-numbered separately, with its front page labeled 1. (Not every organization follows this practice. Although most companies now number appendix pages separately, some continue to number them consecutively as an integral part of the report.) Examples of appendixes appear at the end of the two sample reports in this chapter.

Subsidiary Parts

In addition to the six major parts of a report, there are several additional parts that perform more routine functions. Although I refer to these as "subsidiary," they nevertheless contribute much to the effectiveness of a report. A writer must never assume that these seemingly peripheral details will look after themselves. Because they directly affect the image conveyed of both the writer and the company, they must be prepared no less carefully than the rest of the report.

COVER

Almost every major formal report has a cover. It may be made of cardboard printed in multiple colors and bound with a dressy plastic binding, or it may be only a light cover of colored fiber material stapled on the left-hand side. In any case it has one obvious purpose: to inform the reader of the main topic of the report. Its more subtle purposes are to convey an immediate image of the type of company that has produced the report, and to reflect the type of subject that the report describes. This "matching" of subject matter and company image plays an important part in setting the right tone for a report.

The cover should contain very few words. The report title is an obvious choice, since a good title will invite a prospective reader to open the report and discover what it is about. These words must also stand our clearly on the page. Printing a title like a newspaper banner headline in two-inch high letters may attract attention, but will have a negative effect on the reader's opinion of the author's technical competence. A few words tastefully arranged and well separated from other printed matter will help to present the correct image. The only other words should be the originating company's name and perhaps the name of the author, if he or she is well known enough to lend weight to the technical level of the report and advance the company's image.

Choice of an informative title is specially important. It should be short, yet imply that the report has a worthwhile story to tell. Compare the vague title below with the more informative version written beneath it.

Original vague title

RADOME LEAKAGE

Revised informative title

POROSITY OF FIBERGLAS CAUSES RADOME LEAKS

Technical officers at remote radar sites would glance at the first title with only a muttered, "Looks like somebody else has the same problem we've got." But the second title would encourage them to stop and read the report. They would recognize that someone may have found the cause of (and, hopefully, a remedy for) a trouble spot that has bothered them for some time.

Note how the following titles each *tell* something about the reports they precede:

REDUCING AMBIENT NOISE IN AIR TRAFFIC CONTROL CENTERS

EFFECTS OF DIESEL EXHAUST ON LATEX PAINTS

SALT EROSION OF CONCRETE PAVEMENTS

These titles may not attract every potential reader who comes across them, but they will certainly gain the attention of anyone interested in the topics they describe.

TITLE PAGE

The title page normally carries four main pieces of information: the report title (the same title that appears on the cover); the name of the person, company, or organization to whom the report is directed; the name of the company originating the report (sometimes with the author's name); and the date it was completed. It may also contain the authority or contract under which the report has been prepared, a report number, a security classification such as CONFIDENTIAL or SECRET, and a copy number (important reports given only limited distribution are sometimes assigned copy numbers to control and document their issue). All this information must be tastefully arranged on the page, as has been done in the sample reports later in this chapter.

TABLE OF CONTENTS

All but very short reports contain a Table of Contents (T of C). The T of C not only lists the report's contents, but also shows how the report has been arranged. Just as a prospective book buyer will scan the contents page to find out what a book contains and whether it will be of interest, potential readers will scan a report's T of C to assess how the author has organized the work. They will be searching for a clue to the author's competence to arrange technical information in a logical fashion. If they sense that the report writer knows his or her business both as a technical person and as a writer, then they will delve into the report.

A satisfactory arrangement for a T of C is shown on page 179. This page also contains a list of appendixes attached to the report, with each identified by its full title. In long reports, a list of illustrations with their page numbers may also be helpful. If used, it should be inserted between the T of C and the list of appendixes.

REFERENCES (ENDNOTES), BIBLIOGRAPHY, AND FOOTNOTES

A report writer who refers to another document, such as a textbook, journal article, report, or correspondence, or to other persons' data or even a conversation, must identify the source of this information in the report. To avoid cluttering the report narrative with extensive cross-references, they are placed in a storage area at the end of the report or at the foot of the page. This storage area is known as a List of References, a Bibliography, or Footnotes. Specifically:

A **List of References** is the most convenient and popular way to list source documents. The references are typed as a sequentially numbered list at the end of the narrative sections of the report (usually immediately ahead of the appendix). Such numbered references are frequently referred to as *Endnotes.* For a short example, see page 203.

A **Bibliography** is simply an alphabetical listing of the documents used to research and conduct a project. The documents are listed in alphabetical order of authors' surnames, and the list is placed at the end of the report narrative. A bibliography may list many more documents than are referred to in the report. An example appears on page 164.

Footnotes are like *Endnotes* but are printed at the foot of the page on which the particular reference appears. Footnotes are becoming less and less popular because they are awkward to arrange on the page and, since they are visible, distract reading continuity. Turn to page 165 for an example.

Most industries and engineers prefer to use a List of References for their technical reports.

Preparing a List of References (Endnotes). References should contain certain information, arranged in this order:

(a) Author's name (or authors' names).
(b) Title of document (article, book, paper, report).
(c) Identification details, such as:

For a book: city and state (or country) of publication, publisher's name, and year of publication.

For a magazine article or technical paper: name of magazine or journal, volume and issue number, and date of issue.

For a report: report number, name and location of issuing organization, and date of issue.

For correspondence: name and location of issuing organization, name and location of receiving organization, and the letter's date.

For a conversation or speech: name and location of speaker's organization, name, identification, and location of listener(s), and the date.

(d) The page number (if applicable) on which the referenced item appears or starts.

For example, if your first reference is to an illustration on page 74 of a book by Henrietta J. Tichy, your first entry in your list of references would be like this:

(a) Author's name
(b) Book title (always underlined)

1. H. J. Tichy, Effective Writing for Engineers, Managers, Scientists (New York, NY: John Wiley & Sons, Inc., 1967), p. 74.

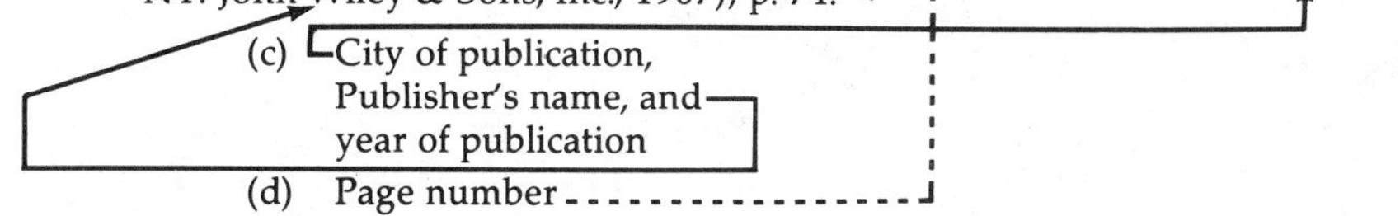

If a book has two authors, both are named:

2. David B. Shaver and John D. Williams, . . . (etc)

But if there are three or more authors, only the first-named author need be listed:

3. Donald R. Kavanagh and others, . . . (etc)

If a book is a second or subsequent edition (as this book is), the edition number is entered immediately after the book title:

4. R. S. Blicq, Technically-Write! 2nd ed (Englewood Cliffs, NJ: Prentice-Hall, Inc., 1981).

Some books contain sections written by several authors, each of whom is named within the book, with the whole book edited by another person. If your reference is to the whole book, then identify it by the editor's name:

5. Robert M. Woelfle, ed, A Guide for Better Technical Presentations (New York, NY: IEEE Press, 1975).

But if your reference is to a particular section of the book, identify it by the specific author, enclose the section title in quotation marks, underline the book title, and then name the editor:

6. J. M. Lufkin, "The Slide Talk: A Tutorial Drama in One Act" in A Guide for Better Technical Presentations, ed. Robert M. Woelfle (New York: NY: IEEE Press, 1975), p. 14.

Similarly, if you are referring to an article in a magazine or journal, and it is your seventh reference, you would list the article like this:

(a) Author's name
(b) Title of article (always in quotation marks, not underlined)

7. William L. Everitt, "The Engineer: a Perennial Student" in IEEE Spectrum, 14:12, December 1977, p. 21.

(c) Title of journal (underlined)
Volume and issue numbers (14:12), and date
(d) Page on which article starts, or of specific quotation referenced

When a magazine article, technical paper, or report is published as one of several documents bound into a volume, then it is listed within quotation marks, and only the title of the volume is underlined, as has been done here. But if the article, technical paper, or report is published as a *separate* document (i.e. not in a magazine or journal), its title is underlined and the quotation marks are omitted:

8. Derek A. Lloyd, Effective Communication and Its Importance in Management Consulting. Report No. 61, Smyrna Development Corporation, Atlanta, GA, February 18, 1980.

If a report or magazine article does not show an author's name, then the first entry for that item will be the title of the report or article:

9. "Continuing Education: Challenge of the 80's" in Engineering Technologist, 8:5, May 1978, p. 113.

For a letter or memorandum, the entry should be like this:

10. Guy Le Page, Robertson Engineering Company, Toronto, Ont. Letter to Ingrid Busch, The Roning Group, Winnipeg, Man., December 12, 1979.

And for a conversation or speech:

11. Elwood R. Phillips, Lakeside Power and Light Company, Montrose, Ohio, in conversation with Anna King, H. L. Winman and Associates, Cleveland, Ohio, March 16, 1980.
12. Emily K. Schlesinger, Baltimore Gas and Electric Company, speaking to the 26th International Technical Communication Conference, Los Angeles, Cal., May 18, 1979.

Every entry in a list of references must have a corresponding reference to it in the Discussion section of your report. At an appropriate place in the narrative you should insert a superscript (raised) number to identify the particular reference. It should look like this:

Earlier tests[3] showed that speeds higher than 2680 rpm were impractical.

(Alternatively, the raised 3 could go here.)

With certain typewriters, it is not possible to type superscript numbers. In such cases you may state the reference number in parentheses:

Earlier tests (3) showed that speeds . . .

If you refer to the same document several times, your list of references need show full details for that document only the first time you refer to it. Subsequent references can be shown in a shortened form containing only the author's name (or authors' names) and the page number. For example, if the first reference you make is to an item on page 48 of the particular book described below, the entry in the list of references would be:

1. Peter F. Drucker, An Introductory View of Management (New York, NY; Harper's College Press, 1977), p. 408.

Now suppose that your second and third sources are other documents but for your fourth source you again refer to *An Introductory View of Management,* this time quoting from page 159. This time you need list only the author's surname and the new page number:

4. Drucker, p. 159.

And you would do the same for each future reference to the same document, simply changing the page number each time. You can even make repeated references to several different documents by the same author by simply inserting the year of publication for the particular document between the author's name and the page number:

9. Drucker, 1977, p. 159.

Preparing a Bibliography. A bibliography lists not only the documents to which you make direct reference, but also many other documents which deal with the topic. The major differences between the elements of the list of references and those of the bibliography are:

1. Bibliography entries are not numbered 1, 2, 3, etc.
2. Bibliography entries are listed in alphabetical order of primary (first-named) authors, with surname given first. That is, a magazine article by Andy **B**urns is listed before a book by Wayne **G**odfrey and Allen **Y**oung, which in turn is listed before a report by Harvey **W**inman.
3. Punctuation is slightly different (see Figure 6-2).
4. The first line of each bibliography entry extends slightly further to the left than subsequent lines for that entry (about five typewritten characters—see Figure 6-2). These extensions serve a practical purpose in that they help readers pick out authors' names easily from a long list (notice how the authors' names stand out clearly in the bibliography in Figure 6-2). This also is why the author's name for each bibliography entry is listed surname first.

Because a bibliography is not numbered, it's not possible to cross-refer directly to it simply by inserting a superscript number in the report narrative, as can be done with a list of references. The most common method is to insert a parenthetical reference in the narrative, and in it include the author's name (or authors' names) and the specific page number:

> Although the tests conducted in Alaska (Faversham, p. 261) showed only moderate decomposition . . .

The full descriptive listing for Faversham's book or report is carried in the bibliography.

If several reports by the same author are listed in the bibliography, then it is normal to include a date in the parenthetical reference to identify which report is being mentioned:

BIBLIOGRAPHY

Agnew, Jeremy, Thick Film Technology (Rochelle Park, NJ: Hayden Book Co., 1973).

Felber, Stanley B. and Arthur Koch, "Group Communication" in What Did You Say? (Englewood Cliffs, NJ: Prentice-Hall, Inc., 1973).

Lovatt, Frederick G. and others, Management is No Mystery, 3rd Edn, ed. Robt B. Arpin (Trent, Arizona: Bonus Books Inc., 1981).

Milnark, S. W., "TACS--A New Approach to Systems Design" in The Airconditioning and Refrigeration Business, 25:2, February 1968, p. 36.

Talbot, J., RATCON Noise Reduction, Report No. 6745-1, CAE Industries Ltd., Electronics Division (Western), Winnipeg, Man., February 1979.

Wood, J., Head, Materials Testing Laboratory, H. L. Winman and Associates, Cleveland, Ohio. Letter to Wayne D. Robertson, President, Robertson Engineering Company, Toronto, Ontario, March 11, 1982.

Figure 6-2. A typical bibliography.

The most significant tests were those conducted 22 kilometers south of Old Crow, Yukon (Crosby, 1976, p. 17), which showed that . . .

The alternative (but less preferred) method is to use footnotes in combination with the bibliography, by inserting superscript numbers into the report narrative and placing footnotes at the bottom of the page. If you use this combination, keep the footnotes short because the documents are described fully in the bibliography. The footnotes should look like this:

[1]Faversham, p. 261.

[2]Crosby, 1976, p. 17.

Preparing Footnotes. Although footnotes once were common, they are now the least-preferred method of source referencing, particularly for technical and business reports. You will still see them used, however, in scholarly works and some scientific papers.

Typical footnotes appear at the foot of this page. The information they contain is the same as that for a list of references. The major differences are:

The first line of each footnote entry is indented slightly and the footnote number is raised slightly.[1]

The footnotes may be numbered consecutively on a page-by-page basis, with each page starting afresh with footnote 1, or they may be numbered consecutively throughout the whole report.

A footnote can be an explanatory note,[2] rather than a reference to another document, although use of footnotes for this purpose is not recommended.

If a document is referenced more than once, it is listed in full only the first time. For subsequent references, only the author's name (or authors' names) and the relevant page number are listed.[3]

The most effective way to refer readers to source material is to insert superscript numbers in the report narrative and to print a numbered list of references at the end of the report. Although footnotes or a bibliography may be used in place of a list of references, they have disadvantages which make them less suitable for engineering and industrial reports.

Source referencing is an essential part of a report. The entries must be complete and accurate, so that readers can identify, and locate or order, every document you list. For that reason, always transcribe numbers, names, titles, dates, and other details extremely carefully from the original document.

DISTRIBUTION LIST

A report sent to several readers will often contain a distribution list that identifies all the persons who are to receive it. The position of the distribution

[1]Charles T. Brusaw and others, Handbook of Technical Writing (New York, NY: St. Martin's Press, 1976), p. 174.

[2]In typeset documents, footnotes normally are printed in a smaller typeface.

[3]Brusaw, p. 106.

list depends on company practice, the preference of the author, and the length of the list. Probably the most suitable position is immediately after the bibliography, although a short distribution list may appear at the foot of the T of C page. A long list may have a page to itself and be inserted as a separate appendix.

ACKNOWLEDGMENTS

Acknowledgments are inserted whenever a report author wishes to acknowlege special help received during the investigation or study which is the topic of the report. Acknowledgments should be brief, simple, and limited only to those persons who have made a significant contribution. They may be inserted on a page by themselves after the summary, after the T of C, as part of the introduction, or immediately after the discussion.

COVER LETTER

The cover letter is a brief courtesy letter that identifies the report and states briefly why it is being forwarded to the addressee. It may be either a letter or an interoffice memorandum and is attached to the outside front of the report. A letter is used for reports distributed outside the originating company, an interoffice memorandum for reports confined to internal distribution. A typical cover letter precedes Report No. 2 in this chapter.

LETTER OF TRANSMITTAL

Occasionally you will find a letter of transmittal inserted as a foreword or preface to a report. Normally written and signed by top management, it reflects company policy and sometimes management's interpretation of the report's findings. Because it repeats much of the information contained in the introduction, conclusions, and recommendations, it usually takes the place of the summary.

The letter of transmittal can also serve a useful purpose when it is used with a technical proposal, which is a form of technical report in which a company describes how it can successfully tackle a proposed task at an economical price for the government or another company. In this case it is used chiefly to discuss financial and legal implications, such as specifications and statements that may be open to more than one interpretation.

An alternative to the letter of transmittal is the *executive summary,* which is an analytical summary of the purpose of the report, its main findings and conclusions, and the author's recommendation. Unlike the normal report summary prepared for all readers, the executive summary can present detailed information on aspects of particular concern to senior executives, and often will introduce financial implications. It may be prepared as a letter or memorandum, or a more formal page titled "Executive Summary."

A letter of transmittal or executive summary is bound inside a report, behind the cover or title page.

The Full Report

THE MAIN PARTS

This description of the full formal report will cover only the ten parts that contribute directly to the report itself. The distribution list, cover letter, letter of transmittal, and acknowledgments are omitted because they add nothing to the technical content. Other parts that sometimes may be omitted are the cover and T of C (if the report is very short), the recommendations (if the writer has none to make), and the bibliography and appendix (if there are no references or supporting data).

THE MAIN PARTS OF A FORMAL REPORT

Cover:	Outside jacket of report; contains title of report and name of originating company; its quality and use of color reflect company "image."
Title Page:	First page of report; contains title of report, name of addressee or recipient, author's name and company, date, and sometimes a report number.
Summary:	An abridged version of whole report, written in nontechnical terms; *very* short and informative; normally describes salient features of report, draws a main conclusion, and makes a recommendation; always written last, after remainder of report has been written.
Table of Contents:	Shows contents and arrangement of report; always includes a list of appendixes and, sometimes, a list of illustrations.
Introduction:	Prepares reader for discussion to come; indicates purpose and scope of report, and provides background information so that reader can read discussion intelligently.
Discussion:	A narrative that provides all the details, evidence, and data needed by the reader to understand what the author was trying to do, what he actually did, what he found out, and what he thinks should be done next.
Conclusions:	A summary of the major conclusions or milestones reached in the discussion; conclusions are only opinions so can never advocate action.
Recommendations:	If the discussion and conclusions suggest that specific action needs to be taken, the recommendations state categorically what must be done.
Bibliography/ References:	A list of reference documents which were used to conduct the project and which the author considers will be useful to the reader; contains sufficient information for the reader to correctly identify and order the documents.
Appendix(es):	A "storage" area at the back of the report that contains supporting data (such as charts, tables, photographs, specifications, and test results) which rightly belong in the discussion but, if included with it, would disrupt and clutter the major narrative.

Figure 6-3. Capsule descriptions of formal report parts.

NORMAL ARRANGEMENT

(Conclusions and Recommendations *after* Discussion)

These ten main parts, complete with capsule descriptions, are listed in Figure 6-3. The order in which they are presented is their order in most formal business and technical reports and is the format illustrated in the mini-report in Figure 6-4. Note that with this arrangement there is a logical flow of information. It starts with an *introduction* that leads into the *discussion,* from which the writer draws *conclusions* and makes *recommendations* (the two latter parts sometimes are referred to jointly as the terminal section). The sample formal report, "Evaluation of Sites for Montrose Residential Teachers' College" (Report No. 1) uses the normal arrangement of report parts.

ALTERNATIVE ARRANGEMENT

(Conclusions and Recommendations *before* Discussion)

In recent years, more and more report writers have altered the organization of their reports to serve the needs of their readers. This alternative arrangement brings the conclusions and recommendations forward, positioning them immediately after the introduction so that the executive reader does not have to leaf through the report to find the terminal section. The advantages of this format are immediately evident: Busy readers have only to read the initial pages to learn the main points contained in the report, and the writer can help them along by gradually increasing the technical content of the report, catering to semitechnical executive readers up to the end of the recommendations, and to fully technical readers in the discussion and appendix. Although the natural flow of information that occurs in the normal arrangement is disrupted, it is replaced by three components, each containing progressively more technical detail:

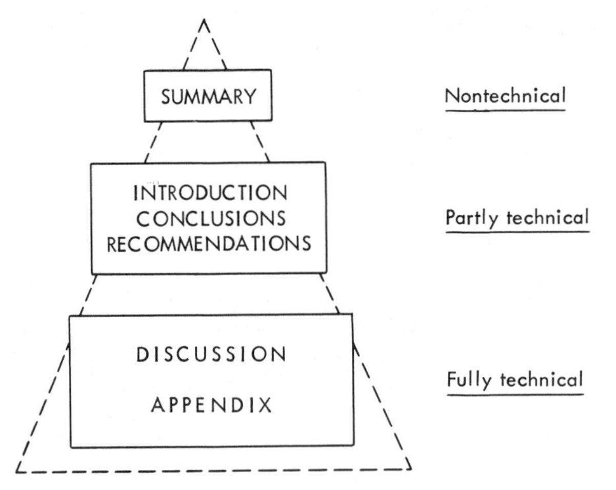

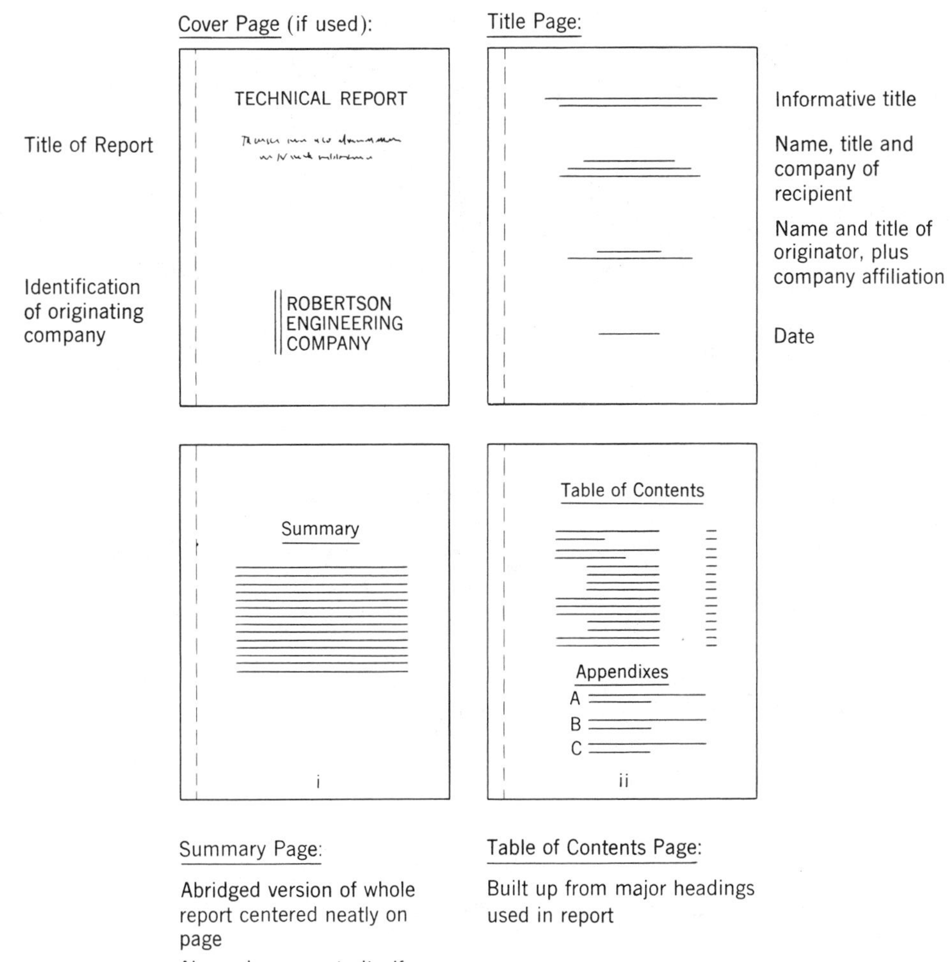

Figure 6-4. Formal report—normal arrangement.

FORMAL REPORT – ARRANGEMENT OF PAGES

(Conclusions and Recommendation following Discussion)

Introduction, Discussion, Conclusions, Recommendation and References

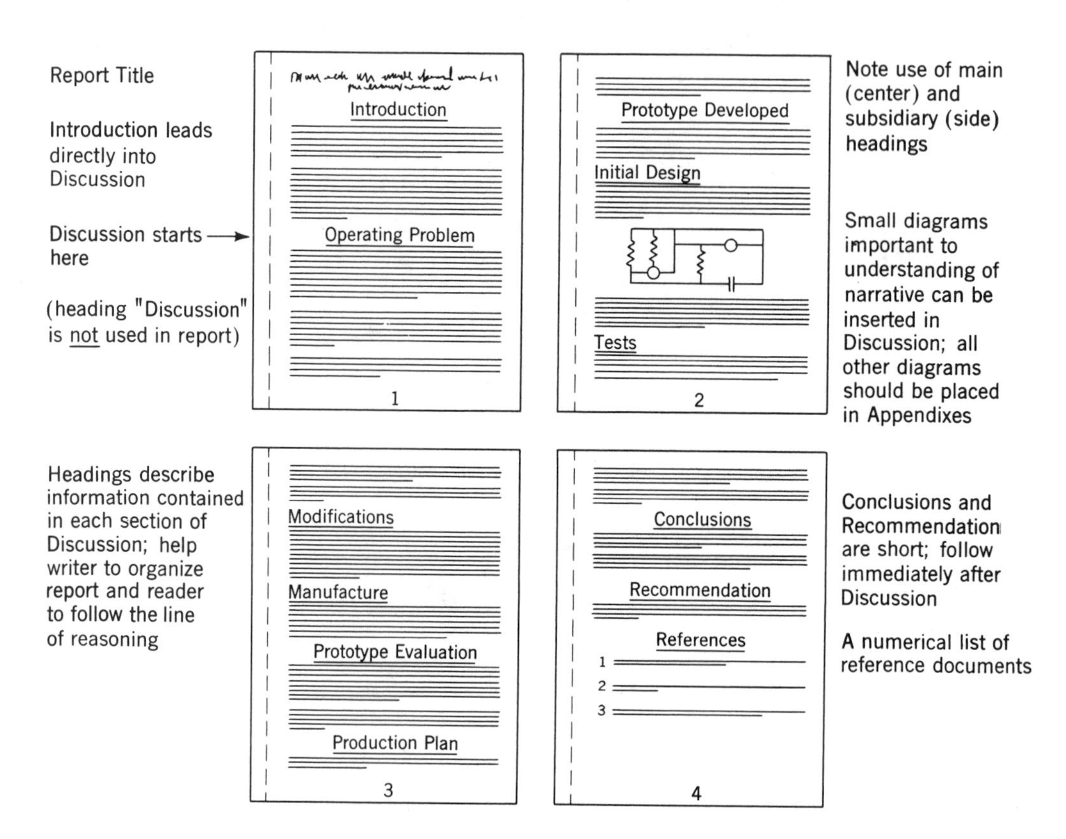

NOTE: Page numbers are arabic numerals either at foot of page as shown here or at top right-hand corner

Left-hand margin is wider to allow for binding edge

FORMAL REPORT – ARRANGEMENT OF PAGES

SUPPORTING DATA (Appendixes)

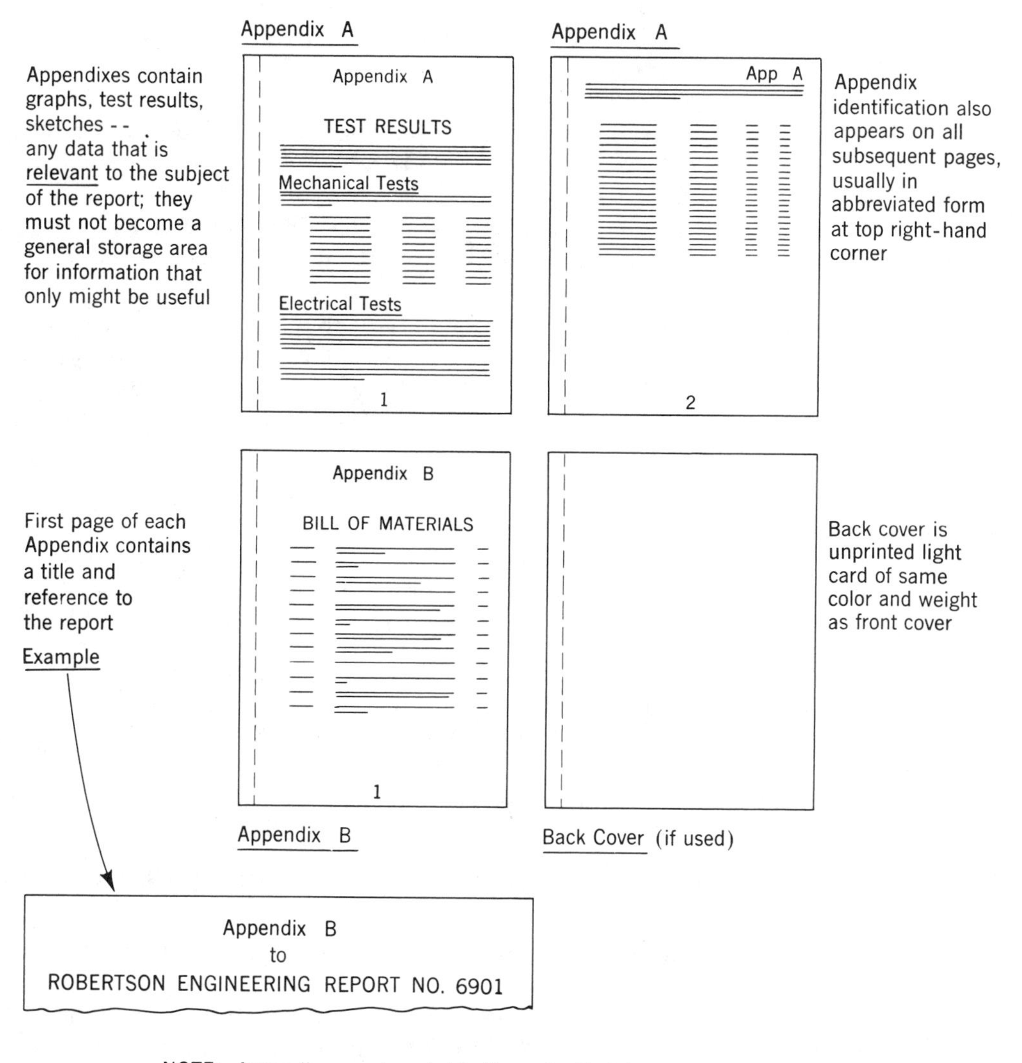

NOTE: Appendixes are inserted in the order that they are referenced in the report

Each Appendix is considered a complete document in itself, thus is numbered separately using arabic numerals

This gradually increasing development of the topic in three separate stages is similar to newspaper technique. If you read any well-written news item, you will find that the first paragraph or two contain a capsule description of the whole story. The next three or four paragraphs contain a slightly more detailed description, and then the newspaper invites you to continue reading on a subsequent page. The final eight or nine paragraphs repeat the same story, but this time with more names, more peripheral information, and more details. Newspapers cater to the busy reader who may not have time to read more than the opening synopsis, and also to the leisurely reader who wants to read all the available information.

A mini-report layout of this alternative arrangement appears in Figure 6-5. This figure contains only the central page of the layout (containing the introduction, conclusions, recommendations, and discussion). The remaining two pages, with the layout for the preliminary information and appendixes, are the same for both arrangements and can be obtained from the first and third pages of Figure 6-4. An example of the alternative arrangement of report parts can be seen in the technical report, "Selecting New Elevators for the Merrywell Building," Report No. 2 in this chapter.

Typical Formal Reports

The two formal reports that follow are typical of the type of work expected of graduate engineers, technologists, and technicians. They comprise:

Report No. 1: *Evaluation of Sites for Montrose Residential Teachers' College*
The writer of this report has placed the conclusions and recommendations after the discussion, following the normal arrangement of report parts, and has used the subject method of development.

Report No. 2: *Selecting New Elevators for the Merrywell Building*
In this report the writer has adopted the alternative arrangement of report parts, placing the introduction, conclusions, and recommendations as a single unit ahead of the discussion. He has used the concept method of development.

On the pages preceding each report I have inserted a few comments, plus some of the data that initiated each project and resulted in the report being written. Both reports have been typed single-spaced. Double-spacing is more common, although some companies prefer to single-space their reports and leave two lines between main paragraphs, as in these two examples.

Formal Report No. 1
Evaluation of Sites for Montrose Residential Teacher's College

This H. L. Winman and Associates' report was prepared originally as a student writing project in which the terms of reference closely paralleled a real situation. The original assignment instructions are printed here so that you may

FORMAL REPORT – ARRANGEMENT OF PAGES

(Conclusions and Recommendation before Discussion)

(ALTERNATIVE ARRANGEMENT)

Introduction, Conclusions, Recommendation, Discussion & References

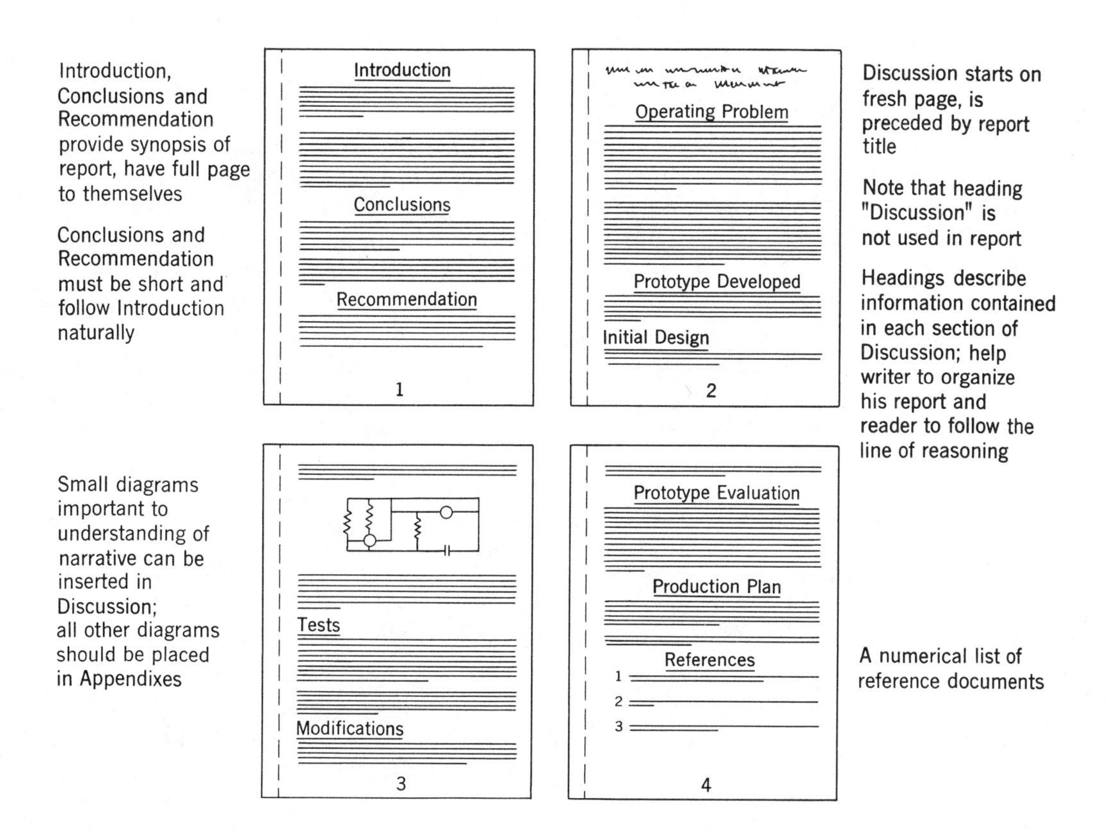

NOTE: Page numbers are arabic numerals either at foot of page as shown here or at top center

Left-hand margin is wider to allow for binding edge

Fgure 6-5. Formal report—alternative arrangement.

see how the writer used the information available to him, plus his own imagination, to write a very realistic report.

ASSIGNMENT DETAILS

You are employed by H. L. Winman and Associates, a Cleveland firm of consulting engineers, for whom you have been conducting a survey to determine the best site for a Residential Teachers' College in the City of Montrose, Ohio (population 275,000). Your task is to prepare a formal report that describes your findings and recommends the best location for the college.

H. L. Winman and Associates was engaged by the College's Board of Governors (composed mainly of local businessmen) to resolve the impasse caused by a three-way split over the choice of a site. Some members of the Board want to build the college at Bluff Heights, so that it will be a prominent feature in an urban redevelopment area; others would like to locate it at Varsity Valley beside the University of Montrose, where it will become part of a proposed city education center; the third group want to place it 22 miles (36 km) out of the city at Hartland Point, where it will develop an entity of its own. The three sites are described briefly below.

> *Bluff Heights.* Part of urban development area; triangular; very few trees; close to expressway; on a medium hill; commands a good view of city. Boundary to the south borders Heritage Industrial Park (light industry); to the east is mainline railway (Northern Railways route from Montrose to Waverly); to the northwest, Highway 79. Favored by three governors because it will rejuvenate a depressed area.
>
> *Varsity Valley.* Shallow valley between Montrose University and Varsity Heights residential area; although new, this residential area promises to become shoddy and run-down (already evidenced by depressed property prices); valley tends to be damp and hold diesel and automotive fumes from expressway. Varsity Valley is a rectangular area bounded by Elm Creek and Varsity Heights to the south and east, and an expressway and U of M to the north; low vegetation and bush with scattered oaks and elms. Favored by three governors because it is close to U of M and so will help form an "Education Complex" for the city.
>
> *Hartland Point.* Superb location; rolling countryside; well-treed (conifers, elm, and poplar); drops gently down from northern boundary to picturesque Hartland River, which bounds south and east edges of area. Boundary to the north is Highway 101, main highway from Montrose to Hartland (population 3265); to the west, Hartland Golf Course. Hartland is three-quarters of a mile east of the site. Favored by three governors because it would truly become a "residential" college; has ample space for expansion; can also provide space for building houses for staff. Disliked by others because it will be too remote from city.

Plans for the college call for immediate acquisition of 80 acres on which the main instructional complex and residence building will be erected. The availability of 80 additional acres adjacent to the first 80 is considered desirable: 40 for future college expansion and 40 for an additional residence building and a social activities center.

(*Note:* A rough sketch of the three sites and a table of comparative data were included with the original assignment. Since these also appear in the report, they are not repeated here.)

READING THE REPORT

Try reading this report in the order in which the members of the Board of Governors would read it:

1. Scan the summary first to get a capsule comment.
   Does it make you want to read further?
2. Read the introduction, then turn to the conclusions and recommendations.
   By now you should have a good grasp of the report's main conclusions. Do you agree with them?
3. Now read the discussion.
   If you previously did not agree with the writer's conclusions, do you now? Has he used a logical and convincing method to develop his discussion?

COMMENTS ON THE REPORT

The writer of this report has been very diplomatic. He has recognized that only three of the nine members of the Board of Governors will agree readily to his selection and that he must persuade the remaining six that he has made a valid choice. By stating in the beginning of the summary that *all three* sites could be used successfully, he is forcing all the members to agree with him and to read further to find out which site he has selected (and also to turn to the discussion to establish why it is better than the other two sites). Then, in the discussion, he writes an unbiased description of each site, presenting facts and figures without trying to persuade the reader that any one has greater advantages than the others (he even refrains from placing them in the normal ascending or descending order of preference). He may not fully persuade all the reluctant members, but at least he will convince them that he has reached an unprejudiced decision.

Some specific comments on the report are:

1. Words like "sociological" and "utilitarian" may make some readers of the summary reach for their dictionaries. Are they too technical? In this context I think not: the Board of Governors will expect to be faced with words that reflect the professional level of the study. They are the right words used in the right place.
2. The subject arrangement used for writing the discussion is clearly evident in the T of C. A reader turning to it from the summary can see immediately how the author has organized his report.
3. The three parts of the introduction are easily recognizable: paragraph 1 is the *purpose,* paragraph 2 contains mostly *background information,* and paragraph 3 identifies the *scope.*
4. The parameters at the end of the introduction establish guidelines for the reader to keep in mind while reading about each site.
5. Two subtle negative phrases in the discussion of character on page 7 provide the first real hint that the author will select Varsity Valley. The first is "a *questionable* combination" at the end of the comment on Bluff Heights; the

second is "activities would be *limited*" at the end of the Hartland Point comment. There is no negative comment concerning Varsity Valley.

6. The rating system used to identify the best site at the end of the discussion may seem artificial, but it provides a practical means for readers to visualize the comparative value of each site.
7. The conclusions provide a convincing terminal summary that follows naturally from the introduction. The first paragraph is the main conclusion, and the remaining paragraphs are three subsidiary conclusions listed in decreasing order of importance.

Formal Report No. 2
Selecting New Elevators
for the Merrywell Building

Barry Kingsley, the writer of this report, has used the concept method of development to guide the reader through the maze of possible elevator combinations to the one choice that offers the most advantages to his client. It is a persuasive report, and quite convincing.

Before reading the report, read the client's letter authorizing H. L. Winman and Associates to initiate an engineering investigation (see p. 190). By comparing it with the report, you can assess how thoroughly Barry has answered the client's requests.

DATA USED BY H. L. WINMAN AND ASSOCIATES

1. The Merrywell Building is owned by Merrywell Enterprises Inc., which occupies the top two floors. It is a nine-story commercial and light manufacturing building located at 617 Carswell Avenue, a busy thoroughfare in the Montrose, Ohio, downtown business district. It has a frontage of 180 feet (55 m), is 130 feet (40 m) deep, and is 71 years old. The present owners purchased it in 1954 and made some moderate renovations to bring it up to date. These included installation of up and down escalators between the ground floor and the second floor, which is occupied by a popular cafeteria patronized by both the building's occupants and the staffs of offices occupying neighboring buildings. Two rather slow manually operated passenger elevators and an even slower freight elevator were left intact, although it was recognized that new elevators eventually would have to be installed.
2. The existing two passenger elevators are slow and cumbersome. They can carry only 11 passengers each, plus the operator, and their speed is only 10 seconds per floor. The freight elevator is 10 feet wide by 8 feet deep (3.1 by 2.45 m), but because of its age, government inspectors have limited its load capacity to 2000 lb (950 kg).
3. Merrywell Enterprises Inc. has set aside a budget appropriation of $500,000 for the purchase and installation of new elevators. They would like the cost to be less than budgeted, but would accept a slight increase above this figure if the right combination of elevators can be found to satisfy the needs of all the companies occupying the building.
4. Top management of Design Consultants Inc. and Vulcan Oil and Fuel Corporation would like an executive elevator for the sole use of executives of each major company in the building. Although such an elevator would be costly and have limited use, David P. Merrywell indicates that he might accept the idea, but hesitates because of the problems it would introduce.

H L Winman and Associates

PROFESSIONAL CONSULTING ENGINEERS
475 Reston Avenue-Cleveland, Ohio, 44104

FEASIBILITY STUDY

EVALUATION OF SITES

FOR

MONTROSE RESIDENTIAL TEACHERS' COLLEGE

Prepared for:

Board of Governors
City of Montrose Residential Teachers' College
Montrose, Ohio

October 10, 19__

SUMMARY

Sound sociological and utilitarian reasons can be found for recommending any one of the three sites proposed as the permanent location of the Montrose Residential Teachers' College. We have therefore analyzed the suitability of each site on tangible factors that would influence the initial construction and eventual growth of the college. By comparing environment, accessibility, servicing, ambient noise, soil condition, and land availability, we found that although the margin of selection between the sites is narrow, Varsity Valley offers the best all-round location with maximum growth potential.

To our assessment we added one other less-tangible factor: the effect that each site would have on the character of the college. Again, Varsity Valley offers the best location. A Teachers' College located here would be able to develop a character of its own, separate from but still within the overall educational environment already established by the University of Montrose.

i

TABLE OF CONTENTS

EVALUATION OF SITES FOR MONTROSE RESIDENTIAL TEACHERS' COLLEGE

INTRODUCTION

This report is the result of our investigation of three possible sites for a Residential Teachers' College at Montrose, Ohio. The study was commissioned by the Board of Governors in May 19__ and was to be completed for their year end meeting on December 10, 19__.

The requirements for a proper site are dictated by the College development plan previously approved by the Board. This plan calls for an immediate land acquisition of 80 acres, with an additional 80 acres to be available for future expansion.

As stipulated in the terms of reference, we limited our study to three sites: Bluff Heights, Varsity Valley, and Hartland Point. Each was evaluated in light of the following requirements, and the main features were compared in the table at Appendix A:

1. Total land cost should not exceed $275,000, of which $135,000 has been budgeted for the initial 80 acres, and $140,000 for the 80 acres to be acquired later.
2. To avoid high development costs the land should be completely serviced, or utilities should be easily accessible to it.
3. The surrounding environment should be compatible with the joint residential/ educational nature of the college.
4. The site should be well drained and offer no foundation problems.
5. Noise levels should be low: preferably no higher than 45 dB by day, and 35 dB at night.
6. The site should be easily accessible by both automobiles and buses.

We also considered the effect that each site would have on the character of the proposed college.

1

BLUFF HEIGHTS

Bluff Heights was selected as a potential site because a Residential Teachers' College located here would contribute much to the redevelopment of the area.

Environment

The site is part of Urban Renewal Area No. 1, which is to begin construction January 1, 19__. The site is elevated 300 feet (92 m) above the general city area and commands an excellent view of the Wabagoon River and the city park to the east. It is a triangular site bounded on the east by the railway to Waverly, on the northwest by Highway 79, and on the south by Heritage Industrial Park.

In spite of being downwind from the industrial park there is no air pollution problem as the industries are of a "light" nature. That is, they are warehouses, assembly plants, printing plants, etc, of a non-smoke or odor-producing variety.

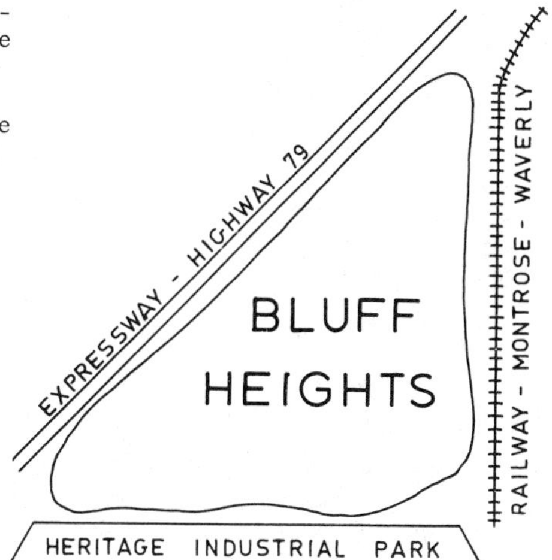

Accessibility

Access to the Bluff Heights site is excellent. Highway 79 Expressway provides a direct route from the Montrose business area, a distance of 3½ miles (5.6 km), and there is a main arterial road in the industrial park that could be extended directly north into the site.

Servicing

All utilities, including sewer, water, power, gas, and telephone services, are installed and available for immediate connection to the college buildings.

Land

Only 120 acres of land are available at this site, with no chance for future expansion beyond this size. Although this would provide the 80 acres required for immediate needs, it could supply only half the land necessary for future expansion. The cost of the site would be $2300 per acre for a total of $276,000.

Soils

The site is dry and well drained. A series of test holes revealed dry sandy clay which should present no particular foundation problems. Bedrock is at an average depth of 24 feet (7.3 m). Poured in place concrete piles could be used but the sandy soil may present some difficulties in boring. Driven timber, steel, or precast concrete piles are recommended.

Noise

The average weekday noise level of 46 decibels at this site is rather high for a residential/educational community. Most of the noise is caused by the high speed traffic on the expressway and by the numerous trains. Traffic volume on both these routes is expected to increase by at least 20% over the next five years, so that the noise level also is likely to increase.

We took a series of noise readings over a seven-day period to determine the variations at night, during the day, and at weekends. These readings are shown in Appendix B for this and the other two sites.

VARSITY VALLEY

Varsity Valley was selected as a possible site for the Residential Teachers' College because of its proximity to the University of Montrose. It was considered that the two seats of higher learning would together provide the start of a comprehensive education complex for the city.

Environment

This is a rectangular site in a shallow valley in the predominantly residential area of Varsity Heights. Montrose University is directly north of the site but separated from it by the Circular Expressway. Elm Creek curves around the east and south sides of the site and runs between it and the Varsity Heights residential area. Although this residential area is only five years old, present decreasing property values may indicate future deterioration into a second class district. The presence of a high class education facility could avert this deterioration.

Because of its proximity to the University full advantage could be taken of all the University facilities, both technological and social. A full rapport could be established between faculties of the two establishments as well as the student bodies.

The valley tends to be damp and as a result holds the exhaust fumes from the expressway. This should be relieved to a large extent when the present drainage program in the valley is completed. The green belt of shrubs and trees already planted along the expressway will trap much of the expressway fumes and noise after two or three years further growth.

3

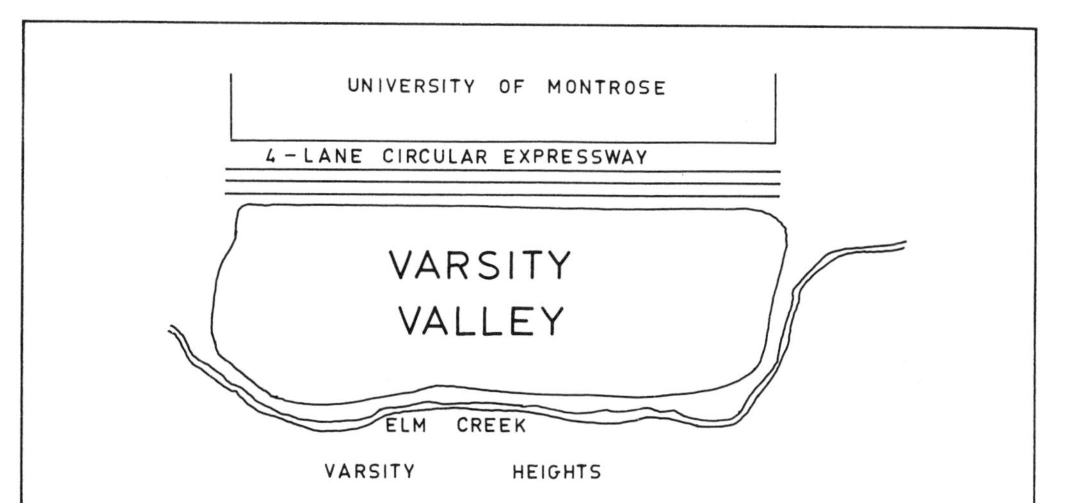

Accessibility

Access to the Varsity Valley site is almost as good as that to the Bluff Heights site. It is six miles (8.9 km) from the city center to Varsity Valley, with good street connections from all areas of the city to the expressway. The University bus line runs to Montrose city center via the expressway from 06:00 to 02:00, and could easily loop through the college site.

If access to the University becomes a requirement, it could be established later by building a footbridge over the expressway.

Servicing

The area is presently 40% serviced, with power, water, and sanitary sewers now available. Gas lines, telephone lines, and storm sewers can be readily extended from either the University or Varsity Heights, and would cause neither high costs nor delays in construction.

Land

There are 150 acres of land available for present and future requirements -- 10 acres less than the desired ultimate area. This need not be a problem since at this site some of the proposed residences and recreation facilities could be eliminated by combining with and making use of the University facilities.

The total cost of the site would be $270,000 at an average price of $1800 per acre. At present there is little demand for land in Varsity Valley and the price is low. But as prices are likely to rise if a large portion of this land is acquired for the college, the full 150 acres would have to be acquired immediately if the land is to be purchased within the prescribed budget.

4

Soils

A recently constructed ditch system now provides surface drainage into Elm Creek. This should relieve the present pooling conditions caused by the relatively impermeable dense, clayey, silt-type soil which covers the area. Bedrock is quite close, at an average depth of 12 feet (3.7 m). Foundation construction therefore would not be costly.

Noise

This site has a lower weekday noise level than Bluff Heights, being in the medium-low range of 40 decibels. The major cause of noise is vehicular traffic on the expressway which, although traffic volume is expected to increase annually, will be screened by the now-growing shelter belt. The present noise level is causing no problems at the University on the north side of the expressway.

HARTLAND POINT

Although Hartland Point lies some distance east of Montrose, it was selected as a potential site for the Teachers' College because it was considered that its remoteness would contribute much to the residential nature of the training establishment.

Environment

The Hartland Point site is a country setting approximately three quarters of a mile (1.2 km) west of Hartland (population 3265). It is a rectangular site adjacent to and south of highway 101 which connects Hartland with Montrose. The Hartland River flows along the south and east sides of the site, while the Hartland Golf Course, which is a private members-only club, acts as the west boundary. The site slopes from the highway to the river in a series of gentle undulations. It is well treed with conifers, elms and poplar.

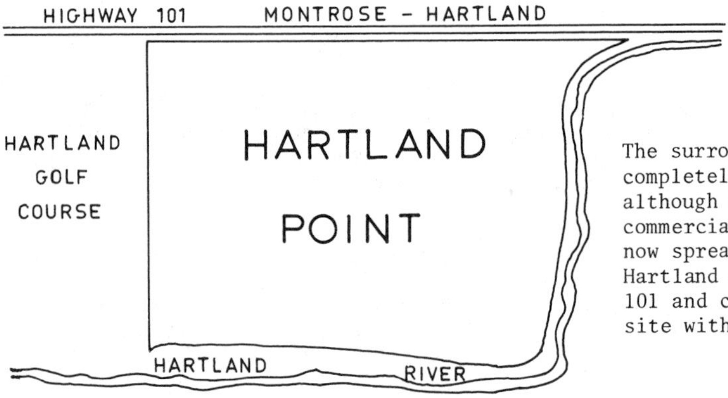

HIGHWAY 101 MONTROSE - HARTLAND

HARTLAND GOLF COURSE

HARTLAND POINT

HARTLAND RIVER

The surrounding area is completely agricultural, although some fringe commercial development is now spreading west from Hartland along highway 101 and could reach the site within five years.

Accessibility

Road access to the site is good via highway 101 from either Hartland or Montrose. However, the latter's city center is 22 miles (36 km) distant, and travel time, especially in the winter, could be a problem.

Bus service is now poor with only the H-B Bus Lines running twice daily past the site. Undoubtedly this service would be improved if the college is situated here, otherwise it will become almost isolated for persons without their own means of transportation. Isolation could also raise overall development costs by causing too high a proportion of the student body to seek on-campus accommodation.

Servicing

At present no services are available at the site, except for a combined power-telephone line along highway 101. All other services such as gas, water, and sanitary sewers would have to be extended from Hartland. This would entail considerable expense which likely would have to be borne by the college. Even then water pressure probably would be insufficient and some type of booster and storage facilities would be required at the site. Surface drainage could be run directly into the Hartland River.

Land

The proposed site comprises approximately 200 acres, which at the present price of $900 per acre would cost $180,000. This would be more than adequate for present and anticipated future needs. If further expansion is necessary, it would have to be north of highway 101.

Soils

The soil, which is of a generally sandy clay underlain with glacial till, provides excellent drainage. Below the till is bedrock at an average depth of 32 feet (8.9 m). Foundation requirements would vary for each building, depending on its loading and excavation depth, with a separate foundation study being necessary in each case.

Noise

Noise at the site is virtually nil, varying from 17 to 30 decibels. The chief cause of noise is traffic on highway 101, which occurs mainly between 08:00 and 18:00 hours.

The site is, however, directly under the aircraft approach line for the east-west runway of the Montrose airport, which is currently being extended. With the expected increase in air traffic, the noise from aircraft could double in five to ten years.

COLLEGE CHARACTER

After investigating each site on the basis of its physical features and the facilities offered, we felt there was one more point that must be considered: this is the college character, or the personality that the college would develop in each location. This character will depend to a great extent on the college site and its immediate surroundings. We believe that:

> The Bluff Heights site will develop no real character of its own but will be a focal point for the urban rejuvenation of the area. It will be an educational complex in an industrial area; a questionable combination.
>
> A college at the Varsity Valley site would fit in with the existing University complex to form a general educational center. It would become another one of the many colleges which now make up the University of Montrose.
>
> An entirely separate college would develop at the Hartland Point site, completely unrelated to any particular city or University. Because of its relative remoteness it would be a residential college in every sense of the word. Outside activities would be limited.

SITE COMPARISONS

Because location of the college at any of the sites offers definite advantages either to the college or to the surrounding area, we decided to run a compara-

SITE RATINGS

Factual Condition	Bluff Heights	Varsity Valley	Hartland Point
Environment	1	2	3
Accessibility	3	2	1
Servicing	3	2	1
Land	1	2	3
Soil	2	3	1
Noise	1	2	3
Total	11	13	12
Character	1	3	2
Total	12	16	14

tive analysis by rating the factual condition for each site as first, second, or third, and awarding 3, 2, or 1 points according to our selection (see table). We totaled the ratings to identify the site with the highest score, and then added one other important but less tangible factor -- college character -- to our analysis.

The site with the highest total score is Varsity Valley, which is third choice in none of the comparison points. In several cases where it is second choice, strong arguments can be made to bring it close to first choice.

CONCLUSIONS

Selection of the best location for the Montrose Residential Teachers' College is difficult because sound reasons can be found for building it at all three sites. Even when the physical advantages of each location are taken into account the margin of selection between the sites remains quite narrow.

At Varsity Valley the college would form part of a comprehensive education complex for the City of Montrose. Since this site offers the best combination of physical features, and also promises to contribute most to the development of college character, it would be our first choice.

At Hartland Point the college would become a truly residential educational institution in a pastoral setting. For some this would represent an ideal location, while for others it would mean isolation. Because its physical features and ability to contribute to college character are slightly less suitable than those at Varsity Valley, we would make this site our second choice.

At Bluff Heights the college would rejuvenate a depressed area of the city. This, we feel, would be the wrong reason for building the college at this site, because the relative isolation of an educational institution in an industrial setting would inhibit development of college character. Although the site offers excellent physical features, it has insufficient land to permit adequate future expansion. We therefore would rate this site as our third choice.

RECOMMENDATIONS

We recommend that Varsity Valley be selected as the permanent site of the Montrose Residential Teachers' College. We further recommend that the entire 150 acres be purchased immediately in readiness for future expansion, to avoid future speculative increases in land prices.

APPENDIX A

to

H L Winman and Associates Report

EVALUATION OF SITES FOR MONTROSE RESIDENTIAL TEACHERS' COLLEGE

ANALYSIS OF FEATURES

	Bluff Heights	Varsity Valley	Hartland Point
Environment	Generally industrial	Generally residential	Countryside
Land Available	120 acres	150 acres	200 acres
Cost per acre (and total cost)	\$2300 (\$276,000)	\$1800 (\$270,000)	\$900 (\$180,000)
Distance from Montrose center	3½ miles; 5.6 km	6 miles; 8.9 km	22 miles; 36 km
Accessibility	Excellent; good roads and bus service	Good; good roads and average bus service	Fair; good roads but poor bus service
Utilities & Services	Installed	40% installed	None; need to be extended from town of Hartland (3/4 mile; 1.2 km)
Soil Conditions	Dry sandy clay; bedrock 24 ft ave (6.7 m)	Dense clayey silt; bedrock 12 ft ave (3.7 m)	Sandy clay underlain with glacial till; bedrock 32 ft ave (8.9 m)
Noise Level (average)	High (46 dB)	Moderate (40 dB)	Low (less than 30 dB)

APPENDIX B

to

H L Winman and Associates Report

EVALUATION OF SITES FOR MONTROSE RESIDENTIAL TEACHERS' COLLEGE

NOISE LEVEL READINGS RECORDED AT EACH SITE

Readings were noted over a 5-minute period each hour and then were averaged for entry in the table below.

	BLUFF HEIGHTS		VARSITY VALLEY		HARTLAND POINT	
Time	Weekdays	Weekend	Weekdays	Weekend	Weekdays	Weekend
0100	38 dB	36 dB	31 dB	30 dB	23 dB	21 dB
0200	36	35	30	30	22	19
0300	37	36	30	29	19	17
0400	34	33	29	30	18	18
0500	35	37	30	29	19	17
0600	39	36	36	31	23	21
0700	48	37	41	34	28	24
0800	53	39	46	34	34	26
0900	54	41	48	35	35	27
1000	52	47	47	36	34	26
1100	53	48	46	37	33	28
1200	53	49	48	36	35	29
1300	54	51	48	38	33	32
1400	53	51	46	39	34	31
1500	52	52	46	36	34	30
1600	54	48	49	37	36	29
1700	55	49	50	39	36	31
1800	47	46	46	36	34	31
1900	46	42	42	34	33	30
2000	43	40	38	35	28	27
2100	41	38	36	33	27	27
2200	39	38	34	32	26	23
2300	40	38	34	32	26	22
2359	39	37	32	31	24	22
Average:	45.6 dB	41.8 dB	40.1 dB	33.9 dB	28.9 dB	25.3 dB

MERRYWELL ENTERPRISES INC

617 Carswell Avenue
Montrose, Ohio 45287

File: WDR/71/007

April 27, 19__

Mr Ian Bailey, PE
Head, Civil Engineering
H L Winman and Associates
Professional Consulting Engineers
475 Reston Avenue
Cleveland, Ohio 44104

Dear Mr Bailey

The elevators in the Merrywell Building are showing their age. Recently we have experienced frequent breakdowns and, even when the elevators are operating properly, it has become increasingly evident that they do not provide adequate service at the start of work, at noon, and at the end of the working day. I have therefore decided to install a complete range of new elevators, with work starting in mid-August.

Before I proceed any further, I would like you to conduct an engineering investigation for me. Specifically, I want you to evaluate the structural condition of my building, assess the elevator requirements of the building's occupants, investigate the types of elevators available, and recommend the best type or combination of elevators that can be purchased and installed within a proposed budget of $500,000.

Please use this letter as your authority to proceed with the investigation. I would appreciate receiving your report by the end of June.

Regards

David P. Merrywell

David P Merrywell, President
Merrywell Enterprises Inc

DPM:tk

COMMENTS ON REPORT

A cover letter travels with this report. It simply refers to the client's letter of request and offers further consulting services.

The summary is short and direct because it is written primarily for one man: the President of Merrywell Enterprises Inc. It encourages him to read the report immediately, and to accept its recommendations, by offering the chance to save $50,000.

Although the background information contained in the first two paragraphs of the introduction seems to repeat details the client already knows, Barry recognizes it is necessary for the benefit of other readers who may not be fully aware of the situation in the Merrywell Building. He then defines the purpose and scope of the investigation by stating the client's terms of reference in paragraph 3 of the introduction. (Note that he has copied them almost verbatim from Mr. Merrywell's letter.)

The conclusions present Barry's answers to Mr. Merrywell's four requests. Their order is different from that in paragraph 3 of the introduction because it is normal to present the main conclusion first (in this case, the best combination of elevators that can be purchased within the stipulated budget), and to follow it with subsidiary conclusions in descending order of importance.

By using the concept method of development for the discussion, Barry permits the reader to follow his line of reasoning. He first evaluates the condition of the building to convince the reader that it is worthwhile to install new elevators. He then establishes the needs of the owner and tenants, and at the foot of page 4 summarizes them as a conclusion to the section. Then in the third section he establishes that it is reasonable to purchase the elevators from only one manufacturer. Having conditioned the reader to accept the factors that will eventually limit his selection to only one combination of elevators, he examines other combinations and demonstrates why each is unsuitable. Gradually he eliminates each combination until he leads the reader to the one that not only fits almost all the requirements tabulated earlier, but also offers a substantial saving in cost.

Conceivably, Barry could have stated in the beginning of the discussion that only one combination would be suitable, and then proved it in the rest of the report. This would have been a very direct method but might not have been nearly as effective. It would not have had dramatic effect because the result would have been known from the beginning, and it would not be nearly as persuasive. A reader likes to be challenged and appreciates a strong, convincing line of reasoning.

Assignments

The ten formal report-writing assignments starting on page 207 require varying levels of technical knowledge. Generally, the technical level increases with each assignment. In projects 1 to 5, all the data required to write a report is supplied. For projects 6 to 10, you will need to spend time researching information, either in a technical library or from resources in your local area.

H L Winman and Associates

PROFESSIONAL CONSULTING ENGINEERS
475 Reston Avenue-Cleveland, Ohio, 44104

June 17, 19__

Mr. David P. Merrywell, President
Merrywell Enterprises Inc
617 Carswell Avenue
Montrose, OH 45287

Dear Mr. Merrywell:

We enclose our report No. 8-23 "Selecting New Elevators for the Merrywell Building," which has been prepared in response to your letter WDR/71/007 dated April 27, 19__.

If you would like us to submit a design for the enlarged elevator shaft, or to manage the installation project on your behalf, we shall be glad to be of service.

Sincerely

Barry V. Kingsley

for Ian Bailey, P.E.
Head, Civil Engineering

BVK:wp
enc

H L Winman and Associates

PROFESSIONAL CONSULTING ENGINEERS

TECHNICAL REPORT

SELECTING NEW ELEVATORS FOR THE MERRYWELL BUILDING

Prepared for:

Mr. David P. Merrywell, President
Merrywell Enterprises Inc
Montrose, Ohio

Prepared by:

Barry V. Kingsley, P.E.
Senior Structural Engineer
H L Winman and Associates

Report No. 8-23

June 17, 19__

SUMMARY

The elevators in the 71-year old Merrywell Building are to be replaced. The new elevators must not only improve the present unsatisfactory elevator service, but must do so within a purchase and installation budget of $500,000.

Of the many types and combinations of elevators considered, the most satisfactory proved to be four 8 ft by 7 ft (2.45 x 2.13 m) deluxe passenger elevators manufactured by the YoYo Elevator Company, one of which will double as a freight elevator during non-peak traffic times. This combination will provide the fast, efficient service requested by the building's tenants for a total price of $450,000, which will be 10% less than the budgeted price.

i

TABLE OF CONTENTS

APPENDIXES

SELECTING NEW ELEVATORS FOR THE MERRYWELL BUILDING

INTRODUCTION

When in 1954 Merrywell Enterprises Inc purchased the Wescon property in Montrose, Ohio, they renamed it "The Merrywell Building" and renovated the entire exterior and part of the interior. The building's two manually operated passenger elevators and a freight elevator were left intact, although it was recognized that eventually they would have to be replaced.

Recently the elevators have been showing their age. There have been frequent breakdowns and passengers have become increasingly dissatisfied with the inadequate service provided at peak traffic hours.

In a letter dated April 27, 19__ to H L Winman and Associates, the President of Merrywell Enterprises Inc stated his company's intention to purchase new elevators. He authorized us to evaluate the structural condition of the building, to assess the elevator requirements of the building's occupants, to investigate the types of elevators available, and to recommend the best type or combination of elevators that can be purchased and installed within the proposed budget of $500,000.

CONCLUSIONS

The best combination of elevators that can be installed in the Merrywell Building will be four deluxe 8 ft by 7 ft (2.45 x 2.13 m) passenger models, one of which will serve as a dual-purpose passenger/freight elevator. This selection will provide the fast, efficient service desired by the building's tenants, and will be able to contend with any foreseeable increase in traffic. Its price at $450,000 will be 10% below the proposed budget.

Installation of special elevators requested by some tenants, such as a full-size freight elevator and a small but speedy executive elevator, would be

1

feasible but costly. A freight elevator would restrict passenger-carrying capability, while an executive elevator would elevate the total price to at least 20% above the proposed budget.

The quality and basic prices of elevators built by the major manufacturers are similar. The YoYo Elevator Company has the most attractive quantity price structure and provides the best maintenance service.

The building is structurally sound, although it will require some minor modifications before the new elevators can be installed.

RECOMMENDATIONS

We recommend that four Model C deluxe 8 ft by 7 ft (2.45 x 2.13 m) passenger elevators manufactured by the YoYo Elevator Company be installed in the Merrywell Building. We further recommend that one of these elevators be programmed to provide express passenger service to the top four floors during peak traffic hours, and to serve as a freight elevator at other times.

SELECTING NEW ELEVATORS FOR THE MERRYWELL BUILDING

Evaluating Building Condition

We have evaluated the condition of the Merrywell Building and find it to be structurally sound. The underpinning done in 1948 by the previous owner was completely successful and there still are no cracks or signs of further settling. Some additional shoring will be required at the head of the elevator shaft immediately above the 9th floor, but this will be routine work that the elevator manufacturer would expect to do in an old building.

The existing elevator shaft is only 24 feet wide by 8 feet deep (7.35 x 2.45 m), which is unlikely to be large enough for the new elevators. We have therefore investigated relocating the elevators to a different part of the building, or enlarging the existing shaft. Relocation, though possible, would entail major structural alterations and would be very expensive. Enlarging the elevator shaft could be done economically by removing a staircase that runs up the center of the building immediately east of the shaft. This staircase is used very little and its removal would not conflict with fire regulations since there are also fire staircases inside the east and west walls of the building. Removal of the staircase will widen the elevator shaft by 11 feet (3.35 m) which will provide just sufficient space for the new elevators.

Establishing Tenants' Needs

To establish the elevator requirements of the building's tenants we asked a senior executive of each company to answer the questionnaire attached as Appendix A. When we had correlated the answers to all the questionnaires, we identified five significant factors that would have to be considered before selecting the new elevators. (There were also several minor suggestions that we did not include in our analysis, either because they were impractical or because they would have been too costly to incorporate.) The five major factors were:

> Every tenant stated that the new elevators must eliminate the lengthy waits that now occur. We carried out a survey at peak travel times and established that passengers waited for their elevators for as much as 70 seconds. Since passengers start becoming impatient after 32 seconds[1], we estimated that at least three, and probably four, faster passenger elevators would have to be installed to contend with peak-hour traffic.

3

Although all tenants occasionally carry light freight up to their offices, only Rad-Art Graphics and Design Consultants Inc considered that a freight elevator was essential. However, both agreed that a separate freight elevator would not be necessary if one of the new passenger elevators was large enough to carry their displays. They initially quoted 9 feet (2.75 m) as the minimum width they would require, but later conceded that with minor modifications they could reduce the length of their displays to 7 feet 6 inches (2.3 m). All tenants agreed that if a passenger elevator doubled as a freight elevator they would restrict freight movements to non-peak travel times.

The three companies occupying the top four floors of the building requested that one elevator be classified as an express elevator serving only the ground floor and floors 6, 7, 8, and 9. Because these companies represent more than 50% of the building's occupants, we consider that their request is justified.

Three companies expressed a preference for deluxe elevators. Rothesay Mutual Insurance Company, Design Consultants Inc, and Rad-Art Graphics all stated that they had to create an impression of business solidarity in the eyes of their customers, and they felt that deluxe elevators would help to convey this image.

The management of Rothesay Mutual Insurance Company and Vulcan Oil and Fuel Corporation requested that a small key-operated executive elevator be included in our selection for the sole use of top executives of the major companies in the building. We asked other companies to express their views but received only marginal interest. The consensus seemed to be that an executive elevator would have only limited use and the privilege could easily be abused. However, we retained the idea for further evaluation, even though we recognized that an executive elevator would prove costly in terms of passenger usage[2].

We decided that the first two of these factors are requirements that must be implemented, while the latter three are preferences that should be incorporated if at all feasible. The controlling influence would be the budget allocation of $500,000 stipulated by the landlord, Merrywell Enterprises Inc. In decreasing order of importance, the requirements are:

1. Passenger waiting time must be no longer than 32 seconds.
2. At least one elevator must be able to accept freight up to 7 ft 6 in. long (2.3 m).
3. An express elevator should serve the top four floors.
4. The elevators should be deluxe models.
5. A small private elevator should be provided for company executives.

Researching Elevator Manufacturers

We asked the three major United States elevator manufacturers to furnish specifications of the elevators they would recommend for the Merrywell Building, plus price quotations and details of their maintenance policies. The catalogs at Appendix B show that except in appearance each company's elevators are basically the same and model for model carry very similar price tags. Significant differences are apparent only in discount policies and maintenance capabilities.

The YoYo Elevator Company of Chicago, Illinois, offers a 10% discount to purchasers contracting for a multiple installation in which all elevators are identical. We queried the other two manufacturers, Jackson Elevators Inc of Detroit, Michigan, and Matson Building Equipment Manufacturers of New York, NY, but neither would agree to incorporate a discount into their contract. Although such a discount agreement might limit the range of elevators from which we could make a selection, we consider that the incentive of a $40,000 to $50,000 reduction in price should not be overlooked.

The YoYo Elevator Company also is the only manufacturer with a service office in Montrose and thus can provide immediate response to calls for emergency maintenance. Both the other manufacturers contract with a local representative to carry out routine maintenance and state that they will fly in a maintenance team within 24 hours if major problems occur. We checked their reputations with the plant superintendents of local buildings using their elevators, and in both cases found that almost invariably delays of three or four days occurred before a maintenance team showed up.

Because of their advantageous pricing policy and superior maintenance service, we consider the new elevators should be selected from the range offered by the YoYo Elevator Company.

Selecting a Suitable Combination of Elevators

There are five basic elevators manufactured by the YoYo Elevator Company which are suitable for installation in the Merrywell Building. These comprise three sizes of passenger elevators that can be supplied in both standard and deluxe versions, a freight elevator, and an executive elevator (see Table 1 for a summary of their specifications, and Appendix C for a more detailed description). From these five basic elevator types we derived numerous combinations of elevators that could be installed in the 35 by 8 ft space (10.7 x 2.45 m), to select a range of elevators which would meet as many as possible of the tenants' requests within the proposed budget of $500,000.

By calculation, we determined that any three of the passenger elevators described in Table 1 could provide an adequate service for the present population of the Merrywell Building, but would be able to handle only a 4%

TABLE 1

Summary of elevators manufactured by the YoYo Elevator Company that are suitable for installation in the Merrywell Building.

MODEL		WIDTH ft (m)	CAPACITY lb (kg)	SPEED (sec/flr)
A	Passenger	6 (1.84)	2000 (910)	5
B	Passenger	7 (2.13)	2600 (1180)	6
C	Passenger	8 (2.45)	3000 (1360)	7
E	Executive Passenger	5 (1.53)	1500 (680)	2
F	Freight	10 (3.05)	5000 (2270)	15

Note: A more detailed description of these elevators is attached as Appendix C.

increase in passenger traffic before again becoming overloaded. Since we estimate that future changes in tenants could possibly increase the population by as much as 23%, we consider that four passenger elevators should be installed. It would be a sounder proposition, both economically and structurally, to anticipate this increase and install the fourth elevator now, than to install it at a later date.

The installation of four passenger elevators would, however, impose a limit on the size of freight elevator that can be installed. Even four of the smallest (model A) passenger elevators would leave only 9 feet (2.75 m) for a freight elevator, thus automatically eliminating the 10-ft wide 5000-lb (3.05 m, 2270 kg) capacity model F freight elevator. The only alternative would be to install one of the largest standard passenger elevators (model C) as a freight elevator, and accept its load restriction of 3000 lb (1360 kg) and maximum interior width of 8 feet (2.45 m). (We have already established that such limitations would be acceptable to the building's tenants.) The price of this combination would, however, exceed the budget by more than $86,000, as shown in Table 2.

To bring the cost down we considered making the model C elevator a dual-purpose unit, serving as a passenger elevator during peak traffic hours, and as a freight elevator at other times. With this arrangement only three regular passenger elevators would have to be purchased, and they could be selected from any of the three models available. Table 2 shows space utilization and the price of the three possible combinations, all of which would cost less than the proposed budget.

TABLE 2

Cost and Space Utilization
for Feasible Elevator Combinations

Combination:	1 "C" Freight PLUS 4 "A" Passenger Units	1 "C" Passenger/Freight Unit PLUS Three Passenger Units: 3 "A"	3 "B"	3 "C"
Standard Passenger Units	$586,400	$470,400	$480,000	$489,600
Deluxe Passenger Units	$596,800	$478,200	$487,800	$497,400
Total Space Occupied (width)	34.5 ft (10.5 m)	28 ft (8.3 m)	31 ft (8.7 m)	34 ft (10.4 m)

The combination proposing three model C passenger elevators and one model C passenger/freight elevator fills the available space most efficiently and has the greatest passenger capacity. It also has a significant advantage in that it qualifies for the 10% discount offered by the manufacturer if all elevators installed are exactly the same model and type. By purchasing four identical model C elevators the total purchase price would be:

	4 Model C Standard Elevators	4 Model C Deluxe Elevators
List price:	$489,600	$500,000
Discount:	48,960	50,000
Purchase price:	$440,640	$450,000

With this combination the need to provide an express elevator serving the ground floor and floors 6, 7, 8, and 9 is easily satisfied. Probably the best elevator to use would be the "freight" elevator, since it could be programmed to stop only at the preselected floors during peak travel hours, and could revert to its normal role as a freight elevator during off-peak hours (thick protective padding could be used to protect the interior paneling when freight is being carried). A second passenger elevator could be similarly programmed if peak-hour passenger loads dictate more than one express elevator is needed. We do not consider that an express elevator is necessary during non-peak travel times.

The combination of four deluxe model C elevators would therefore satisfy all but one of the tenants' requirements established at the beginning of this report. It reduces waiting time to less than 32 seconds; it provides a freight elevator able to handle articles up to 8 ft (2.45 m) long; it provides an express elevator serving the top four floors; and it offers prestige. Not only does it meet the budget appropriation of $500,000 established by Merrywell Enterprises Inc, but it also qualifies for a discount of $50,000 which will bring the total purchase price down to $450,000.

This arrangement does not, however, satisfy the request for a small private elevator for company executives. The only combination in Table 2 that leaves enough room in the elevator shaft to install an executive elevator consists of 3 model A passenger units and 1 model C passenger/freight unit. With the addition of a model E executive elevator, the occupied space would increase to 33.5 feet (10.3 m), and the cost would rise to $613,000 or $620,800, depending on whether standard or deluxe passenger elevators are installed. Although this combination would provide superior service for all of the building's occupants, we consider that its extra cost of at least $113,000 more than the budget, and $163,000 more than the cost of four model C deluxe passenger elevators, is too expensive to warrant the limited extra convenience it offers.

REFERENCES

1. Byron Johnson, "Don't Keep Passengers Waiting" in Elevators in the Industrial Complex (New York: Antrim Book Company, 1971), p. 75.

2. K. K. Krauston, "The Executive's Private Elevator" in Business and Industry, 27:5, May 1980, p. 43.

8

APPENDIX A

to

H L WINMAN AND ASSOCIATES REPORT No. 8-23

QUESTIONNAIRE

To: All Occupants of the Merrywell Building

To assist us in selecting a new range of elevators for the Merrywell Building, please complete the following questionnaire and return it to Barry V Kingsley, H L Winman and Associates, 475 Reston Avenue, Cleveland OH, 44104.

Company Name:...

Floor:...............   Total No. of employees:..........................

1. Do you consider the present elevator service satisfactory? ....

   ....
   (please elaborate)

2. What is the average length of time your employees have to wait for an elevator? -- At ground level:......secs. At your floor:......secs.

3. Do you need a freight elevator? .... (If so, please state dimensions and weight of largest loads carried) ....

4. What percentage of your employees' travel is between your floor and the ground floor? ......%; between your floor and other floors? ......%

5. Would a small private elevator be of use to your company's executives? .... If so, how many members of your staff would use it? ....

6. If you have additional comments or suggestions, please write them here:

Thank you for your cooperation.

APPENDIX B

to

H L WINMAN AND ASSOCIATES REPORT No. 8-23

ELEVATOR MANUFACTURERS' CATALOGS

This appendix contains catalogs supplied by the three major North American elevator manufacturers:

1. YoYo Elevator Company, Chicago, Ohio — Page 2

2. Jackson Elevators Inc, Detroit, Michigan — Page 13

3. Matson Building Equipment Manufacturers, New York — Page 27

Price quotations and maintenance policies appear on the final page of each catalog.

NOTE:

Because of their bulk, the catalogs have been omitted from this copy of the report.

1

APPENDIX C

to

H L WINMAN AND ASSOCIATES REPORT No. 8-23

DESCRIPTIONS OF SUITABLE ELEVATORS

The new elevators will be selected from the following range offered by the YoYo Elevator Manufacturing Company. Standard elevators have brown metal paneling, vinyl-tiled floors, and plain trim on doors and exterior; deluxe units have deep-pile carpeting, mahogany paneling, fancy trim inside and out, and an emergency telephone. All passenger elevators are fully automatic.

TYPE A

Size 6 ft (1.84 m) square; capacity 2000 lb (910 kg); speed 5 seconds per floor. Price: standard unit $116,000; deluxe unit $118,600.

TYPE B

Size 7 ft (2.13 m) square; capacity 2600 lb (1180 kg); speed 6 seconds per floor. Price: standard unit $119,200; deluxe unit $121,800.

TYPE C

Size 8 ft wide x 7 ft deep (2.45 x 2.13 m); capacity 3000 lb (1360 kg); speed 7 seconds per floor. Price: standard unit $122,400; deluxe unit $125,000.

TYPE E (Executive)

Deluxe unit 5 ft (1.53 m) square; capacity 1500 lb (680 kg); very high speed of 2 seconds per floor (on non-stop runs of 6 floors or more). Price: $142,600. Two versions available:

Type EB - push-button operated
Type EK - key operated

TYPE F (Freight)

Rugged unit 10 ft wide x 7 ft 6 in. deep (3.05 x 2.3 m); manually operated; capacity 5000 lb (2270 kg); slow speed (15 seconds per floor). Price: $130,000.

Elevator sizes quoted are exterior dimensions. An additional clearance of 6 in. (15.2 cm) is required between each elevator, and 3 in. (7.6 cm) between each end elevator and the wall. Prices include purchase, installation, and one year's free maintenance.

PROJECT NO. 1: BUILDING A TOOTHPICK TOWER

You are to assume that a local organization known as "The Association of Structural Engineers Inc" (ASEI) has offered an award to students of the course you are taking. The President of ASEI (Frederick K. Bartley, P.E.) has written to the head of your English/Communications Department to announce the award. His letter says (in part):

> . . .it is to be known as the ASEI Communication's Award and it will be presented annually. It will be an engraved plaque displaying the award winner's name, the name of the educational institution, and the year. The winner will be the student who shows the most ingenuity in completing a project we assign and concurrently displays the greatest competence in technical communication and report writing.

You are to compete for this year's award, for which ASEI has published the following requirements:

You are to design, build, and test to destruction a structure that supports the maximum weight for the lowest possible price. Specifications for the structure are:

1. It is to be built solely of wooden toothpicks and glue. Only flat style toothpicks are to be used and they are to be referred to as "structural members." The type of glue to be used is at the student's discretion.
2. The structure must support a horizontal 5 × 3 × ¼ in. wooden platform (128 × 76 × 65 mm), on which the load will be exerted.
3. The top surface of the platform must be no less than 10 in. (255 mm), and no more than 12 in. (305 mm), above the base of the structure.
4. There must be no protrusions above the platform's upper surface, to permit a load to be placed or exerted on it.
5. The cost of the structure is to be calculated according to the number of structural members used. You are to assume that each structural member costs $60. (Short portions of structural members that are half or less the length of a full member may be assumed to cost $30.)
6. An economy vs. load factor is to be calculated in dollars per pound (or metric equivalent) of maximum load supported. For example, if 200 structural members are used to build a structure that supports 18 pounds before it fails, the structure can be rated at $666.67 per pound of load. This figure is derived as follows:

   No. of members × cost per member ÷ maximum load supported
   or: 200 × $60 ÷ 18 = $666.67/lb.
7. Since the objective is to build a structure which will withstand the highest possible load for the lowest possible price, the winning design will be that which achieves the lowest cost per pound of load.

At the end of the project you are to prepare a formal report that describes the project and the results achieved, and suggests what you could have done to obtain a better cost/load ratio.

Some of the topics you should consider when planning and writing your report are:

Why the project was undertaken.

Your terms of reference.

The different types of structure that could be used, and the advantages and disadvantages of each.

Why you chose a particular design for your structure.

The effect that striving to achieve an optimum strength/cost ratio had on your design decision.

Whether it is better to aim for high strength at high cost, or low strength at low cost, to achieve this optimum ratio.

The types of glue available, their influence on structure strength, and why you selected a particular type of glue.

How the structure was built.

How the structure was tested.

How and why the structure failed.

The results of the test, and how the cost/load factor calculations were made.

How the structure's strength could have been increased without an equivalent increase in cost.

Your conclusions on the success achieved by your structure, and some recommendations for improving it.

Your instructor will inform you of any other requirements and where,

when, and how your structure is to be tested. The photograph shows a previous (not very successful) design.

PROJECT NO. 2: MOVING A BUS STOP

H. L. Winman and Associates is the engineering consultant for Montrose (Ohio) Community College. In response to a telephone call from Darryl Kane, the college's Technical Services Administrator, Karen Woodford assigns you to a new project at the college. She tells you that Mr. Kane will give you all the details and that your time is to be charged against Montrose Community College Project No. M71-C.

When you visit Darryl Kane the following morning, he tells you that he has received a series of noisy telephone calls from an indignant cafe owner, Georges V. Picolino, who operates a restaurant at 227 Wallace Avenue, just south of Main Street. Yesterday, Mr. Picolino visited the college.

"Indignant" hardly described Mr. Picolino adequately. He was bristling with anger. He had been to the Transit Office; he had been to the Montrose Streets and Traffic Department; he had even called at the home of his City Councillor; and to all he woefully related his tale. They all did the same thing: told him they could not help him and that he should see someone else. Angry at being shuffled from person to person, none of whom showed any real interest in his problem, Mr. Picolino finally turned to the college, because it is college students who are the source of his annoyance.

With some difficulty, Mr. Kane calmed him down and elicited a collection of details from which he pieced together the cause of Mr. Picolino's indignation. Apparently, college students traveling from east and west Main Street disembark from their buses at Wallace Avenue and wait for the college bus at a special stop on Wallace Avenue immediately in front of Picolino's Pantry. On cold or wet mornings the students crowd into the Pantry to keep warm while they wait. Georges Picolino's complaint is: (1) they don't buy anything (other than an occasional cup of coffee); and (2) they discourage other customers from coming in. The cafe becomes so crowded that passers-by think it is full and turn elsewhere to seek their breakfast. This is the time of day when Mr. Picolino used to do a lot of business, but since the bus stop was put in five months ago his early morning business has dropped to virtually nil.

Mr. Kane soothed Mr. Picolino's ruffled surface by promising that the problem would be resolved shortly, and then placed the telephone call to Karen Woodford that resulted in your visiting the college this morning. He asks you to evaluate Mr. Picolino's complaint and, if it is valid, to investigate ways to resolve it. He asks for a report of your findings that he can present to city and college officials.

In your investigation you take the following approach, asking yourself pertinent questions as you go along:

1. Is Mr. Picolino's complaint warranted?
   Yes, it seems to be. You check on two consecutive mornings:

Time	*No. of Students in Cafe on Wednesday*	*No. of Students in Cafe on Thursday*
7:00	0	0
7:10	5	0
7:20	8	7
7:30	13	10
7:40	15	18
7:50	27	36
8:00	42	39
8:10	23	31
8:20	6	5
8:30	1	0

a. Mr. Picolino informs you that these are quieter mornings than usual. You check up and find that this week 35% of the college's student population are writing special tests in the afternoon and are unlikely to be attending classes in the morning.

b. There are seats for 36 customers in Picolino's Pantry.

2. How frequent are the buses?
A bus loads passengers at the bus stop at 7:15, 7:32, 7:45, 7:55, 8:05, 8:15, and 8:28.

3. What is the bus route in this area?
The bus travels north along Wallace Avenue (a one-way street), then turns left to travel west along Main Street.

4. Could the bus stop be moved?
There are four possible directions:
   a. South: Yes. But it would have to be at least 120 yards (110 m) south; there is a crosswalk immediately south of the present stop.
   b. North: No. No room between the existing stop and Main Street, and Wallace Avenue north of Main Street is not on the bus route.
   c. East: No. Not on bus route.
   d. West: Yes. By Caernarvon Hotel.

5. What factors affect the location of the bus stop?
   a. Connecting bus routes drop passengers at points A and B on Main Street (refer to diagram).
   b. Montrose Rapid Transit System insists that the connecting stop must be within 100 yards (91 m) of the intersection of Main Street and Wallace Avenue.
   c. There is a cab rank outside the Caernarvon Hotel, but Montrose Streets and Traffic Department is adamant: It will not authorize a bus stop in that block.
   d. The bus stop could be moved directly across the street, to the west side of Wallace Avenue, where there are no stopping restrictions. But this poses two problems: Bus doors would open onto the traffic side of the bus rather than onto the sidewalk, and it is still too close to Picolino's Pantry.

6. You note that the lot at the southwest corner of the Main and Wallace intersection is vacant. Previously a Walston Oil service station, it is now abandoned and has been for ten months; the building is boarded up. You investigate ownership. The land was purchased by the City of Montrose six months ago. The city plans to build a car park on the lot, but not for five years, when it plans to acquire adjoining lots.

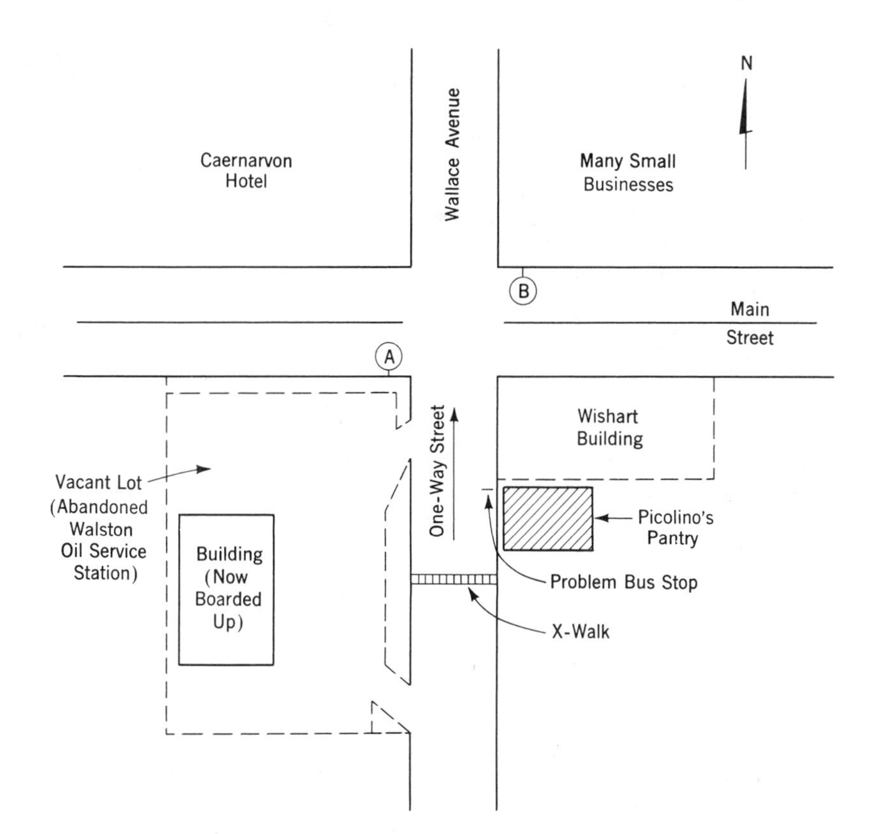

INTERSECTION OF MAIN STREET AND WALLACE AVENUE, MONTROSE, OHIO

7. An idea begins to form: Why not build a bus shelter on the vacant lot and route the bus through the lot? (It could use the old entry and exit ramps built for the service station.)
8. You discuss your idea with the Traffic Engineer at Montrose Streets and Traffic Department (D.V. Botting), who says his department would not object providing the shelter conforms to City of Montrose, Transit Division, Building Construction Specification BCS 232. You also check with Carol Whiteside, Traffic Superintendent of the Montrose Rapid Transit System, who says her department will go along with it. (She also says Transit will pay the cost of building a shelter on the lot, providing the cost is no greater than $5000.) Then you check with the City of Montrose (Planning Division) and get approval to use the land for five years. The city will also authorize expenditure of $600 a year for general upkeep of the lot. You ask all three parties to state their agreement with your plan in writing, and you receive letters from them indicating their willingness to participate.
9. You write to three contractors asking them to give you a price quotation, telling them that the shelter must be able to hold at least 40 people, that the price quotation must include an adequate heating/cooling system (temperature range: 10° to 27°C), seats, and lighting, and that the work must be completed within three weeks of receipt of an order. You also inform them that the work must conform to City of Montrose Bylaw 61A.

10. The lowest price quotation you receive is submitted by Donovan Construction Company, 2821 Girton Boulevard, Montrose, whose price of $5945 is still $945 above the maximum stipulated by Montrose Transit. (Other quotations were $6667 and $6788.) You visit Gavin M. Donovan, owner-manager of Donovan Construction Company, to see if he can bring his price down to within Montrose Transit's limit.

    "Impossible" he says. "No way! Nobody could do it."

    Then, as an afterthought, he adds: "But tell you what: We could do as good a job by renovating the service station."

    He digs out some files from his desk. "See here," he says, "we ran an estimate on renovating the existing building and installing seats. Electric heat and light are already in it. The price would be $4565—and I'll throw in some landscaping."

Since this resolves the problem, you return to your office and prepare a report for Darryl Kane. You decide to prepare a formal investigation report, since you want to submit copies to all parties involved and to use it as the document for implementing conversion of the Walston Oil building and moving the bus stop.

PROJECT NO. 3: INDIVIDUAL VERSUS BULK METERING OF APARTMENT BLOCKS

Your branch manager, Vern Rogers, asks you to accompany him to a project start-up meeting with Hugh Rempel, your local power utility's Chief of Consumer Services. On the way, he tells you that Hugh has commissioned H. L. Winman and Associates to determine whether individually metered apartments are better conservers of electricity than bulk-metered apartments. Vern says he is assigning the project to you and that when your research is complete you are to prepare a formal report of your findings which he will forward to Hugh Rempel.

"We believe," Hugh Rempel says when you and Vern have seated yourselves in his office, "we believe that apartment dwellers who pay their own electricity bills are more energy-conscious than those whose electricity is included in their rent."

"You've got data to substantiate that?" Vern asks.

"No, we haven't." Hugh replies. "Or, rather," he corrects himself, "we haven't any refined data."

He turns to you: "I can give you all the consumption data you want on every house and apartment block in the city," he says. "But it's all raw data. You are going to have to sort through it."

Hugh Rempel says he will find an office for you at City Hall (where the consumption records are kept and bills are paid), so you will have easy access to the information you need.

The first thing you do after moving into City Hall is to jot down six questions:

1. *Where do I start?* (There are hundreds of apartment blocks.)

2. *How many apartment blocks should I survey?* (It's impractical to survey more than a small proportion.)
3. *Which apartment blocks should I select?* (What criteria should be used to select them?)
4. *What factors should I use to make the comparisons?* (Cost? Consumption?)
5. *What time span should my survey cover?* (Six months? One year? Two years?)
6. *How much time do I have for the study?*

Question 6 is answered the following morning by Vern Rogers, who calls you to say that Hugh Rempel wants your report to be on his desk in three months. Vern adds a caution: "That may seem like a long time, but you'll need every minute. Don't make your sample too large."

His warning helps you answer question 2, which in turn answers question 1. You set yourself an arbitrary figure of 2400 apartment units, one-half of which should be individually metered and one-half bulk metered.

You have more difficulty answering question 3. Obviously, the apartment blocks on average should be roughly the same size (that is, there should be the same number of large, medium, and small blocks in the two groups); they also should have similar facilities and services that draw electricity (underground parking, elevators, air-conditioning, saunas, heated swimming pools, etc); and they should be of comparable age and construction.

Then Records Clerk Rachel Wendover introduces a further complication: "There are two kinds of apartment blocks," she says, "those heated by electricity, and those heated by other means such as gas or oil." (Suddenly you realize you have a *four-part* comparison.) "And that's not all," she goes on to say. "There aren't all that many individually metered apartment blocks. Individual apartment meters, because they are more expensive to install, have been only a recent development. Builders don't like to put them in, but they are being pressured by the government."

From your discussion with Rachel, you discover that most individual metering has been done in smaller apartment blocks. So, to simplify your survey, you decide to:

Examine the records of 80 apartment blocks, 40 of which should be individually metered.

Divide each group of 40 into 20 electrically heated and 20 heated by other means.

Eliminate apartment blocks with underground parking, saunas, elevators, and central air-conditioning.

Limit your choice to three-story walk-up apartment blocks.

Select blocks which, on average, have 30 individual units.

In the next two months you:

1. Visit each apartment block to find out:
   - The number of units.
   - The gross square footage (exterior of building).
   - The fuel used for heating the building.
2. Eliminate the type of construction as a criterion.

3. Find only 16 electrically heated apartment blocks which are individually metered, so also reduce the bulk-metered group to 16 blocks.
4. Examine one year's (last year's) consumption records. (To do more than one year is impossible, because 13 of the individually metered blocks are less than two years old.)
5. Combine the records onto four tables (see pages 215–16).
6. Analyze the results.
7. Recheck the figures for apartment blocks showing abnormally low and high figures:

   401 (high); 219 and 416 (low); 120 (exceptionally low).

   All are correct.
8. Can find no obvious explanation for the abnormal "Annual kWh per sq ft" results for apartment blocks 108 and 119 (particularly).
9. Check populations and special features of those that showed abnormal figures, as well as any blocks showing generally low consumption rates.
10. Discover that:

    Apartment block 120 contains only bachelor apartments.

    Apartment blocks 219 and 416 contain 60% and 50% bachelor apartments

    A conserve-energy awareness program has been conducted and posters displayed in blocks 118, 218, 312, 315, 316, and 411.

    The populations of blocks 118, 220, 315, 316, and 413 are predominantly "professional" people such as lawyers, university professors, engineers, and executives. (Rents are higher than average, and individual apartments are larger.)

    Construction of all apartment blocks is generally similar.

    Apartment block 401 is north of the airport on a slight rise; it's rather exposed.

You check all 72 apartment blocks to identify in which ones conservation-awareness programs have been conducted. You find only three more blocks than previously identified; they are 217, 310, and 410.

In your report, discuss your results, analyze the statistics, comment on abnormalities, draw a conclusion, and make a recommendation. Consider also: (1) whether a further study should be done after another one or two years; (2) if any of the factors that seem to be contributing to lower energy consumption could be implemented in all apartment blocks; (3) why professional people should contribute to generally lower energy consumption; and (4) whether anticipated increases in electricity rates will influence consumption in apartment blocks (if so, which type of blocks?). Use appendices, and insert at least one illustration or chart in the Discussion section.

PROJECT NO. 4: EVALUATING EMPLOYMENT POSSIBILITIES

H. L. Winman and Associates has been engaged by your local Department of Education to study employment patterns for technical graduates of junior colleges and technological institutes over the past six years. The intent is to assess which fields offer the best employment opportunities for graduates. The project

Group 1: Individually Metered—Not Electrically Heated

Apt Block No.	No. of Units	Total Annual kWh	Annual kWh per Unit	Building Size (sq ft)	Annual kWh per sq ft
101	12	68,267	5,689	12,495	5.46
102	48	261,292	5,444	42,044	6.22
103	12	65,210	5,434	11,319	5.76
104	30	159,520	5,317	37,704	4.23
105	41	198,990	4,853	38,760	5.13
106	12	57,230	4,769	12,495	4.58
107	12	55,897	4,658	11,319	4.94
108	31	143,121	4,617	19,524	7.33
109	12	55,391	4,616	12,495	4.43
110	13	57,438	4,418	12,495	4.60
111	59	259,150	4,392	52,098	4.97
112	12	50,092	4,174	12,495	4.01
113	32	131,363	4,105	27,784	4.73
114	12	47,430	3,953	12,495	3.80
115	12	47,354	3,946	12,495	3.79
116	48	182,773	3,808	41,925	4.36
117	32	113,079	3,534	27,968	4.04
118	25	88,165	3,527	23,292	3.79
119	29	94,965	3,275	13,254	7.17
120	49	97,537	1,991	33,184	2.94
	533	2,234,264	4,192 (Ave)	467,460	4.78 (Ave)

Group 2: Bulk Metered—Not Electrically Heated

Apt Block No.	No of Units	Total Annual kWh	Annual kWh per Unit	Building Size (sq ft)	Annual kWh per sq ft
201	18	161,510	8,973	25,272	6.39
202	19	165,280	8,699	19,032	8.68
203	26	179,280	6,895	28,017	6.40
204	24	156,420	6,518	22,500	6.95
205	18	109,920	6,107	19,032	5.78
206	24	145,680	6,070	25,272	5.76
207	51	291,360	5,713	48,171	6.05
208	53	301,760	5,694	44,712	6.75
209	23	122,640	5,332	19,950	6.15
210	24	126,890	5,287	25,272	5.02
211	39	202,080	5,182	36,750	5.50
212	51	259,040	5,079	45,000	5.76
213	48	242,640	5,055	46,665	5.20
214	46	210,880	4,584	43,680	4.83
215	8	36,290	4,536	6,552	5.54
216	45	201,120	4,469	38,574	5.21
217	24	102,540	4,273	22,950	4.47
218	59	237,680	4,028	57,792	4.11
219	27	108,200	4,007	33,909	3.19
220	21	83,700	3,986	18,834	4.44
	648	3,444,910	5,316 (Ave)	627,936	5.49 (Ave)

Group 3: Individually Metered—Electrically Heated

Apt Block No.	No. of Units	Total Annual kWh	Annual kWh per Unit	Building Size (sq ft)	Annual kWh per sq ft
301	12	263,530	21,961	11,592	22.73
302	12	258,500	21,542	11,448	22.58
303	12	255,410	21,284	12,888	19.82
304	12	249,760	20,813	13,248	18.58
305	12	223,080	18,590	9,882	22.57
306	24	419,430	17,476	26,930	15.58
307	24	405,050	16,877	22,928	17.67
308	24	396,480	16,520	26,928	14.72
309	24	383,920	15,997	22,928	16.74
310	24	377,230	15,718	26,760	14.10
311	24	373,985	15,583	26,930	13.89
312	24	373,580	15,566	26,930	13.87
313	24	363,200	15,133	22,928	15.84
314	24	362,312	15,096	22,928	15.80
315	24	368,584	14,941	26,930	13.69
316	24	307,250	12,802	22,928	13.40
	324	5,381,301	16,609 (Ave)	335,106	16.06 (Ave)

Group 4: Bulk Metered—Electrically Heated

Apt Block No.	No. of Units	Total Annual kWh	Annual kWh per Unit	Building Size (sq ft)	Annual kWh per sq ft
401	29	643,840	22,201	25,092	25.66
402	58	1,194,400	20,593	51,288	23.29
403	30	613,440	20,448	28,650	21.41
404	41	830,400	20,254	40,941	20.28
405	29	582,080	20,072	25,092	23.20
406	54	998,400	18,489	51,282	19.47
407	54	997,200	18,467	51,282	19.45
408	18	321,760	17,876	13,504	23.83
409	18	310,000	17,222	13,504	22.96
410	33	522,120	15,822	29,760	17.54
411	42	660,800	15,733	39,933	16.55
412	20	308,960	15,448	13,504	22.88
413	15	227,880	15,192	13,416	16.99
414	18	268,080	14,893	13,504	19.85
415	18	254,280	14,127	14,592	17.43
416	18	243,600	13,533	16,896	14.42
	495	8,977,240	18,136 (Ave)	442,240	20.30 (Ave)

is authorized by Dennis M. Carstairs, Coordinator of Special Services in the Department of Education.

At H. L. Winman and Associates the project has been directed by Vern Rogers, the local branch manager. Several months have been spent conducting a survey of a random selection of graduates. Results of the survey have been tabulated, and Vern has assigned you to analyze them and prepare a report of your findings. Factors you should consider are:

Employment patterns and trends.

Probable future trends for particular fields.

Fields that offer the best and worst employment opportunities.

Which groups of employers offer the best opportunities from the point of view of:

1. Company growth.
2. Salary.
3. Stability.

The number of graduates likely to be hired in the next two years.

Whether colleges are churning out more graduates than the employment market can bear.

The survey divided employers of graduates into the following general categories:

1. Engineering and Research Organizations
2. Project Management Organizations
3. Manufacturing Organizations
4. Education
5. Government
6. Service Organizations
7. Miscellaneous Technical Employers

In the survey of each year's graduates in the table on page 218, the following coding was used:

T—Total number of graduates surveyed

1 to 7—Fields in which employed

0—Graduates employed in fields other than that for which they were trained

U—Graduates currently unemployed

Two figures appear against each coding in the table:
The number of graduates *currently* employed in that category
The average monthly salary *currently* earned by these graduates

A factor not evident in the table applies to employment category 6—Service Organizations. A breakdown of this category shows that in one group of employers (Computer Companies), the statistics contradict the results for the other employee groups: (see table on page 219).

Survey of Technical Graduates by Employment Categories

Employment Category	*Total No. Currently Employed in Category*	*Number Currently Employed in Each Category (according to year of graduation) Plus Current Average Monthly Salary*					
		Number of Years Since Graduation					
		6 years	*5 years*	*4 years*	*3 years*	*2 years*	*1 year*
T	1666	140	195	241	323	368	399
1	274	31 ($1775)	42 ($1700)	45 ($1625)	48 ($1540)	57 ($1450)	51 ($1375)
2	181	2 ($1675)	5 ($1680)	12 ($1530)	32 ($1450)	58 ($1370)	72 ($1290)
3	115	7 ($1650)	21 ($1580)	30 ($1515)	25 ($1425)	18 ($1345)	14 ($1270)
4	174	23 ($1640)	28 ($1590)	26 ($1530)	30 ($1460)	32 ($1385)	35 ($1315)
5	109	18 ($1625)	11 ($1545)	15 ($1495)	28 ($1420)	16 ($1350)	21 ($1270)
6	219	36 ($1745)	38 ($1670)	32 ($1595)	47 ($1535)	35 ($1430)	31 ($1350)
7	205	15 ($1660)	29 ($1585)	34 ($1510)	51 ($1435)	37 ($1370)	39 ($1300)
0	291	8 ($1470)	20 ($1430)	39 ($1375)	41 ($1310)	86 ($1235)	97 ($1155)
U	98	0	1	8	21	29	39

Graduates Currently Employed by Service Organizations—Category 6

No. Years Since Graduation	Category 6 Total	Computer Companies	Other Companies
6	36	4	32
5	38	7	31
4	32	15	17
3	47	29	18
2	35	26	9
1	31	28	3

Analyze the results of the survey and then write a formal report of your findings for Mr. Carstairs. Prepare charts and graphs to illustrate your report.

PROJECT NO. 5: STORMWATER DISPOSAL, CAYMAN FLATS

The Fairview Development Company wants to develop an area of virgin land known as Cayman Flats on the southern perimeter of Montrose, Ohio. The City of Montrose displays interest in the proposal and asks for a formal presentation of the company's development plans.

The land is flat, and Fairview Development Company soon realizes that it has a stormwater drainage problem to overcome before it can complete its presentation. It calls in H. L. Winman and Associates to resolve the problem (see letter). The project is assigned to you.

You start your investigation by examining Cayman Flats. It is generally flat, low-lying, and frequently waterlogged. The maximum variation in height is 9 feet (2.75 m); its area is 1594 acres (646 hectares). To the east is a railway line (Northern Railways) into Montrose center, to the north a residential area, and to the west and south lie arable land 70% cultivated (see map).

You calculate the maximum stormwater runoff for the land in its present condition. (Stormwater is rainwater that must be drained from the land quickly to prevent flooding of low-lying areas and basements. Maximum stormwater runoff is the largest amount of water likely to occur; it is calculated on past records of maximum precipitation accumulated from the heaviest rainstorms.) Using the Rational Formula, you calculate that a 21-foot (6.4 m) diameter culvert would be required to handle the heaviest peaks.

Since such a large culvert would offer construction problems, use a lot of property, represent an eyesore, and be a source of danger to small children, you look for other methods. The most obvious is to construct storm sewers throughout Cayman Flats before development starts. Because the runoff would then be channeled, the stormwater would be "staged" (that is, reach the outfall, or disposal sewer system, as a series of peaks). You estimate that the largest peak could be handled by a 96 inch (2.44 m) storm sewer.

Your next step is to examine the territory between Cayman Flats and the

FAIRVIEW DEVELOPMENT COMPANY
212 Bligh Street
Montrose, Ohio
45287

Harvey L. Winman, President
H. L. Winman and Associates
475 Reston Avenue
Cleveland, Ohio 44104

Dear Mr. Winman:

We are preparing a feasibility study for the City of Montrose, in which we are proposing to develop the Cayman Flats area to the south of the city as a new residential district. This low, flat land offers drainage problems because of its distance from the Wabagoon River. The storm sewers of the intervening developed areas cannot be used since they have insufficient capacity to handle the additional stormwater runoff that Cayman Flats will generate.

We are asking you to conduct an engineering investigation into this stormwater disposal problem and to recommend an economical method that we can include in our presentation to the City of Montrose.

Yours sincerely,

Frederick C. Magnusson,
President

FCM/jms

Wabagoon River. The whole area has been developed; to build a sewer of the required size directly to the river would be phenomenally expensive, and you would likely have difficulty in getting municipal approval. The distance is too great to build it around the southwest perimeter of the city. But there is a 40 foot (12 m) wide belt of open land on the west side of the railway line. It is owned by Northern Railways, who agree to lease it to the city for $1200 per year and will guarantee a 20 year lease.

Some other factors you discover are:

1. Storm sewers of the residential areas north of Cayman Flats are of only 18 inch (0.46 m) diameter in zones S-4 and S-6, and 24 inch (0.60 m) diameter in zone S-5. Those in S-4 feed into the 30-inch (0.76 m) sewers in S-2; those in S-6 run directly down to the river, following the boundary line between S-2 and S-3; those from S-5 feed into the 36-inch (0.91 m) sewers in S-3.
2. The municipalities view with alarm your inquiries to find an easy route to the river for a 96 inch (2.44 m) storm sewer through zones S-3, S-5, and S-6. They fear (rightly) that you could not excavate deeply enough to avoid interfering with the existing services. (Too deep an excavation would foil the natural flow into Wabagoon River.)
3. Cayman Flats slopes very slightly downhill from the southwest corner to the northeast corner. Total drop is 9 feet 3 inches (2.83 m).
4. Although zones S-15, S-17, and S-19 are all classed as "light industry," S-17 has never been developed. It is a small zone of approximately 95 acres (38 ha), lower in elevation than the surrounding zones, and hence rather swampy. Undoubtedly the amount of fill needed to build it up has hindered its development.

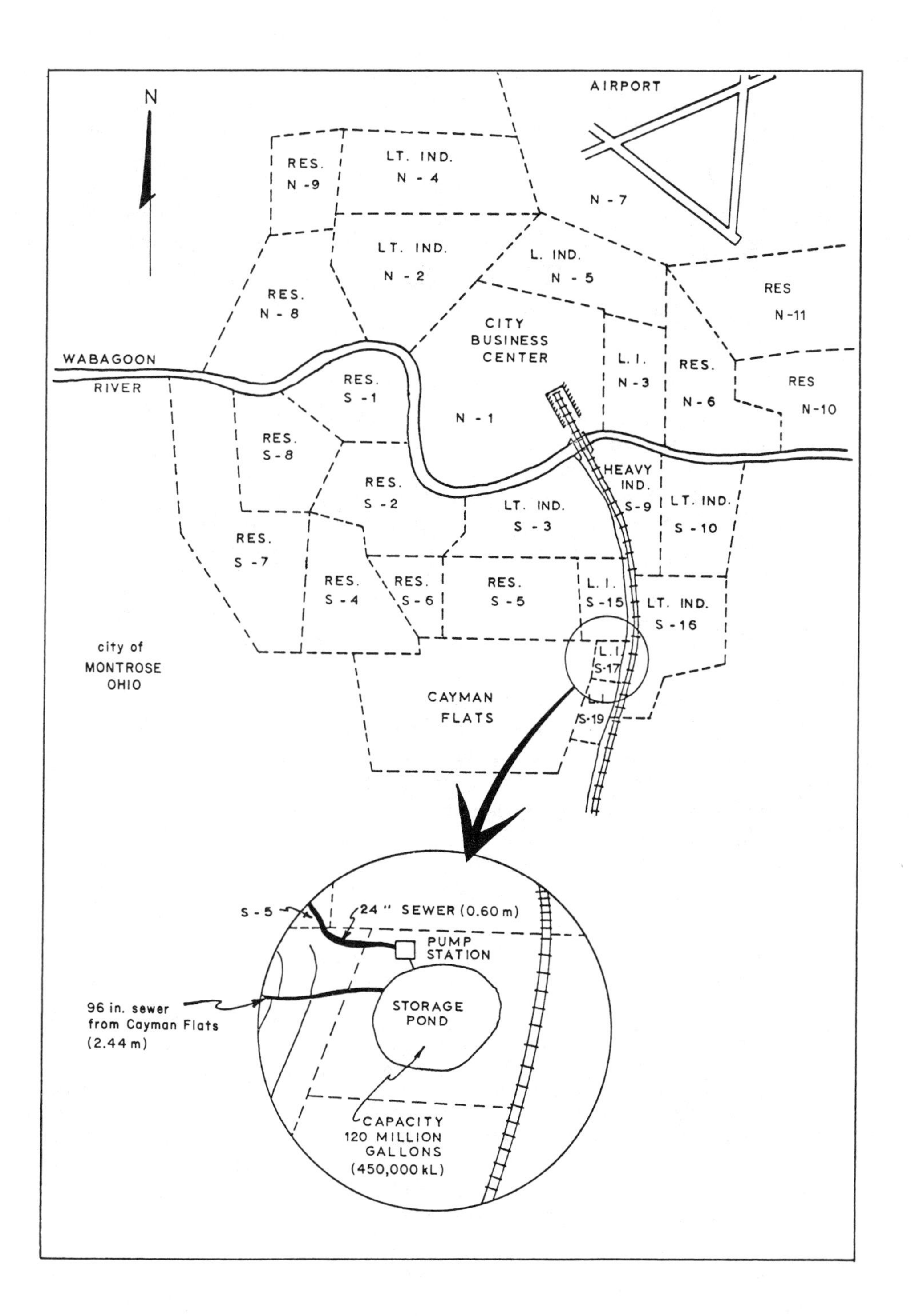
N
AIRPORT
RES.
N -9
LT. IND.
N - 4
N - 7
LT. IND.
N - 2
L. IND.
N - 5
RES.
N - 8
RES
N-11
CITY
BUSINESS
CENTER
WABAGOON
RIVER
RES.
S - 1
L. I.
N - 3
RES.
N - 6
RES
N-10
N - 1
RES.
S - 8
RES.
S - 2
HEAVY
IND.
S - 9
LT. IND.
S - 10
LT. IND.
S - 3
RES.
S - 7
RES.
S - 4
RES.
S - 6
RES.
S - 5
L. I.
S - 15
LT. IND.
S - 16
L. I.
S-17
L. I
S-19
city of
MONTROSE
OHIO
CAYMAN
FLATS
S - 5
24" SEWER (0.60 m)
PUMP
STATION
96 in. sewer
from Cayman Flats
(2.44 m)
STORAGE
POND
CAPACITY
120 MILLION
GALLONS
(450,000 kL)

At this point an idea occurs to you. If zone S-17 is already acting as a collection point for some of the stormwater from the surrounding zones (principally Cayman Flats), why not excavate it even further and use it as a quick runoff storage pond? If it were big enough, you could hold all the stormwater runoff from Cayman Flats and then pump it at a controlled rate into a smaller diameter outfall sewer from S-17, along the railway line, to the river. It would mean building a pump station (it would have to be automatic), and fencing the storage pond, but it would be feasible. You make some quick calculations:

Size of storage pond required: 120,000,000 gallons (450,000 kL).
Size of outfall sewer to river: 24 inch (0.60 m) diameter.

To evaluate the ideas so far you work up some cost estimates:

1.	Cost of installing storm sewer system throughout Cayman Flats.	$987,000
2.	Cost of 96-inch diameter storm sewer along railway line from Cayman Flats to Wabagoon River.	$1,272,000
3.	Cost of 24 inch (0.60 m) diameter storm sewer over same route.	$827,000
4.	Cost of purchasing zone S-17, excavating the storage pond and fencing it, and building an automatic pumphouse and pump system.	$621,500
5.	Pumphouse annual maintenance and operating costs.	$ 6,500

At this stage you feel your problem has been resolved and you have a reasonable proposal to offer Fairview Development Company. Your report is almost finished before another obvious and logical idea occurs to you: If you are going to *control* the flow of water into the outfall storm sewer, why not control it even further and pump the stored stormwater into the *existing* 24 inch (0.60 m) storm sewers of zone S-5? You would have to wait until the stormwaters from S-5 had been fed into the river, but this would be no problem because you have planned an oversize storage pond that could hold the water from the heaviest storms recorded during the past 30 years. All you would have to do is obtain municipal approval and build a 24 inch (0.60 m) sewer from the pumphouse beside the pond to one of the storm sewers in Zone S-5 (see sketch).

You approach the City of Montrose with your unusual idea. The City Engineer hesitates briefly (for three weeks), and then gives approval in principal. You then calculate one more cost:

6.	Cost of building a 24 inch (0.60 m) storm sewer from the pumphouse in zone S-17 to southeast tip of zone S-5 and connecting it to the storm sewer in zone S-5.	$162,000

When you are preparing your report for Fairview Development Company, you should bear in mind how they intend to use it. They will probably attach it to their proposal to the City of Montrose as evidence that they have researched and resolved the stormwater disposal problem. Hence you must write with the knowledge that although you are addressing it to Fairview Development Company, the ultimate readers are likely to be the City Councillors of the City of Montrose.

PROJECT NO. 6: EXAMINING THE EFFECTS OF AIRCRAFT NOISE

You are employed in the Environmental Studies Department of H. L. Winman and Associates. Your department head, Vic Braun, assigns you to investigate an aircraft noise problem affecting two communities in the city of Montrose, Ohio. He suggests that you drive over to Montrose and arrange to meet Mr. Lorne J. Carson, who is vice-president of Manston Machinery Company. Mr. Carson will give you the details and you are then to undertake the project.

As well as being an executive of a manufacturing company, Mr. Carson is an active member of the Montrose Junior Chamber of Commerce (JCC). Currently he is chairman of the JCC Environment Committee, and it is in this capacity that he has engaged H. L. Winman and Associates to study the aircraft noise problem.

He explains that the JCC has been approached by a citizen's action group representing residents of the Silver Heights and Fairleigh Estates residential communities. These communities lie under the flight path of the approach to the main (northwest-southeast) runway of Montrose International Airport, and the residents are complaining bitterly of aircraft noise. They claim that overflying aircraft disturb both their sleep and peace of mind, awaken sleeping babies, rattle the china in cupboards and cabinets, cause cracks to appear in walls, and generally lower property values.

The action group has even raised a petition calling for a ban on all takeoffs and landings between 11:00 P.M. and 7:00 A.M., and response has been high, particularly from residents of Silver Heights. The action group already has over 9000 signatures, is threatening legal action, and has asked the JCC Environment Committee for support if it takes the case to court.

"If their case is valid," Mr. Carson explains, "then I would like to support them. But if it is not, then I would like to advise them accordingly."

He and his committee have carried out a preliminary survey but have not been able to come up with any hard facts. So far all the objections have been based on the subjective opinions of individuals—and not everyone seems to have a complaint.

"Noise is such an odd thing," he says. "What to one person is hardly noticeable background noise, to another is irritating, peace-destroying drumming. It's entirely personal."

Mr. Carson says he wants specific facts on which to assess whether the JCC Environment Committee should support the action group in their quest, or whether he should persuade the action group not to proceed with expensive legal proceedings if their case is weak or invalid. He asks you to measure the sound levels in the two communities, compare them against accepted standards for noise in residential areas, and identify whether or not the sound levels really produce the detrimental effects claimed by the action group.

"I want you to prepare a report which I can show to my committee and the action group, and which, if necessary, can be used as objective evidence. In it I want you to present your findings, discuss noise 'standards,' compare your sound level readings against the standards, draw a conclusion, and if possible make a recommendation."

During the next three weeks you:

Research information on sound and its effects on people.

Discover few firm guidelines which you can accept as standards, and certainly none for the City of Montrose (the problem has not been addressed by the city).

Talk to people in the community.

Identify that Silver Heights and Fairleigh Estates are relatively new residential developments, both built in the last 10 to 15 years.

Measure sound levels at five points under the flight path, using a GenRad 1551-C sound-level meter (see Figure 7-1 in Chapter 7).

You start by consulting a city map and identifying five locations to take sound level readings. You plot them onto a smaller map (see sketch) and identify them as positions 1 to 5. Three of them are in the Silver Heights area:

1. On Airport Road, at the junction of Moss Grove.
2. On Marjorie Street at Wilhelmina Crescent.
3. At the intersection of Grassy Boulevard and Craven Avenue.

One is in Fairleigh Estates:

4. On Kississimee Drive, midway between Wall and Denver Avenues.

For comparison, you also identify one location further southeast along the flight path, from which there have been no objections to date:

5. On the roof of the Faculty of Architecture, in the University of Montrose.

You calculate the distances from the southeast end of the runway to these five positions:

	Distance From Runway			
Position	*(yards)*	*(meters)*	*(km)*	*(miles)*
1	850	810	0.8	0.5
2	1500	1380	1.4	0.83
3	2200	2000	2.0	1.24
4	3500	3200	3.2	2.0
5	5500	5000	5.0	3.1

Over a two-week period you take sound-level readings at various times of the day and night, and on different types of aircraft. You then consolidate your results into a comparison chart (see table). While taking the measurements you note that flying tactics differ between pure jet and propeller-driven aircraft:

1. Propellor aircraft (particularly piston-powered aircraft) tend to approach more steeply and take most of the power off on the landing approach.
2. Pure jets make a long, low approach with power on all the way in.
3. Pure jets climb away *much* more steeply than propellor-driven aircraft.

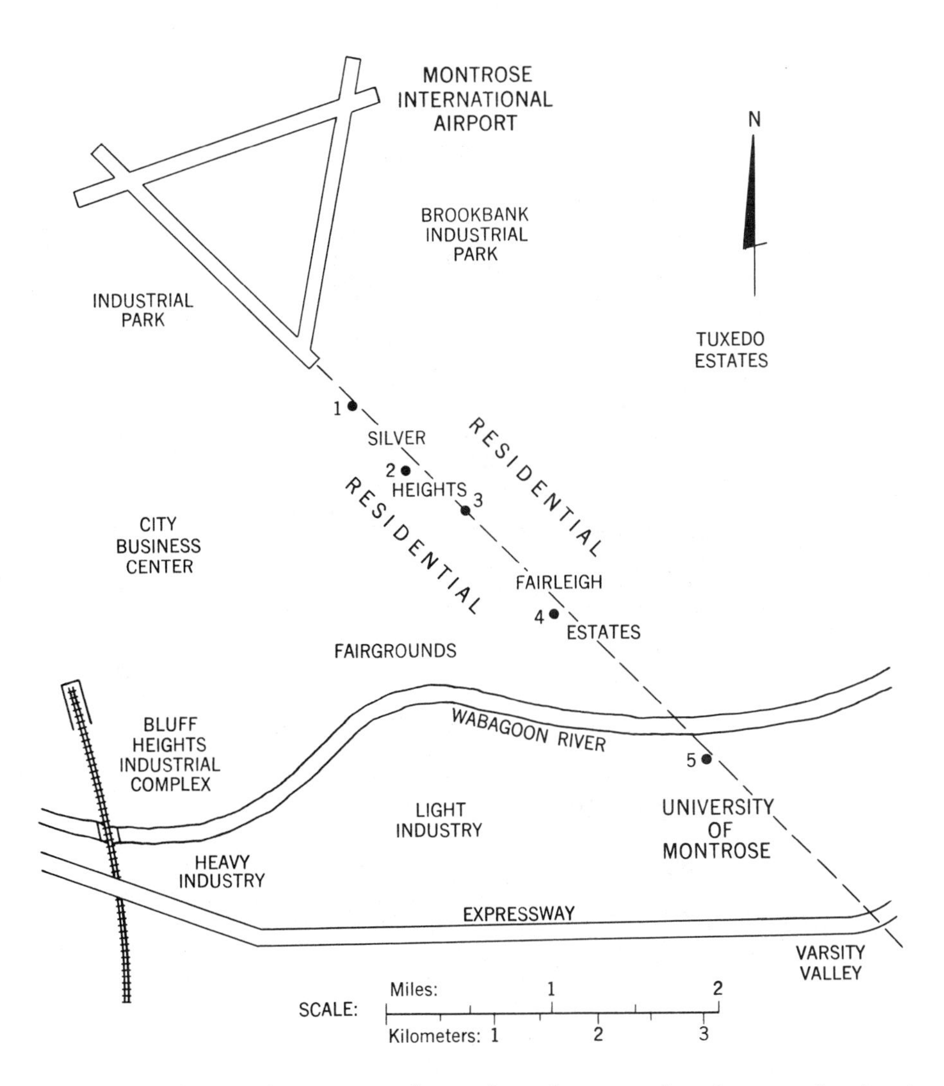

4. Light aircraft turn into and away from the approach path sooner than heavier aircraft, and so do not always overfly location 5, and sometimes even location 4.

Senior air traffic control officer Phil Tyson tells you that prevailing winds are from the northwest, so that the northwest-southeast runway is most frequently used at Montrose International Airport. Of all landings and takeoffs:

52% are to the northwest.

12% are to the southeast.

36% use the other two runways.

You interview airport manager Vince Roberts, who tells you:

On average, 820 aircraft movements (takeoffs or landings) occur during a 24-hour day.

Only 7% of the traffic movements occur between midnight and 6:00 A.M.

*Average Sound Levels Measured**
(Aircraft Immediately Overhead)

Aircraft Type	Takeoff or Landing	Location and Distance (Yards/Meters) From Runway				
		1 850 yd 810 m	2 1500 yd 1380 m	3 2200 yd 2000 m	4 3500 yd 3200 m	5 5500 yd 5000 m
Heavy Jet	Takeoff	88 dB	86 dB	78 dB	74 dB	73 dB
	Landing	96 dB	92 dB	89 dB	87 dB	86 dB
Medium Jet	Takeoff	87 dB	84 dB	82 dB	72 dB	71 dB
	Landing	92 dB	90 dB	87 dB	85 dB	83 dB
Wide-Body Jet	Takeoff	84 dB	82 dB	78 dB	72 dB	70 dB
	Landing	87 dB	83 dB	78 dB	75 dB	72 dB
Medium Prop-Jet	Takeoff	84 dB	83 dB	82 dB	77 dB	69 dB
	Landing	77 dB	79 dB	78 dB	74 dB	71 dB
Light Piston	Takeoff	83 dB	81 dB	78 dB	72 dB†	63 db†
	Landing	64 dB	66 dB	65 dB	61 dB†	58 db†

*Like the City of Montrose, these figures are hypothetical.
†Not all light piston aircraft overflow these points

Movements by aircraft types can be divided roughly into:

8% wide-body jets, such as Boeing 747, DC-10, L-1011.
7% heavy jets, such as DC-8, Boeing 707.
15% medium jets, such as DC-9, Boeing 727 and 737.
15% medium prop-jets.
55% piston aircraft.

Over the next eight years the heavy jets and some of the medium jets will be gradually replaced by a new generation of wide-body jets, such as the Boeing 767, which are predicted to be as quiet or quieter than the current wide-body jets.

Piston aircraft are generally single- or twin-engine light business planes; large four-engine piston aircraft amount to less than 1% of total traffic.

Because complaints of noise generally are based on subjective opinions, you try to identify what harmful physical effects of noise have been documented. The most obvious is permanent hearing loss. Lesser known are a reduction in gastric activity, a rise in blood pressure, a temporary rise in breathing and heart rate, and psychological impact. You also consider what factors affect the extent of physical damage that may occur; for example:

Individual susceptibility.

Noise intensity and frequency.

Length and continuity of exposure.

You use your research to establish "permissible" levels of exposure to noise and the effects that mild, moderate, and extensive exposure can have on an individual. From these you develop criteria against which you compare the sound-level readings measured at the five locations.

You then write a formal report and submit it to Lorne Carson.

PROJECT NO. 7: INSTALLING DIAL-A-WASH IN THE UNITED STATES AND CANADA

You are to assume that when Mr. Ralph Rosenkratz, President of Dial-a-Wash Inc., 3300 Mountain Drive, Topeka, Kansas, visited H. L. Winman and Associates earlier this month, he discussed his company's plans to build a series of semiautomatic dial-operated car wash systems in major centers in the United States and Canada, with the first to be installed in your city. He was looking not only for a suitable site but also for a firm of consulting engineers to research other sites and manage the construction of Dial-a-Wash systems.

Dial-a-Wash systems have been very successful in Great Britain. They are built to a standard design, each installation having five bays. For one dollar, a customer washes his or her own car with a high-pressure hose and brush, using a special type of detergent patented by the Dial-a-Wash company. The remainder of the operation is automatic, the customer dialing the pre-rinse, after-rinse, dry and wax operations from inside a telephone-type booth.

Location of the Dial-a-Wash systems is important. They must be easily accessible from a major trunk road, have space for entry and exit ramps, and be away from competition. Each bay is 14 feet wide by 25 feet long (4.27 by 7.6 m), and the whole unit measures 90 feet wide by 25 feet deep (27.4 by 7.6 m).

Mr. Rosenkratz asked H. L. Winman and Associates to assess sites in your area and to recommend the one most suitable for development. He requested a proposal that justifies the choice of site, indicates its availability for either purchase or lease, and quotes a firm price for building the Dial-a-Wash unit.

Vern Rogers assigns the project to you, with instructions to do a thorough research job. He stresses that your proposal must convince Mr. Rosenkratz that he should appoint H. L. Winman and Associates as project engineers, since there will be many more Dial-a-Wash units to follow.

This assignment will require you to identify several sites and to assess their suitability with regard to traffic movement, competition, parking, means of exit and entry, availability, zoning, and price (either for purchase or lease).

PROJECT NO. 8: WATER ANALYSIS FOR A STEAM-GENERATING PLANT

Wakeling Processors Inc is to build a food-processing plant in Market Drayton, some 40 miles (64 km) from your office. As it will require large supplies of hot water and steam, it will have its own steam-generating plant.

Oliver R. Wakeling, the company's President and General Manager, is concerned that local water may cause problems in the recirculating system of this steam-generating plant. He writes to Vern Rogers, manager of the local H. L. Winman and Associates branch where you are employed, and requests that a chemical technologist study the situation.

Vern Rogers assigns the problem to you as project No. 28. He asks you to investigate the implications of using local water in the recirculating system of the proposed plant, consider some of the factors that will influence operation, and identify aspects that Wakeling Processors should know before they design or select a system.

When your studies are complete, you are to prepare a formal report of your findings, which both you and Vern Rogers will sign. Vern comments that Mr. Wakeling would like you not only to draw conclusions but also to make any necessary recommendations.

You drive to Market Drayton and discuss the situation with Frank Hedges, Market Drayton's town engineer. He tells you that Market Drayton pumps all its water from your city, so any problems will be identical to those for a local plant. You check whether any chemicals are added in Market Drayton, but Frank says no: "Our water is exactly the same as the water you drink and use to wash dishes and take showers."

You return home, analyze the local water supply, and consider factors affecting the proposed steam-generating plant. For example:

1. The ionic and biological content of the local water supply.
2. The implications of seasonal changes on this content.
3. The maximum concentrations of ionic and biological content that a recirculating system can accept or operate with.
4. Ways for maintaining these concentrations within acceptable limits.
5. A means for detecting changes in concentrations of the feed water (what methods or monitoring equipment can be used?).
6. Operating conditions (temperature, pressure, etc) of the system, and whether they might cause internal production of contaminants.
7. Materials that can be used in the system (and those that cannot, because of the effects of contaminants).
8. Preventive maintenance measures, to keep the recirculating system operational.

When your analysis is complete, you prepare your report for Mr. Wakeling.

PROJECT NO. 9: BUILDING AND TESTING A MEETING TIMER

To your surprise, Jim Perchanski obtains company approval for you to design, build, and test a prototype of the electronic meeting timer you proposed in Project No. 6 of Chapter 5. Approval comes in a memo to you from H. L. Winman and Associates Executive Vice-President Martin Dawes:

> Your proposal of (date) to design and build a meeting timer has been approved. When your design has been finalized, and testing is complete, please provide me with six copies of a report which I can submit to a marketing consultant and

possible manufacturers. Please coordinate your progress with A. Rittman, Department Head, Special Projects.

Andy Rittman assigns work order No. 6763 to the project. He suggests that your report be comprehensive, since its readers will range from technically knowledgeable people to marketing specialists who have not previously heard of your concept.

Note: This project assumes that you really build a timer, either as a hobby project or as part of a technology course. Ideally, you will test it during a real meeting, either at your college or privately in industry. Ask meeting participants to comment on its effectiveness and to state whether they would regard it as a useful management aid or merely as a gimmick.

PROJECT NO. 10: A MINICOMPUTER FOR WESMAN DISTRIBUTORS?

You are a technologist employed by the local branch of H. L. Winman and Associates. Privately, you are interested in minicomputers and "hobby" computers. Knowing of your interest in this field, Branch Manager Vern Rogers calls you into his office and introduces you to Wesley Baird and Manfred Jaeger, who jointly own a local wholesale firm known as Wesman Distributors.

"We've heard a lot lately about microcomputers and minicomputers," Wesley explains. "Several salesmen have been in to see us, but we're not sure we're big enough to make it worthwhile having one."

Manfred interrupts to describe Wesman Distributors. Their firm handles machine parts and replacement equipment for the agricultural implement market. They supply parts to sales outlets and repair shops mainly in rural communities. Their product line is varied, ranging from a tiny thumbscrew for a milling machine to a large drum for a grain cleaner. "Our product line lists about 14,000 different items," he explains, "but we carry only about 800 of them: those items most often in demand."

"Then you are particularly interested in inventory control and accounting functions?" you ask.

They agree, and explain that they would like H. L. Winman and Associates to do a study for them. Vern Rogers suggests you assess whether their volume of business warrants using a computer and, if so, whether a minicomputer would meet their needs. Wesley adds, "We'd like a specific recommendation so we'll know what to do or buy."

During your study you learn these facts about Wesman Distributors.

Wesley and Manfred have 9 employees, in addition to themselves.

They processed 8620 orders last year, for a total of 25,760 items.

The total volume of orders received last year amounted to $1,136,000.

They have been in business for nine years.

For the past three years their business has shown a continual increase, both in volume and dollar value:

Last year: 18%
The year before: 11%
The year before that: 16%

Most standard items are taken directly from their warehouse (20,430 last year); special or very large items are ordered from the manufacturer and delivered direct to destination (5330 last year).

In discussion with Wesley and Manfred you identify the following activities they would like to handle by computer, or information they would like to store and have available for recall:

Inventory control
Accounts receivable and payable
Payroll records
Check issuing
Automatic production of a machine-printed purchase order when the stock of a particular item drops to a predetermined level
Form-type correspondence
Profit and loss statements
Detection of slow-moving items
Manufacturer and client address lists
Taxation data
Stock levels
Monitoring of outstanding orders from suppliers.

Wesley and Manfred also have some questions they want answered:

1. How much is a computer going to cost?
2. How difficult is it to train someone to use it? (Could it be one of their regular clerks?)
3. How reliable are the computers?
4. Is there a danger of losing stored information? (Can it be "wiped out" by, say, a power cut?)
5. Will the chosen system be able to expand with the company (presuming the present expansion rate continues)?

To evaluate Wesman Distributors' needs against existing micro/mini computer systems, you consider:

Storage size and methods, and means for interfacing with mass storage.
Speed of input and information retrieval.
Types of input and output.
Display methods.
Attachments and "peripherals."
System cost, flexibility, and adaptability.

You then decide which would be the best system (or combination) to use, or whether it would be better for Wesman Distributors to rent time on a large computer, or even to use none at all. Finally, you write your report.

Note: You are free to increase the scope of the report or introduce additional features, if to do so will increase the quality of your report.

• • •

The following projects in previous chapters are also suitable for presentation as formal reports:

Chapter 4, Project 11
Chapter 5, Projects 1, 2, 3.

7

OTHER TECHNICAL DOCUMENTS

In our technological era, with its increasingly complex range of equipment and processes, there is a growing need for manufacturers to write clear technical manuals to accompany their products. Most equipment is issued with a book of operating instructions that contains: (1) a brief description of the equipment; (2) instructions on how to use it; and (3) a list of likely replacement parts. Also available, but usually only to qualified repair specialists, is a set of maintenance instructions with detailed maintenance and repair procedures plus a comprehensive parts list describing every item in the equipment. Both publications perform the same task for a different type of reader. Operating instructions assume that the reader has only slight technical knowledge, whereas maintenance instructions assume that the reader is fully competent.

This chapter describes how to write a technical description and instruction and prepare a parts list. It assumes that you will know your topic thoroughly and consequently may find it difficult to write in simple terms for someone who does not. It also offers suggestions on how to convert your knowledge of a process, equipment, or new technique into an interesting magazine article or technical paper.

Technical Description

The suggestions that follow apply to any type of technical description. Whether you are describing a piece of heavy construction machinery or a delicate instrument, a site plan for a building or street intersection, a photograph of faulty machining or a schematic of an electronic circuit, a chart depicting output, a new batching process, or a revised assembly method, you must organize your description so that it follows a distinguishable, coherent pattern.

SPATIAL ARRANGEMENT

Suppose Harvey Winman asks you to write a description of a new business complex known as "Tower Twenty-One," which he can give to visitors to the construction site. The bottom five floors are ready for occupancy, while the remainder are in varying stages of completion, ranging from nearly ready at the sixth floor to the final concrete-pouring stages on the twenty-first. A logical approach would be to describe the floors in sequence from the basement up, or, alternatively, from the penthouse down. Readers would immediately understand your approach because it would be natural for them to think of a vertical building as a series of floors arranged in sequence from 1 to 21. But if you started by describing the bottom five floors, jumped to the top, then back to floors 10, 11, and 12, then up to 17 and 18, and so on, you would confuse them. They would find your description difficult to follow because it would be *incoherent.*

Coherence depends on arranging information in such an order that the logic is evident to the readers. There can be many patterns. A spatial arrangement (*spatial* means "arranged in space") depends on the shape of the subject:

Vertical Subjects As indicated by the "Tower Twenty-One" example, tall, narrow subjects lend themselves to a vertical description, each item being described in order from top to bottom, or bottom to top. Vertically arranged control panels and electronic equipment racks fall into this category. The sound-level meter in Figure 7-1 is a typical example.

Horizontal Subjects Long, relatively flat subjects demand a horizontal description, with each part being described in order from left to right, or right to left.

Circular Subjects Round subjects, such as a clock, a revolution counter, a pilot's altitude indicator, or a circular slide rule, suit a circular arrangement, with the description of markings starting at a specific point (e.g. "12" on a clock) and continuing in a clockwise direction until the entire circle has been described. In effect, this is the same as the horizontal description, if you consider that the circle can be broken at one point and unwound into a straight line. Alternatively, if the subject has a series of items that are arranged more or less concentrically (such as the scales of a circular slide rule), they can be described in sequence starting at the center and working outwards, or from the outermost item inwards.

Regardless of the shape of the subject and the arrangement of its parts, there is almost always some pattern that can describe it. If it is not clearly left-to-right, top-to-bottom, or center-to-circumference, you may have to search for a logical arrangement. The timer control box in Figure 7-2 would pose such a problem. Probably it would be best to start with the control knob at the center and work outwards, even though the positions of the electrical outlets and the slide switch are diametrically opposite. A top left to bottom right basic arrangement might also be practical, combined with a concentric description of the items forming the central cluster. Note how the following description guides the reader to the correct part of the control box by unobtrusively introducing gentle directions ("on top," "in the center," "around," and "at the bottom right"):

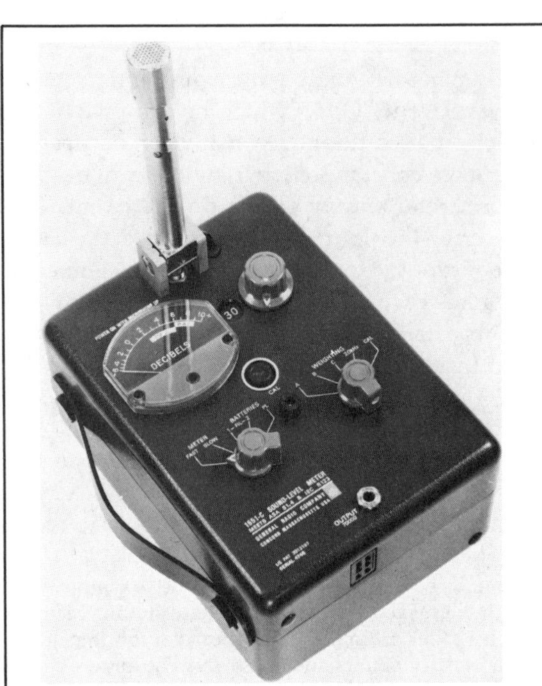

The GenRad 1551-C Sound-Level Meter measures sound levels from 25 dB to 150 dB. The microphone at the top doubles as an on-off switch: when the microphone is raised as shown, the unit is "on"; when the microphone is lowered, so that it rests downward along the front of the meter, the unit is "off." The DECIBELS scale at the upper left registers how much the recorded sound level deviates (from −6 to +10 dB) below or above a preset level visible in the small circular window beside the scale. This preset level is adjusted in 10-degree increments, from 30 dB to 140 dB, by the control knob at the upper right.

At the center is a pushbutton for calibrating the meter scale. Below it, in the lower half of the unit, are two control knobs for selecting various operating modes. The left-hand knob selects either fast or slow meter needle response, or tests the unit's batteries. The right-hand knob selects a weighting network suitable for the range of sound levels being measured, or the calibration function. At the bottom is a 7000 ohm output jack for connecting the sound-level meter to other instruments, or to a recorder.

Figure 7-1. Technical description: GenRad 1551-C Sound-Level Meter. An overall vertical (top to bottom) pattern has been used to describe the meter, with an occasional horizontal (left to right) pattern within it. (Photo and description courtesy GenRad Inc., Concord MA.)

Spatial Description

Two electrical outlets on top of the control box connect the timer to the exposure lamp and the heater in the developing tray. The group of controls in the center contains a control knob with a pointer for setting exposure time, a time setting ring with a raised flange that acts as a stop for the pointer, and a knurled clamp screw that when tightened holds the time setting ring firmly in position. Around these controls is a dial that indicates exposure time (in seconds) and "activate," or developing, time. At the bottom right is a three-position

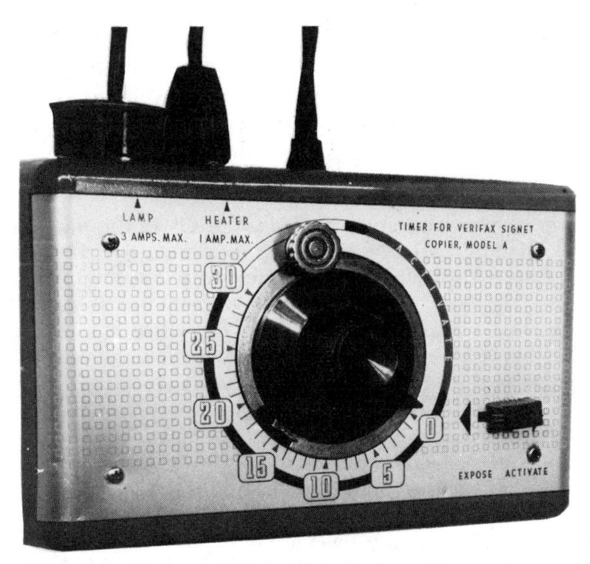

Figure 7-2. Timer control box. Photo: Anthony Simmonds.

slide switch that starts the timer for either the expose or activate cycles.

SEQUENTIAL ARRANGEMENT

Sometimes it is more suitable to describe a subject sequentially, introducing items in a natural order that does not depend on physical layout. This method is particularly suitable for describing processes, techniques, or equipment that has to be operated. The two most common methods are operating order and cause to effect.

Operating Order. The items are introduced in the order in which they are employed when the equipment is in use. This makes a very natural description which prepares the reader for the operating instructions that follow. Its one major disadvantage is that the writer has to find a way to mention static items that the operator does not use. In the case of the timer control box, a sequential description would differ significantly from the spatial description mentioned earlier:

Sequential Description (Operating Order) The raised flange on the time setting ring rotates freely around the control knob at the center of the control box. When locked in position by the knurled clamp screw, it presets the desired exposure time and acts as a stop for the control knob and pointer. The scale around the central cluster indicates exposure time (in seconds) and "activate," or developing, time. Operation is controlled by a three-position slide switch at the bottom right, which starts the timer for either the expose or activate cycles. Two electrical outlets on top of the control box connect the timer to the exposure lamp and the heater in the developing tray.

Cause to Effect. This method is used mainly to describe processes or operations which lead to a direct result. Each step in the process is described in sequence. It can be a simple description of the action that takes place when I depress a key on my typewriter to cause an imprint on the paper, or it can be a complex description of the effect of soil disturbance on permafrost in the discontinuous zones of Alaska and Canada.

Sequential Description (Cause to Effect)

When exploration crews in search of oil and mineral deposits first cleared long narrow strips of undergrowth for their seismic lines, they little knew the damage they were causing. Beneath its thin protective covering of vegetation, the soil had remained frozen for aeons at a temperature only a degree or two below freezing. Then, with the stubby spruce gone and the insulating layer of moss torn up and thrown aside, the ice in the soil began to melt. Small pools of water formed which spread into shallow ponds that in succeeding summers became a narrow stretch of muskeg. The short summer months of the subarctic offer little time for vegetation to grow, so that now, 30 years later, pools of muskeg are still visible between the thin cover of poplar and coarse grasses that have slowly replaced the spruce and moss. The seismic lines have become a permanent scar upon the landscape.

Which method should you use? In some cases the physical shape of the subject, or arrangement of its parts, will dictate the most suitable method. In other cases you may have a choice. When you do, try to find the method most natural for the specific situation, always trying to see the description through the eyes of a reader unfamiliar with the equipment or process. You can help your reader by inserting a photograph or drawing beside your description.

Technical Instruction

When Harvey Winman's special projects engineer, Andy Rittman, wants a job done, he issues instructions in clear, concise terms: "Take your crew over to the east end of the bridge and lay down control points 3, 4, and 7," he may say to the survey crew chief. If he fails to make himself clear, the chief has only to walk back across the bridge to ask questions. But Fred Stokes, chief engineer at Robertson Engineering Company, seldom gives spoken instructions to his electrical crews. Most of the time they work at remote sites and follow printed instructions, with no opportunity to walk across a project site to clarify an ambiguous order.

A technical instruction tells somebody to do something. It may be a simple one-sentence statement that defines what has to be done but leaves the time and the method to the reader. Or it may be a step-by-step procedure that describes exactly what has to be done and tells when and how. It is the latter type of technical instruction that will be described here.

Before attempting to write an instruction, you must first define your readers, or at least establish their level of technical knowledge and familiarity with your subject. Only then can you decide the depth of detail you must provide. If they are familiar with a piece of equipment, you may assume that the simple statement "Open the cover plate" will not pose a problem. But if the equipment is new to them, you may have to broaden the statement to help them first identify and open the cover plate:

> Find the hinged cover plate at the bottom rear of the cabinet. Open it by inserting a Robertson No. 2 screwdriver into the narrow slot just above the hinge and then rotating the screwdriver half a turn counterclockwise.

GIVE YOUR READER CONFIDENCE

A well-written technical instruction automatically instills confidence in its readers. They feel they have the ability to do the work even though it may be highly complex and quite new to them. Consider these examples:

Vague Before the trap is set, it is a good idea to place a small piece of cheese on the bait pan. If it is too small it may fall off and if it is too big it might not fit under the serrated edge, so make sure you get the right size.

Clear and Concise Cut a 17-inch (0.62 m) length of 10-gauge wire and strip 1 inch (20 mm) of insulation from each end. Solder one end of the wire to terminal 7 and the other end to pin 49.

The first excerpt is much too ambiguous. It only suggests what should be done, it hints where it should instruct (almost inviting readers to nip their fingers), and in 31 explanatory words it fails to define the size of a "small" piece of cheese. The second excerpt is assertive and keeps strictly to the point. Such clear and authoritative writing immediately convinces its readers of the accuracy and validity of the steps they have to perform.

The best way to be authoritative is to write in the imperative mood. This means you should begin each step with an active verb, so that your instructions are commands:

Ignite the mixture. . . .
Mount the transit on its tripod. . . .
Apply the voltage to. . . .
Cut a 2 inch wide strip of. . . .
Connect the green wire. . . .
Excavate 3 feet down. . . .
Measure the current at. . . .
Count the number of blips. . . .

The imperative mood in the clear, concise excerpt quoted above keeps the instruction taut and definite. Notice how the active verbs (*cut, strip,* and *solder*) make readers feel they have no alternative but to follow the instructions. The vague excerpt would have been equally effective (and much shorter) if it had been written in the imperative mood and if the vague verb "place" had been replaced by an image-conveying verb-adverb combination such as "wedge firmly":

> Before setting the trap, wedge a 3/8 inch cube of cheese firmly under the serrated edge of the bait pan.

The following two statements clearly show the difference between an instruction written in the imperative mood and one that is not:

Disengage the gear, then start the engine.
(*Definite; uses active verbs*)

The gear should be disengaged before starting the engine.
(*Indefinite; uses passive verbs*)

The first statement is strong because it tells readers to do something. The second is weak because it neither instructs them to do anything nor insists it is really necessary that anything be done ("should" implies it is only *preferable* that the gear be disengaged before the engine is started).

Although an active verb will most often be the first word in a sentence, sometimes it may be preceded by an introductory or conditioning clause:

Before connecting the meter to the power source, *set* all the switches to "zero."

The imperative mood is maintained here because the active verb starts the statement's primary clause (in this case the clause that describes the action to be taken).

If you want to check whether a sentence you have written is in the imperative mood, ask yourself whether it *tells* the reader to *do* something. If it does, then you have written an *instruction.*

AVOID AMBIGUITY

There is no room for ambiguity in technical instructions. You have to assume that the person following your instructions cannot ask questions, and so you must never write anything that could be interpreted more than one way. This statement is wide open to misinterpretation:

Align the trace so that it is inclined approximately 30° to the horizontal.

Each technician will align the trace with a different degree of accuracy, depending on his or her interpretation of "approximately." How accurate does "approximately" require the technician to be? Within 5°? Within ½°? It may be that even 10° either side of 30° is acceptable, but the reader does not know this and is left with a feeling of doubt. Worse still, the reader's confidence in the technical validity of the whole instruction will be undermined. Vague references like this must be replaced by clearly stated tolerances:

Align the trace so that it is inclined 30° (±5°) to the horizontal.

More subtle, but equally open to misinterpretation, is this statement:

Adjust the capstan handle until the rotating head is close to the base.

Here the offending word is "close," which needs to be replaced by a specific distance:

Adjust the capstan handle until the distance between the rotating head and the base is 25 mm.

Similarly, such vague references as "*relatively* high," "*near* the top," and "*an adequate* supply" must be replaced by clearly stated measurements, tolerances, and quantities.

Specifying Tolerances. Many instructions call for readers to take a series of readings and then record their results in special places provided either on the instruction sheet itself or on a separate data sheet. This is particularly common in field testing and troubleshooting of electronic equipment, where a typical entry might read like this:

> Connect the voltmeter to test points
> 8 and 17, then note the reading   5.5 V ________V

The figure 5.5 V is the measurement that the technician should obtain, and the short line is provided for entering the voltage actually recorded. Since it is very difficult to obtain an exact voltage, a slight variation from the specified voltage is permissible. This tolerance can be stated:

As a percentage:	5.5 V (±6%)
As a specific voltage:	5.5 V (±0.33V)
As a voltage range:	5.17 to 5.83 V

The voltage range offers an advantage, since the technician does not have to make a calculation.

Indefinite Words. Weak words such as "should," "could," "would," "might," and "may" so weaken the authority of an instruction that they reduce the reader's confidence in the writer. Though their meaning is clear, they sound indefinite:

> Set the meter to the +300 V range. The needle should indicate 120 volts (±2 volts).

In this example, "should" implies it is not really essential that the needle indicate 120 volts (although it would be nice if it did). No doubt the writer meant it *must* read the specified voltage, but he or she has failed to say so. Nor has the reader been told to note the reading. The writer has forgotten the cardinal rule of instruction writing: *Tell* the reader to *do* something. To be authoritative, the instruction needs very few changes:

> Set the meter to the + 300 V range, then check that its needle indicates 120 volts (±2 volts).

Notice how the steps in the sample instructions in Figure 7-3 are clear, concise, and definite. You need not be a specialist in the subjects to recognize that they would be easy to follow; you feel that the writers know exactly what they are doing, and if you abide by their instructions you will not run into difficulty.

WRITE BITE-SIZE STEPS

Technicians working on complex equipment in cramped conditions need easy-to-follow instructions. You can help them by writing short paragraphs, each containing only one main step. If a step is complicated and its paragraph grows unwieldy, divide it into a major step and a series of substeps. Instruction writing lends itself to subparagraphing like this:

3. List the documentary evidence in block J of Form 658. Check that blocks A to

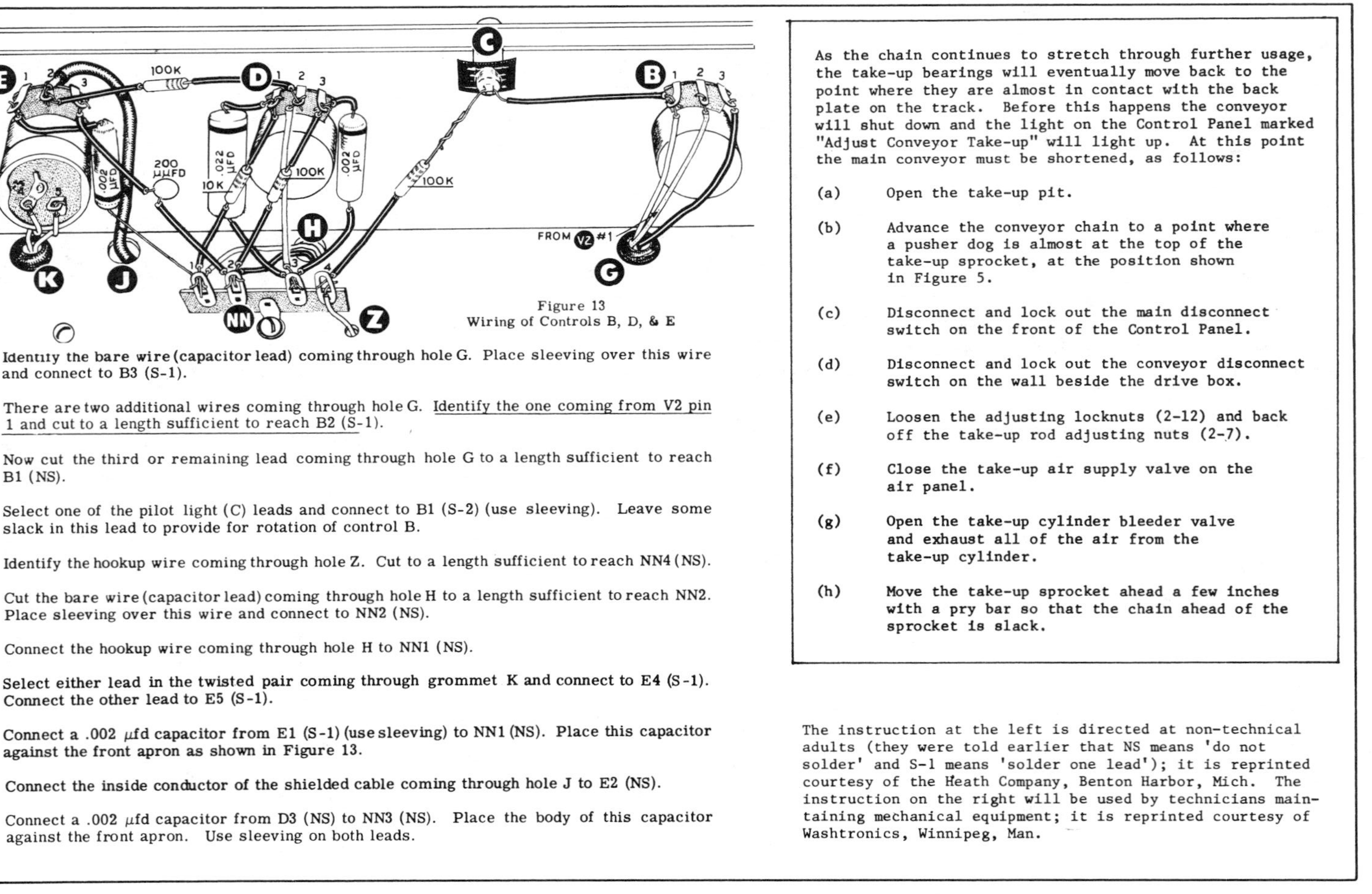

Figure 13
Wiring of Controls B, D, & E

Identify the bare wire (capacitor lead) coming through hole G. Place sleeving over this wire and connect to B3 (S-1).

There are two additional wires coming through hole G. Identify the one coming from V2 pin 1 and cut to a length sufficient to reach B2 (S-1).

Now cut the third or remaining lead coming through hole G to a length sufficient to reach B1 (NS).

Select one of the pilot light (C) leads and connect to B1 (S-2) (use sleeving). Leave some slack in this lead to provide for rotation of control B.

Identify the hookup wire coming through hole Z. Cut to a length sufficient to reach NN4 (NS).

Cut the bare wire (capacitor lead) coming through hole H to a length sufficient to reach NN2. Place sleeving over this wire and connect to NN2 (NS).

Connect the hookup wire coming through hole H to NN1 (NS).

Select either lead in the twisted pair coming through grommet K and connect to E4 (S-1). Connect the other lead to E5 (S-1).

Connect a .002 μfd capacitor from E1 (S-1) (use sleeving) to NN1 (NS). Place this capacitor against the front apron as shown in Figure 13.

Connect the inside conductor of the shielded cable coming through hole J to E2 (NS).

Connect a .002 μfd capacitor from D3 (NS) to NN3 (NS). Place the body of this capacitor against the front apron. Use sleeving on both leads.

As the chain continues to stretch through further usage, the take-up bearings will eventually move back to the point where they are almost in contact with the back plate on the track. Before this happens the conveyor will shut down and the light on the Control Panel marked "Adjust Conveyor Take-up" will light up. At this point the main conveyor must be shortened, as follows:

(a) Open the take-up pit.

(b) Advance the conveyor chain to a point where a pusher dog is almost at the top of the take-up sprocket, at the position shown in Figure 5.

(c) Disconnect and lock out the main disconnect switch on the front of the Control Panel.

(d) Disconnect and lock out the conveyor disconnect switch on the wall beside the drive box.

(e) Loosen the adjusting locknuts (2-12) and back off the take-up rod adjusting nuts (2-7).

(f) Close the take-up air supply valve on the air panel.

(g) Open the take-up cylinder bleeder valve and exhaust all of the air from the take-up cylinder.

(h) Move the take-up sprocket ahead a few inches with a pry bar so that the chain ahead of the sprocket is slack.

The instruction at the left is directed at non-technical adults (they were told earlier that NS means 'do not solder' and S-1 means 'solder one lead'); it is reprinted courtesy of the Heath Company, Benton Harbor, Mich. The instruction on the right will be used by technicians maintaining mechanical equipment; it is reprinted courtesy of Washtronics, Winnipeg, Man.

Figure 7-3. Excerpts from typical instruction manuals.

G have been completed correctly, then sign the form and distribute copies as follows:

3.1 Attach the documentary evidence to Copies 1 and 2 and mail them to the Chief Recording Clerk, Room 217, Civic Center, Montrose, Ohio.

3.2 Mail Copy 3 to the Computer Data Center, using one of the special preaddressed envelopes.

3.3 File Copy 4 in the "Hold—Pending Receipt" file.

3.4 When Copy 2 is returned by the Chief Recording Clerk, attach it to Copy 4 and file them both in the "Action Complete" file.

To avoid ambiguity, use a simple paragraph numbering system. Unless your instruction is going to be very long, there is nothing better than a straightforward system that starts at 1. Subparagraphs can be assigned decimal subparagraph numbers, such as 7.1, 7.2, and 7.3. This system also builds in a means for easy cross-referencing between steps; for example:

18. Reassemble the unit, reversing the procedure of steps 6 to 9.

IDENTIFY SWITCHES AND CONTROLS CLEARLY

Opinions differ as to the best method for identifying switches, controls, and operating positions marked on equipment. Some authorities recommend using capital letters throughout, while others prefer to enclose some of the words within quotation marks. I suggest you follow these two rules:

1. Identify switches, controls, and selector positions exactly as they appear on the equipment. The loudspeaker selector switch shown in the diagram would thus be described as the SPKR switch. This helps the reader to locate the correct control on a panel containing many dials and switches. As most equipment labeling is in capital letters, much of the time you will have to use capital letters.

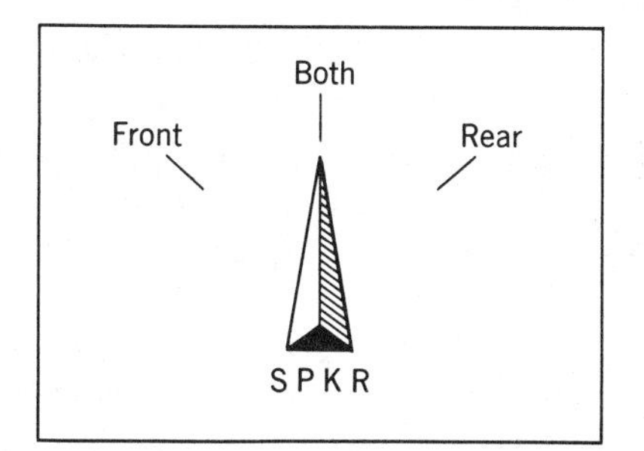

2. Differentiate between switches, controls, handles, and valves, and the positions to which they can be set, by enclosing the position settings within quotation marks. The loudspeaker selector switch in the diagram would be described as being set to the "BOTH" position. If you wanted to tell the readers to select only the front loudspeaker, you would instruct them to do so like this:

Set the SPKR switch to the "FRONT" position

or

Set the SPKR switch to "FRONT."

INSERT FAIL-SAFE PRECAUTIONS

Precautionary comments are inserted in instructions to warn readers of dangerous conditions or damage that can occur if they do not exercise care. There are three types:

Warnings: To alert readers of an element of personal danger (such as the presence of unprotected high voltage terminals).

Cautions: To tell them when care is needed to prevent equipment damage.

Notes: To make general comments (e.g. "on some older models the valve is at the rear of the unit").

Draw attention to a precautionary comment by placing it in the middle of the text, identing it on both sides, and preceding it with the single word WARNING, CAUTION, or NOTE. Draw a box around the words WARNING and CAUTION to give them extra prominence:

WARNING

Disconnect the power source before removing the cover plate.

A precautionary comment must always *precede* the step to which it refers. This will prevent an absorbed reader who concentrates on only one step at a time from acting before reading the warning. Never assume that mechanical devices, such as indentation and the box drawn around the precautionary word, are enough to catch a reader's attention.

Warnings and cautions must be used sparingly. A single warning will catch a reader's attention. Too many of them will cause a reader to treat them all as comments rather than as important protective devices.

INSIST ON AN OPERATIONAL CHECK

The final test for any technical instruction is the reader's ease in following it. Since you cannot always peer over a reader's shoulder to correct mistakes, you should find out whether users are likely to run into difficulty *before* you send it out. To obtain an objective check, give the draft instruction to a technician of roughly equal competence to those who eventually will be using it, and observe how well the task is performed.

As you watch (and you must watch with a zipped lip!), note every time the user hesitates or has difficulty. When the task is finished, ask if any parts need clarification. Then rewrite ambiguities, and recheck your instruction with another technician. Repeat this procedure until you are confident that it can be followed easily.

Parts List

A parts list helps equipment owners identify and describe the parts they need when ordering replacements. If they select the correct items from the parts list and describe them clearly, they are likely to receive what they want without delay. But if the parts list is ambiguous and they misinterpret it, or if they describe the items only vaguely, then they may receive the wrong parts or have to wait while the manufacturer writes for clarification.

Since neither the manufacturer (who wants to keep on good terms with equipment owners) nor the user (who wants to get the equipment working again as quickly as possible) wishes to cloud the ordering procedure, both need to communicate their information clearly. There is a method to listing and ordering parts which, if adopted by both parties, will prevent supply errors from occurring.

MANUFACTURER'S RESPONSIBILITY

To help equipment users identify the parts they require, manufacturers endeavor to list parts in a logical order that will be self-evident. Common methods are to group the items according to physical location, as in the list of front panel components for the amplifier in Figure 7-4, or to group them by class, e.g. all resistors, all hardware, all switches and controls, and so on. Within the groups, the items are described in three ways:

1. *By item number.* Each entry is assigned an item number, usually starting at 1 for the first item in the entire list and continuing in sequential order to the end of the list. Item numbers are sometimes accompanied by an illustration, which is keyed to each item in the parts list (see Figure 7-4). In large systems each major assembly may be assigned a whole number, and its parts numbered as subsections. For example, part 27 of assembly 14 would be identified as item 14–27.
2. *By part number.* Every item has a separate number to identify it. Quoting this number is the first step toward ensuring delivery of the correct part.
3. *By description.* Each part is described in noun-adjective order. This means identifying the noun that best describes the item and stating it first, then adding descriptive adjectives that will pin down its specific characteristics. The amber pilot lamp identified as item 26 in Figure 7-4 would be described in this order:

   Lamp, pilot, amber

   noun *adjectives*

   Similarly, the description for a cadmium plated hexagonal head No. 7 wood screw, 32.5 mm long, would be:

   Screw, hex hd, cad pl, No. 7, 32.5 mm

   noun *adjectives*

   The order in which the descriptive adjectives appear is not too important (although normally the words appear first, followed by specific dimensions and tolerances). However, the order should be consistent throughout a parts list, just as the abbreviations should be consistent. If, for instance, the first resistor in the parts list is described as "Resistor, 22 kohm, ½ watt," then the descriptions of all other resistors should state the resistive value before the wattage rating. If a dimension is quoted in decimal form, with the unit of measurement abbreviated in a particular manner (e.g. 27.5 mm), then all dimensions should adopt the same form; occasional use of a fraction and a different unit of measurement (e.g. 4¼ in.) would be inconsistent. This is no more than a technical application of parallelism, discussed in Chapter 10.

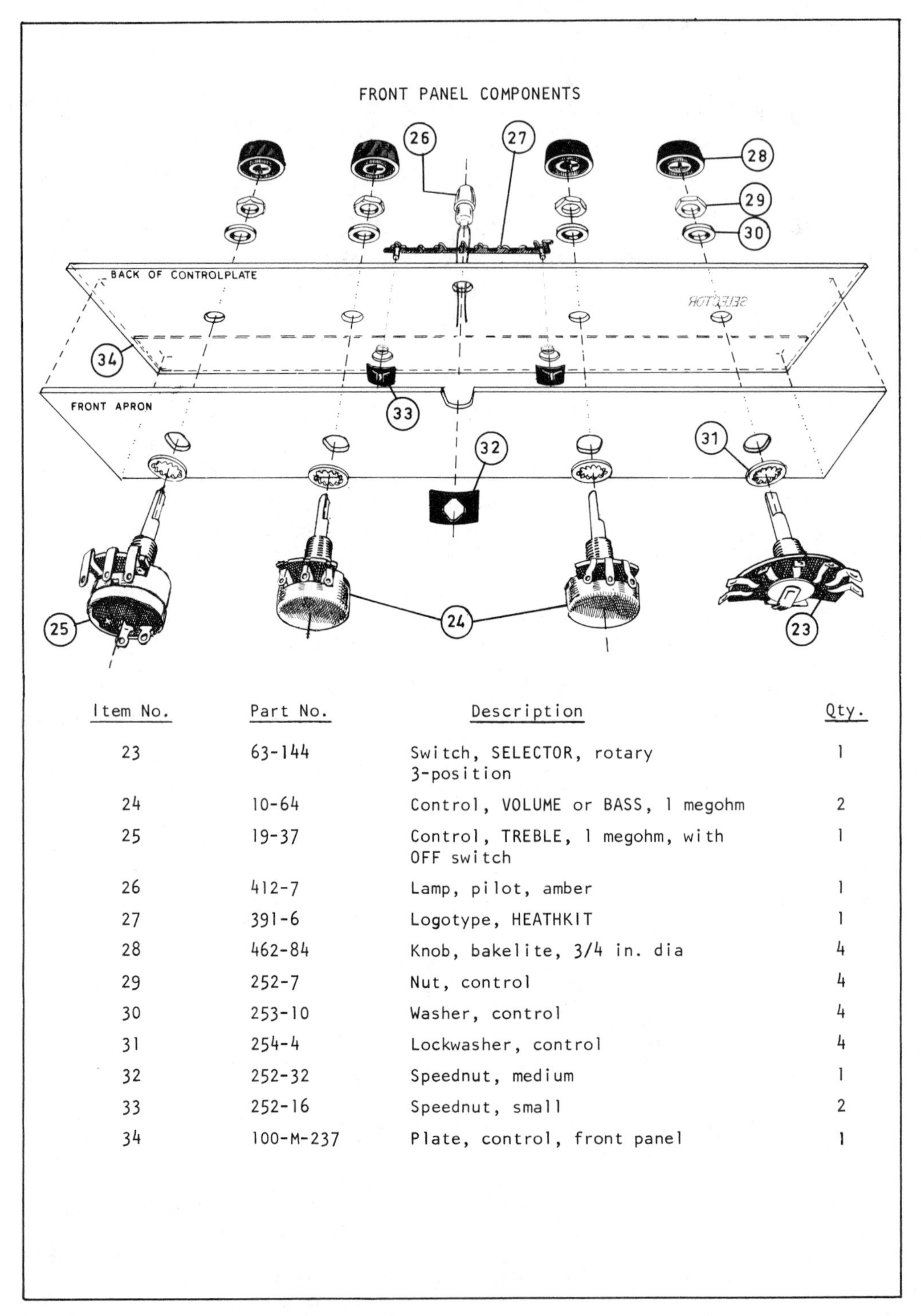

Item No.	Part No.	Description	Qty.
23	63-144	Switch, SELECTOR, rotary 3-position	1
24	10-64	Control, VOLUME or BASS, 1 megohm	2
25	19-37	Control, TREBLE, 1 megohm, with OFF switch	1
26	412-7	Lamp, pilot, amber	1
27	391-6	Logotype, HEATHKIT	1
28	462-84	Knob, bakelite, 3/4 in. dia	4
29	252-7	Nut, control	4
30	253-10	Washer, control	4
31	254-4	Lockwasher, control	4
32	252-32	Speednut, medium	1
33	252-16	Speednut, small	2
34	100-M-237	Plate, control, front panel	1

Figure 7-4. Excerpt from a typical parts list (front panel controls for Heathkit EA-3 amplifier; courtesy the Heath Company, Benton Harbor, Mich.).

The "Qty" column in the parts list shows the number of like items present in the equipment. If a unit of measurement is necessary, the unit is abbreviated and entered immediately after the quantity:

3 kg	30 cm^3
28 ft	7 doz
14 mm	8 gal

USER'S RESPONSIBILITY

Manufacturers who issue thoroughly detailed parts lists have gone as far as they can in helping users of their products order replacements. Individuals needing spare parts must do an equally thorough job when describing their needs. If a request is made on a company purchase order, spaces usually are available for copying the appropriate entries from the parts list. If done by letter, the request must be extremely explicit:

Dear Sir:

Please ship the following replacements for our Heathkit EA-3 amplifier:

Part No.	*Description*	*Qty Reqd*
19–37	Control, TREBLE, 1 megohm, with OFF switch	1
3E–10	Resistor, wirewound, 110 ohm, 5 watt, 10%	3

The need to specify clearly exactly what parts one requires is especially important when sending in a request from the field. Too often a brief note on a scrap of paper like this is attached to a progress report:

I need 13 rolls tarpaper and some bitumastic compound for sealing joints type ML-3. Please ship by air express special delivery.

Jack Ogilvie

If you received this note, you would have to decipher it before you could fill Jack Ogilvie's request. And you may easily ship the wrong materials, or too little or too much. A properly written request (even if it is on a scrap of paper) will always speed up delivery of the spare parts:

Please send me the following items
by air express:

13 rolls Tarpaper, 50 m by 1 m.
1 doz Sealing compound, bitumastic,
type ML-3 (1 kg tubes,
for calking gun).

Technical Papers and Articles

WHY WRITE FOR PUBLICATION?

The likelihood that one day you may be asked to write a technical paper for publication, or even want to do so, may seem so remote to you now that you might be justified in skipping this section. Yet this is something you should think about, for getting one's name into print is one of the fastest ways to obtain recognition. Suddenly you become an expert in your field and are of more value to your employer, who is happy because the company's name appears in print beneath yours. You become of more value to prospective employers, who rate authors of technical papers more highly than equally qualified persons who have not published. And you now have positive proof of your competence, and sometimes a few extra dollars from the publisher.

Mickey Wendell has an interesting topic to write about: As a senior technician in H. L. Winman and Associates' Materials Testing Laboratory, he has been testing concretes with various additives to find a grout that can be installed in frozen soil during the Alaskan winter. One mixture that he analyzed but discarded contained a new product known as Aluminum KL. As a by-product of his tests he has discovered that mixing Aluminum KL with cement in the right proportions results in a concrete with very high salt resistance. He reasons that such concrete could prove invaluable to builders of concrete pavements in snow-affected areas of the United States and Canada, where salt mixtures are applied in winter to melt the snow.

Mickey has been thinking about publishing this particular aspect of his findings and has jotted down a few headings as a preliminary outline. Here are the four steps he must take before his ideas appear in print.

STEP 1: SOLICIT COMPANY APPROVAL

Most companies encourage their employees to write for publication, and some even offer incentives such as cash awards to those who do get into print. However, they expect prospective authors to ask for permission before they submit their manuscripts.

To obtain permission, Mickey must write a brief memorandum outlining his ideas to John Wood, his department head. He should ask for approval to submit a paper, explain what he wants to write about and why he thinks the information should be published, and outline where he intends to send it. His proposal is shown in Figure 7-5. John Wood will discuss the matter at management level, then signify the company's approval or denial in writing. It is important for Mickey to have written consent to publish his findings. It sometimes takes months for an article to appear in print, and if memories are short or there have been personnel changes, it will help him to prove he has obtained company approval.

STEP 2: CONSIDER THE MARKET

Mickey must decide very early where he will try to place his paper. If he prefers to present his findings as a technical paper before a society meeting, as suggested by John Wood (see his comment in Figure 7-5), he will be writing for a limited audience with specialized interests. If he decides to publish in the journal of a technical society, he will be writing for a larger audience, but still within a limited field. If he plans to publish in a technical magazine, he will be appealing to a wide readership with a broad range of technical knowledge. His approach must therefore differ, depending on the type of publication and level of reader.

A guiding factor may be Mickey's writing capability. A paper to be published by a technical society requires high quality writing. The editor of a society journal normally does not do much prepublication editing, other than making minor changes to suit the format and style of the society's publications. A technical magazine article, however, will be edited—sometimes quite fiercely—by a professional editor who knows the exact style that readers expect. Such an editor prefers authors to approximate that style and expects them to organize their work well and to write coherently; but he or she is always ready to prune or graft, and sometimes even completely rewrite portions of a manuscript. Hence, the pressure on the authors of magazine articles is not so great.

A secondary consideration may be the state of Mickey's wallet: if it is thin and he needs a new set of tires for his car, he may choose to write a magazine

H L Winman and Associates

INTER - OFFICE MEMORANDUM

From: M Wendell
Date: June 24, 19__

To: J Wood
Subject: Approval for Proposed Technical Article

May I have company approval to write an article on concrete additives for publication in a technical journal? Specifically, I want to describe our experiments with Aluminum KL, and the salt-corrosion resistance it imparted to the concrete samples we tested for the Alaska transmission tower project. I believe that our findings will be of interest to many municipal engineers in the northern United States, who for years have been trying to combat pavement erosion caused by the application of salt during snow removal.

I was thinking of submitting the article to the editor of "Municipal Engineering", but I'm open to suggestions if you can think of a more suitable magazine.

MW

MW:kr

Approval granted. Let me see an outline and the first draft before you submit them. Suggest you also consider preparing a technical paper for presentation at the Combined Conference on Concrete to be held in Chicago next March.

John Wood
6 July 19–

Figure 7-3. Soliciting company approval.

article. Normally, there is no pay for writers of technical papers, other than recognition by one's peers. Perhaps the most important factor is for Mickey to be able to identify a potential audience for his information. Readers of society journals and technical magazines may be the same people, but they expect different information coverage in a technical paper than they do in a magazine article.

Technical Paper. Readers of society journals are looking for facts. They neither expect nor want explanations of basic theories, and they can accept a strongly technical vocabulary. A technical paper can be very specific. It can describe a minute aspect of a large project without seeming incomplete, or it can outline in bold terms the findings of a major experiment. No topic is too large or too small, too specialized or too complete, to be published as a technical paper.

Technical Article. Most readers of magazine articles are looking for information that will keep them up to date on new developments. Some will have definite interest in a specific topic and would welcome a lot of technical details. Others will be looking mainly for general information, with no more than just the highlights of a new idea. Magazine authors must therefore appeal to a maximum number of readers. Their articles should be of general interest; their style can be brief and informal; their vocabulary must be understandable; and they should sketch in background details for readers whose technical knowledge is only marginal.

STEP 3: WRITE AN ABSTRACT AND OUTLINE

Most editors prefer to read either a summary of a proposed paper or an abstract and outline before the author submits the complete manuscript. They may want to suggest a change in emphasis to suit editorial policy, or even decline to print an interesting paper because someone else is working on a similar topic.

This type of summary is much longer than the summary at the head of a technical report; the abstract, however, usually is quite short. The summary contains a condensed version of the full paper in about 500 to 1000 words. An abstract contains only very brief highlights and the main conclusion (rather like the summary of a report), since it is supported by a comprehensive topic outline.

Some authors write the complete first draft of the paper before attempting to write a summary or abstract, then leave the revising and final polishing until after the paper has been accepted by an editor. Others prepare a fairly comprehensive outline, often using the freewheeling approach suggested in Chapter 1, and leave the writing until after acceptance. Both methods leave room for the author to incorporate changes before the final manuscript is written.

Since the summary or abstract and outline have to "sell" an editor on the newsworthiness of his topic, Mickey Wendell must make sure that the material he submits is complete and informative. In addition he must indicate clearly:

1. Why the topic will be of interest to readers.
2. How deeply the topic will be covered.
3. How the article or paper will be organized.

4. How long it will be (in words).
5. His capability to write it.

Mickey can cover the first four items in a single paragraph. The fifth he will have to prove in two ways. He can prove his technical capability by mentioning his involvement in the topic and experience in similar projects. His ability to write well he can demonstrate by submitting a clear, well-written summary or abstract.

If Mickey later decides to prepare his paper for presentation before the Combined Conference on Concrete, as John Wood has suggested, he will have to prepare a summary in response to a "call for papers" letter sent out by the society, and submit it to the chairman of the papers selection committee. If his paper is to be accepted, he has to convince the committee that the subject is original, topical, and interesting, and that he has the technical capability to prepare it. If he has presented papers previously, he can use this fact to demonstrate his experience in oral reporting; if he has not, he will have to prove his capability during the conference. This aspect of oral reporting is covered in Chapter 8.

STEP 4: WRITE THE ARTICLE OR PAPER

A good technical paper is written in an interesting narrative style that combines storytelling with factual reporting. Articles published in general interest magazines tend to be written like feature newspaper stories, whereas technical papers more nearly resemble formal reports. If the article deals with a factual or established topic, the writing is likely to be crisp, definite, and authoritative. If it deals with development of a new idea or concept, the narrative will generally be more persuasive, since the writer is trying to convince the reader of the logic of his or her argument.

The parts of an article or paper are very similar to those of a report:

Summary A synopsis that tells very briefly what the article is about. It should summarize the three major sections that follow. Like the summary of a report, it should catch and hold the reader's interest.

Introduction Circumstances that led up to the event, discovery, or concept that is to be described. It should contain all the facts readers will need if they are to understand the discussion that follows.

Discussion How the author went about the project, what he found out, and what inferences he drew from his findings. The topic can be described chronologically (for a series of events that led up to a result), by subject (for descriptive analyses of experiments, processes, equipments, or methods), or by concept (for the development of an idea from concept to fruition). The methods are very similar to those used for writing the discussion of the formal report (see p. 149).

Conclusion A summing-up in which the writer draws conclusions from and discusses the implications of his major findings. Although he will not normally make recommendations, he may suggest what he feels needs to be done in the future, or outline work that he or others have already started if there is a subsequent stage to the project.

Illustrations are a useful way to convey ideas quickly, to draw attention to an article, and to break up heavy blocks of type. They should be instantly clear and usefully *supplement* the narrative. They should never be inserted simply to save writing time; neither should they convey exactly the same message as the written words. For examples of effective illustrating, turn to any major publication in your technical field and study how its authors have used charts, graphs, sketches, and photographs as part of the story. For further suggestions on how to prepare illustrative material, see Chapter 9.

Since an article or paper is to be read by many persons, considerably more revision time is required than for in-plant reports. Mickey Wendell should work closely with technical editor Anna King and should give himself time between major revisions to put the article aside so that he can return to it with a fresh mind. He will also be wise to call on at least one independent reviewer to read the paper when it is nearly ready for submission. He should respect his reviewer's comments, because they are likely to be similar to those of his eventual readers.

Finally, a comment on the editor's role: Mickey should not be surprised or disturbed if the editor who handles his manuscript makes some changes. These changes will affect only the arrangement of the information and will seldom alter the technical content (if they do, Mickey will have good reason for raising his voice). The editor knows the particular readers well. If the editor feels the material is too lengthy, too detailed, or wrongly emphasized, he or she will make changes to bring it up to the expected standard. Mickey may feel that the alterations have ruined his carefully chosen phrases, but readers will not even be aware that changes have been made. They will simply recognize a well-written paper, for which Mickey rather than the editor will reap the compliments.

Assignments
Description Writing

PROJECT NO. 1: BLUEPRINTING PROCESS

Write a description of the blueprinting process, based on the information provided below. Assume that your readers will be drafting students who have not yet used a blueprinter.

This is how a technician told me he made some blueprints:

> The blueprinter was relatively simple to operate. Since the power was already on and the machine was operating, the steps involved in making my 20 blueprints were easy to follow. In the first place, for each print I took a sheet of blueprint paper and placed it on the feed table with its face side up. (The face side is the yellow side.) The original tracing, right way up, was then placed onto the blueprint paper. Before pushing the two sheets of paper into the machine, I had to set the speed selector to the correct speed, which is 15 for medium speed blueprint paper. (Mine was medium speed.) The two sheets were then pushed toward the machine until they were grasped between the belts and the exposure roller. When the two sheets emerged I had to separate them and fold back the blueprint paper and feed it up and over a developing roller. The original tracing came right out and I removed it in readiness for making another blueprint. This printing procedure was repeated until

I had made all the copies I needed. At the top of the machine there was a guide bar that directed the prints to one of two delivery trays. The front delivery tray is at the front of the machine immediately in front of the operator; the back delivery tray is used when you want to stack a succession of prints without removing them. Because I had 20 prints to make, I selected "rear delivery." When I had finished printing I had to remember to turn the speed selector back to 5.

The speed selector looks like this:

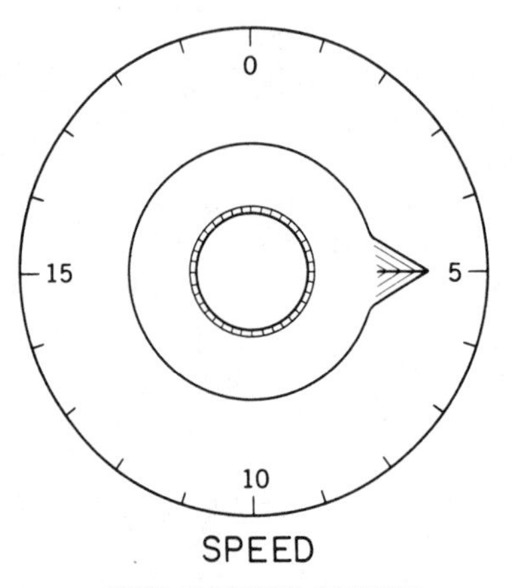

SPEED SELECTOR CONTROL

The blueprinter viewed from the side looks like this:

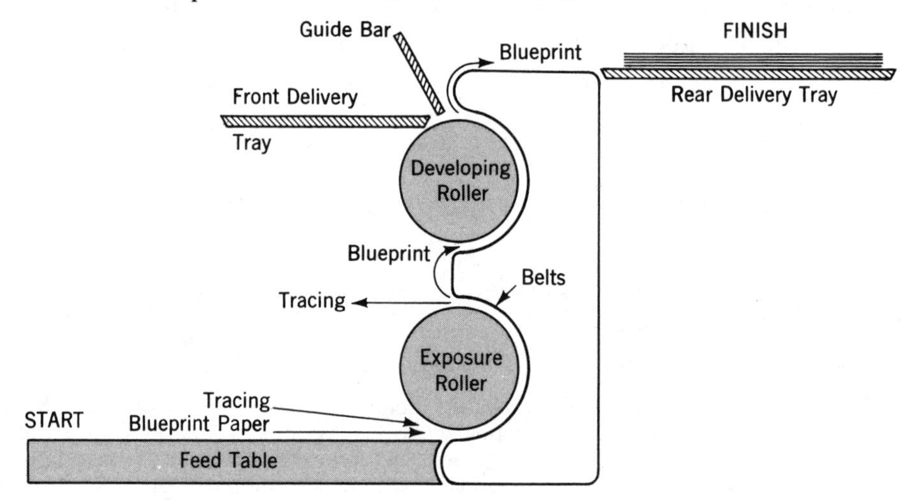

BLUEPRINTER (VIEWED FROM SIDE)

1. The exposure roller is made of glass and has a high intensity light inside it.
2. The light burns off the yellow coating on the blueprint paper (except where the image of the tracing prevents light from reaching it).
3. The developing roller forces the paper through ammonia fumes (produced by heating liquid ammonia).

4. The blueprinter has a chimney to exhaust the fumes (because they are toxic).
5. Fumes "develop" the blueprint (i.e. the yellow areas turn blue).
6. The number set on the speed selector is the number of feet of paper the blueprinter produces in a minute.
7. The type of paper described here is more correctly known as "whiteprint" or "blueline" paper. In true blueprint paper the image appears white against a dark blue background.

PROJECT NO. 2: RESISTORS

Mr. Wayne D. Robertson, president of Robertson Engineering Company, receives the following letter from his friend, Dave Kostyn, who owns a wholesale grocery company:

> Dear Wayne,
>
> Can you help me? I have a dozen high school students in the Junior Achievement group I'm working with and they have planned a project to sell packages of "resistors" to radio amateurs and hobbyists around the city. I cannot help them very much because at the moment I have no idea what a resistor is or what it looks like! All I know is that a bunch of resistors will be coming to us loose and the youngsters will have to divide them into groups of 25, package them, and arrange to sell them.
>
> What I need right now is a paragraph or two from you describing what a resistor looks like and what it's made of (plus anything else you think I should know).
>
> Can you do that for me? I certainly would appreciate it!
>
> Regards,
>
> Dave

Mr. Robertson passes the letter to the chief engineer, who passes it down to your supervisor, who in turn passes it to you to write the description. You may assume that Dave Kostyn will post your description on the wall, where it can be read by all the Junior Achievers.

PROJECT NO. 3: LAC LE ROULET AREA

Some H. L. Winman and Associates personnel will shortly be assigned to Lac le Roulet for one year. As you have been there previously, technical editor Anna King asks you to write a 200- to 250-word description of the area which she can insert into the company magazine. Use the sketch and legend as a guide, and find a pattern for organizing your description.

Legend: Map of Lac le Roulet and Surrounding Area

(1) Hillock; top 837 ft (255 m) above mean sea level (AMSL); has a gravel base and a park-like appearance with scattered evergreens growing to a height of 40 ft (12 m); also has low underbrush.

(2) Proposed airstrip; 4000 ft (1200 m) long; 780 ft (238 m) AMSL.

(3) Site for permanent camp.

(4) Site of temporary camp.

(5) Existing seaplane dock; deteriorating badly; needs complete rebuilding.

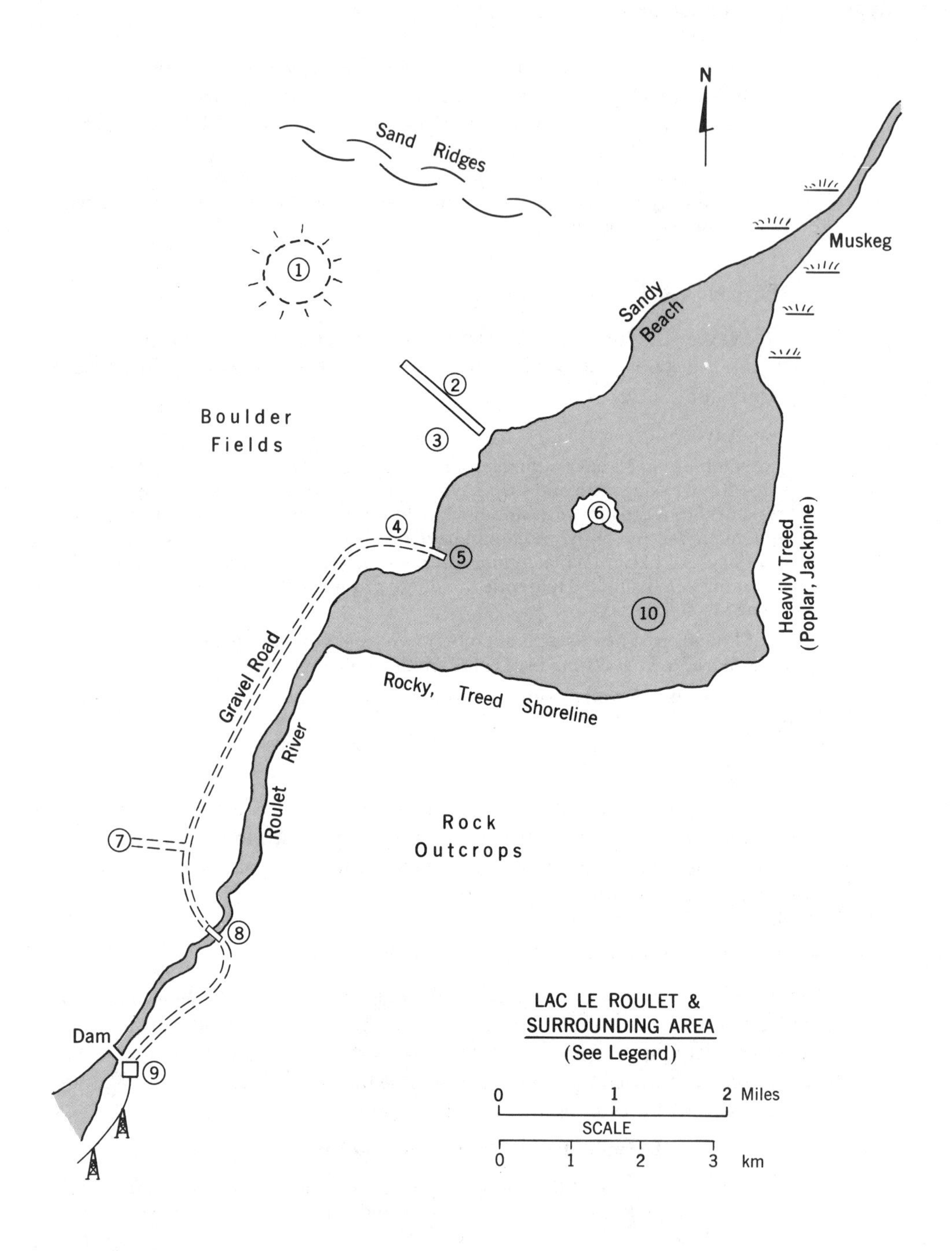
N
Sand Ridges
1
Muskeg
Sandy Beach
2
Boulder Fields
3
6
4
5
Heavily Treed (Poplar, Jackpine)
10
Gravel Road
Rocky, Treed Shoreline
Roulet River
Rock Outcrops
7
8
Dam
9
LAC LE ROULET & SURROUNDING AREA
(See Legend)
0 1 2 Miles
SCALE
0 1 2 3 km

(6) Island; top 791 ft (241 m) AMSL plus height of trees (approx 30 ft; 9 m).
(7) Site for radio transmission tower.
(8) Bailey bridge over rapids; water level drops 8 ft (2.44 m) through rapids.
(9) Roulet hydroelectric power generating station; water levels: 806 ft (245.67 m) AMSL above dam; 774 ft (235.9 m) AMSL below dam.
(10) Lac le Roulet; water level 765 ft (231.1 m) AMSL; level varies no more than ±2 ft (0.6 m); contains excellent stock of whitefish and northern pike.

PROJECT NO. 4: AUTO-MART SITES

Part 1. Assume that you are part way through your evaluation of possible Auto-Mart sites (see Project No. 2 in Chapter 5). Vern Rogers asks you to write him a 100- to 150-word description of each site, giving its proximity to the downtown and residential areas.

Part 2. Vern also asks you to write a 150-word comparison of the major differences (advantages/disadvantages) between the two sites, with particular reference to traffic patterns and accessibility.

PROJECT NO. 5: MEETING TIMER

Part 1. Describe the meeting timer illustrated on page 142 (Project 6 in Chapter 5). Assume that the description is to be included with a press release to trade magazines.

Part 2. If you have designed a new timer (see Project 9 in Chapter 6), describe it for a press release. Include:

The technology you employed.
The circuit.
How the timer will be used.

PROJECT NO. 6: TOOTHPICK TOWER

Describe the "toothpick tower" on page 208. Read Project No. 1 of Chapter 6 so you will understand why the tower was built and how it was used. Assume that your description will be given to students at other colleges who have not heard of the ASEI award.

Note: If you have built a toothpick tower, then you may describe it instead of the tower illustrated in Chapter 6.

PROJECT NO. 7: INTERSECTION AND BUS STOP

Mr. D. V. Botting, traffic engineer for Montrose, Ohio, Streets and Traffic Department, is aware that you have been doing an investigation into a bus stop problem at the corner of Main Street and Wallace Avenue. He writes you a note asking for recent photographs showing the properties at all four corners of the intersection. Since no photographs are available, you describe the intersection and the four properties in about 200 to 250 words. Assume that the Walston Oil Service Station building has been converted into a bus shelter and that a bus stop is now located on the lot.

For more information turn to Project No. 2 of Chapter 6.

PROJECT NO. 8: CAYMAN FLATS AREA

Describe the general area of Cayman Flats in relation to major features of the City of Montrose (there is a map of the city on p. 221). Assume that your readers are construction companies who will be submitting bids on:

1. Installing a storm sewer system throughout Cayman Flats.
2. Excavating zone S-17 and building a pump station on it.
3. Installing a storm sewer between the pump station and the existing storm sewers of zone S-5.

PROJECT NO. 9: MONTROSE AIRPORT FLIGHT PATH

Write a 150-word description of the areas under and on each side of the flight path as aircraft approach Montrose International Airport from the southeast (see Project No. 6 of Chapter 6).

PROJECT NO. 10. WATER CONSUMPTION OF MONTROSE

In Project No. 1 of Chapter 9 you are asked to prepare illustrations depicting quantities of water sold by the City of Montrose, Ohio, to various groups of consumers during the previous year. When this has been done, prepare an analysis of water consumption for the City Engineer, who will mail it with one of the illustrations to all water consumers. In your analysis identify when each segment of the community draws most and least water, and suggest reasons for these highs and lows.

PROJECT NO. 11: RESEARCHING A NEW MANUFACTURING MATERIAL OR PROCESS

You are to research information on a topic allied to your technology and then prepare it for both written and oral presentation. The topic may be a new manufacturing material, method, or process. The written and spoken presentations must:

1. Introduce the topic.
2. State why it is worth evaluating.
3. Describe the material, method, or process.
4. Discuss its uniqueness and usefulness.
5. Show how it can be applied in your particular field.

To obtain data for your topic, you will have to research current literature and probably talk to industrial users, manufacturers, and suppliers. Typical examples of topics are: a new oil that can be used at very low temperatures; a method for supporting the deck of a bridge during concrete-pouring by building up a base on compacted fill; a new paint for use on concrete surfaces; and a new materials-handling system.

You may assume that both your readers and your audience are technicians to whom the topic will be entirely new.

Instruction Writing

PROJECT NO. 12: TESTING SPLIT-BOLT CONNECTORS

Write an instruction to all installation supervisors at sites M1 through M18 telling them to test all split-bolt connectors (SBCs) on site (see Project No. 6 of Chapter 4).

Additional information:

1. Background to this project can be found on page 100.
2. The instruction is to be written as an interoffice memorandum.
3. Don Gibbon, electrical engineering coordinator at H. L. Winman and Associates, will sign it.
4. Tell the site installation supervisors that they are to report their findings to Don Gibbon.
5. Tell them to test the SBCs using procedure PR27-7.
6. All SBCs on site are to be tested. (It would be best to test those in stock first, then use tested ones to replace those in use.)
7. Inform them that faulty SBCs are to be sent to the contractor with a note that they are to be held for analysis under project HW44.
8. Their tests are to be completed within 7 days.

PROJECT NO. 13: USING THE MEETING TIMER

Write an instruction for using the meeting timer installed in H. L. Winman and Associates' conference room. Assume that the timer is wall-mounted. (See Project No. 6 of Chapter 5 for details and an illustration of the timer.)

PROJECT NO. 14: USING A BLUEPRINT MACHINE

Write an instruction for users of a blueprint machine that is accessible to you. Assume that the machine will be running, so that users do not have to switch it on or off. Also assume that the instructions will be pinned to the wall beside the machine.

If a blueprint machine is not available to you, base your instruction on the description and illustration included with Project No. 1 of this chapter.

Include a warning of the dangers of inhaling ammonia fumes.

Technical Article

PROJECT NO. 15

This project assumes that you have built and tested either the toothpick tower (Project 1, Chapter 6) or the electronic meeting timer (Project 6, Chapter 5; and Project 9, Chapter 6). You believe the topic would make an interesting article for a journal, magazine, or newspaper to publish.

Part 1. Select a publication which you feel would be interested in printing your article. Describe why you have selected this particular magazine.

Part 2. Prepare an outline of your proposed article. Describe your planned audience.

Part 3. Before submitting an article for publication, you should always obtain management permission (a "release") to write and print it. Write a memorandum either to your department head at the college you are attending, or to one of the H. L. Winman and Associates' department heads, requesting approval to submit an article. Outline briefly what the article will contain and where you plan to submit it.

Part 4. Write a 500-word summary of the article, plus a letter introducing it to the editor of the magazine.

8

TECHNICALLY—SPEAK!

This chapter covers two facets of public speaking, both concerned with the oral presentation of technical information. The first is the oral report, sometimes called the technical briefing, delivered to a client or one's colleagues. The second is the technical paper presented before a meeting of scientific or engineering-oriented persons. Both depend on public speaking techniques for their effectiveness, although neither requires vast experience or knowledge in this field. Also included in this chapter are some suggestions on how to contribute properly to meetings you attend, and how to prepare for and present yourself well at an interview.

The Technical Briefing

Your department head approaches your desk, a letter in hand, and says:

> Mr. Winman has had a letter from the RAFAC Corporation. They're sending in some representatives next Tuesday. I'd like you to give them a rundown on the project you're working on.

Every day visitors are being shown around industrial organizations, and every day engineers and technicians are being called upon to stand up and say a few words about their work. On paper, this sounds straightforward enough, but to those who have to make the oral presentation it can be a traumatic experience. Much of their nervousness can be reduced (it can seldom be entirely eliminated, as any experienced speaker will tell you) if they are given some hints on public speaking. The best training, of course, is practical experience, which can be gained only by standing up and doing the job.

ESTABLISH THE CIRCUMSTANCES

As soon as you have been informed that you are going to deliver an oral report, you should establish some of the circumstances surrounding the visit—

factors that will have a bearing on your approach. Jot down what you need to know, then ask your department head questions like these:

Who are the visitors?
You will probably be introduced to them, but at the critical moment before you speak you don't want to be concentrating on names. If you have heard them before they will be easy to remember.

How much will they know already?
You don't want to bore your visitors by repeating unnecessary details. Find out if they will come to you with no knowledge of your project, or whether management will have given them some preliminary information before you speak.

How long do you want me to talk?
Find out if management wants you to describe the project in detail, or simply touch on the highlights. It could be that the stop in your area is only a two-minute pause on a plant tour, or it may be a specific visit to study your project. The intent will directly influence your subject coverage.

Where is the briefing to take place?
Are you to address the visitors in the board room? Or will they be coming down to the project area? Availability of equipment may dictate how you tackle your subject and whether you need to make drawings to illustrate your talk.

Only when these factors have been firmly established can you start making notes. Jot down the topics you intend to discuss, and arrange them in an interesting, logical order. Think of an unenlightened person sitting in front of the equipment, and try to look at in the same way that he or she will. Don't let your familiarity with the project blind you to characteristics that are unimportant to you but would be interesting to the observer.

FIND A PATTERN

The best technical briefings follow an identifiable pattern, just as written formal reports do. You can establish a pattern for your briefing by answering three questions that you would be likely to ask if you were a visitor to another plant.

1. What Are You Trying To Do? Use the answer to this question to build an *Introduction*, as you would for a formal report. Offer your listeners some background information, which may comprise:

How your company became involved in the project (with, perhaps, a comment on your own involvement, to add a personal touch).

Exactly what you are attempting to do (in more formal terms, your objectives).

The extent or depth of the project (i.e. the scope).

2. What Have You Done So Far? This would be the *Discussion* section of the formal report. The answers to this question should cover:

How you set about tackling the project.

What you have accomplished to date (work done, objectives achieved, results obtained, and so on).

Preliminary conclusions you have reached as a result of the work done (if the work is complete, these will be the final conclusions).

3. What Remains to Be Done? (or **What Do You Plan to Do Next?**) This question is relevant only if the project is still in progress, in which case it is equivalent to the *Future Plans* section of a written progress report. Answers to this question should cover:

The scope of future work.

Results you hope to achieve.

A time schedule for reaching specific targets and final completion.

If the project is complete, this question is not relevant and is replaced by an alternative question: **What Are the Results of Your Project?** The answers would then be combined with the final answer to question 2 and would be equivalent to the *Conclusions* section of a written report.

Now we have a pattern for the main part of your briefing. But you still have to give it a beginning and an ending. The beginning should be a quick synopsis of the project in easy-to-understand terms—the equivalent of a report Summary. The ending can be a quick summing up (a Terminal Summary) that leads into an opportunity for your listeners to ask questions. The complete pattern is shown as a flow diagram in Figure 8-1.

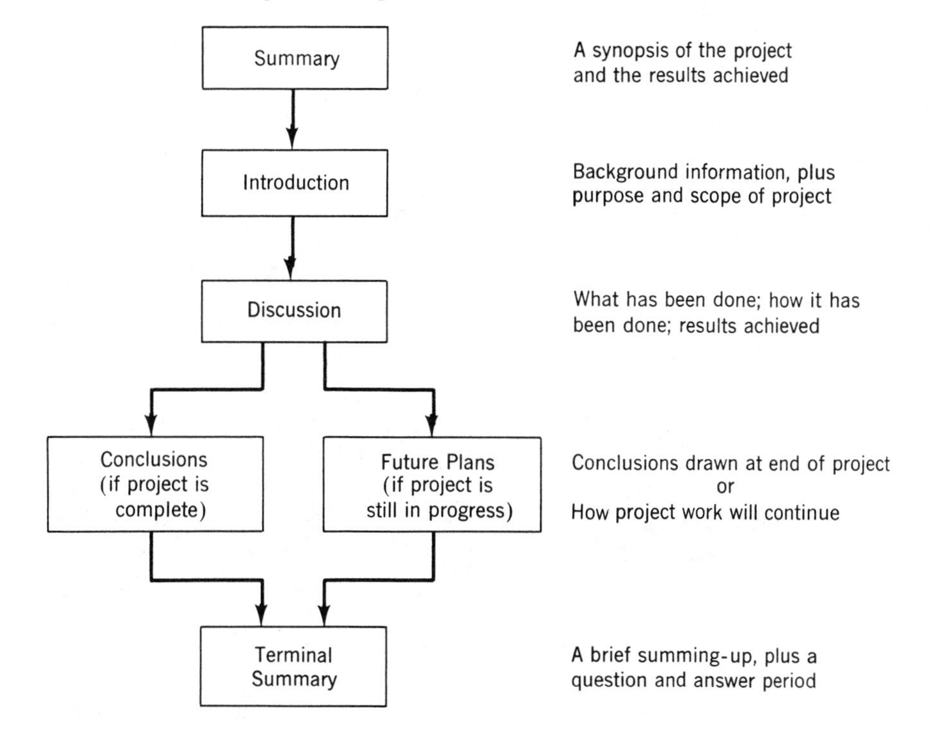

Figure 8-1. Flow diagram for a technical briefing.

PREPARE TO SPEAK

Make brief speaking notes on prompt cards no smaller than 6 × 4 inches (15 × 10 cm). If it is not convenient to hold the cards, place them on a makeshift speaker's stand, or even mount them on the back of a piece of equipment where you can read them easily without straining. Write in large, bold letters that you can see at a glance, using a series of brief headings to develop the information in sufficient detail. A specimen card is shown in Figure 8-2. Prompt cards like this are scaled-down versions of the speaking notes I recommend for technical paper presentation. For comparison, a typical page of notes is illustrated in Figure 8-3.

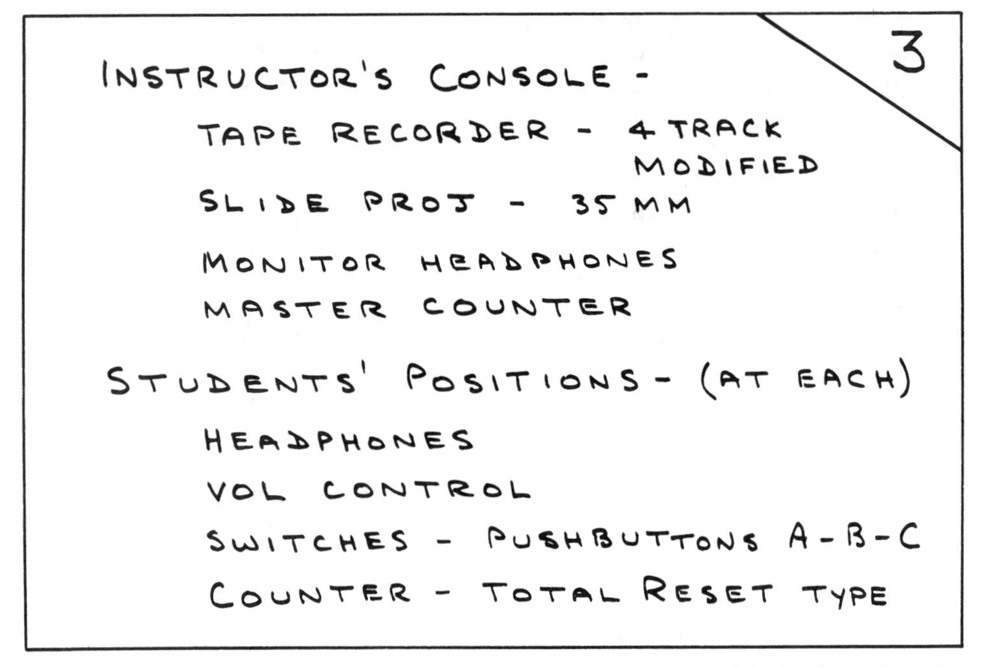

Figure 8-2. Prompt card for an oral report. (This is one of several prompt cards made by Ron Brophy for a talk he presented on the APL System described in Figure 5-3).

Don't overlook the practical aspects of the briefing. If you have equipment to demonstrate, consider its layout in relation to a logically organized description. Try to arrange the briefing so that you will move progressively from one side of the display area to the other, instead of jumping back and forth in a disorganized way. If the display is large and easy to see, let it remain unobtrusively at the back of the area. If it is small, consider moving it forward and talking from beside or behind it.

If visual aids will help you give a clearer, more readily understood briefing, then prepare as many as you will need. They may range from a series of steps listed as headings on a flip chart to a working model that demonstrates a complex process. Strive for simplicity; let the visual aid support your commentary rather than make the commentary explain an overly complex aid. Some suggestions for preparing graphs, charts, and diagrams are contained in Chapter 9.

Take a leaf from the experienced technical speakers' notebook and practice your briefing. Run through it several times, working entirely from your prompt cards, until you can do so without undue hesitation or stumbling over awkward words. If the cards are too hard to follow, or contain too much detail, amend them. Then ask a colleague to sit through your demonstration and give critical comments.

The time and effort you invest in preparing for a briefing will depend on your confidence as a speaker and your familiarity with the topic. The more confident you are, the less time you will need. No one will expect you to give a professional briefing, but everyone—visitors and management alike—will appreciate a carefully prepared talk presented in an interesting manner.

The Technical Paper

Chapter 7 discussed the steps Mickey Wendell would have to take to publish a magazine article or a technical paper. (He is a senior technician in H. L. Winman and Associates' Materials Testing Laboratory, and he has discovered that an additive called Aluminum KL mixed with cement in the right proportions produces a concrete with high salt resistance.) This chapter assumes that the papers committee of the Combined Conference on Concrete liked Mickey's abstract and summary, and the chairman of the committee has notified him that his paper has been selected for presentation at the forthcoming Chicago conference. Mickey has four months to prepare for it.

Mickey must recognize immediately that presenting a paper before a society meeting is much more demanding than delivering the same information at a technical briefing to company visitors. The occasion is more formal, the audience is much larger, and the speaker is working in unfamiliar surroundings. Many experienced engineers and scientists duck their responsibility to the audience when faced with such a situation and simply read their papers verbatim. This can result in a dull, monotonous delivery that would turn even a superior technical paper into a dreary, uninteresting recital. If Mickey is to avoid this trap, he must start preparing early.

PREPARATION

The key to an effective oral presentation is to have good speaker's notes, and to practice with them. This takes time. Mickey must write the publication version of his paper soon after he hears that it has been accepted (not leave it to the very last week, as so often happens), because he will need it to prepare his speaking notes.

The spoken version of a technical paper does not have to cover every point encompassed by the written version. In the 20 to 30 minutes allotted to speakers at many society meetings, there is time to present only the highlights—to trigger interest in the listeners so that they will want to read the published version. Mickey has to consider how he is to stimulate and hold this interest.

Selecting Topic Headings. His speaking notes will consist mainly of topic headings extracted from the written version of his paper. He should jot these headings onto a sheet of paper and then study them with four questions in mind:

1. Which points will prove of most interest to the audience?
2. Which are the most important points?
3. How many can I discuss in the limited time available?
4. In what order should I present them?

When Mickey was writing his paper, he was preparing information for a reader. Now he is preparing the same information for a listener, and the rules that guided him before may not apply. The logical and orderly arrangement of material prepared for publication is not necessarily that which an audience will find either interesting or easy to digest.

It is reasonable to assume that the audience at a society meeting is technically knowledgeable, has some background information in the subject area, and is interested in the topic. In Mickey's case most of the audience will be civil engineers and technologists, with a sprinkling of sales, construction, and management people. He must keep this in mind as he examines his list of headings, identifies which points he intends to talk about, and arranges them in the order he feels will most suit his listeners.

Preparing Speaking Notes. There are many ways in which Mickey can prepare his speaking notes. He can type them onto prompt cards similar to those illustrated in Figure 8-2, print them in bold letters on 8½ × 11 sheets, or enter them in a notebook. But he should never take the shortcut of simply entering the headings in the margin of the typed copy of his paper. The temptation to read the paper may become too great if he is very nervous, and once he starts reading he will find it difficult to return to extemporaneous speech.

Anna King suggested that Mickey use a notebook because its pages are bound together. It would be reassuring to know that if he inadvertently drops his notes, he only has to turn to the correct page to continue speaking. It could be a catastrophe to find his carefully prepared prompt cards scattered around his feet!

The notebook she recommended should be about 9 × 7 inches (25 × 18 cm) when closed (slightly wider than this textbook), should lie flat when opened, and should have wide-spaced horizontal lines. Its left-hand page should carry speaking notes, and its right-hand page demonstration notes (see Figure 8-3). The left-hand page is divided into four columns, the first for "time elapsed" and the remaining three each containing progressively more information. The right-hand page is a storage area for notes indicating when demonstrations are to be carried out and slides or diagrams are to be presented. It also carries comments and excerpts to be read to the audience.

In effect, Mickey Wendell would be wise to have two notebooks: one for

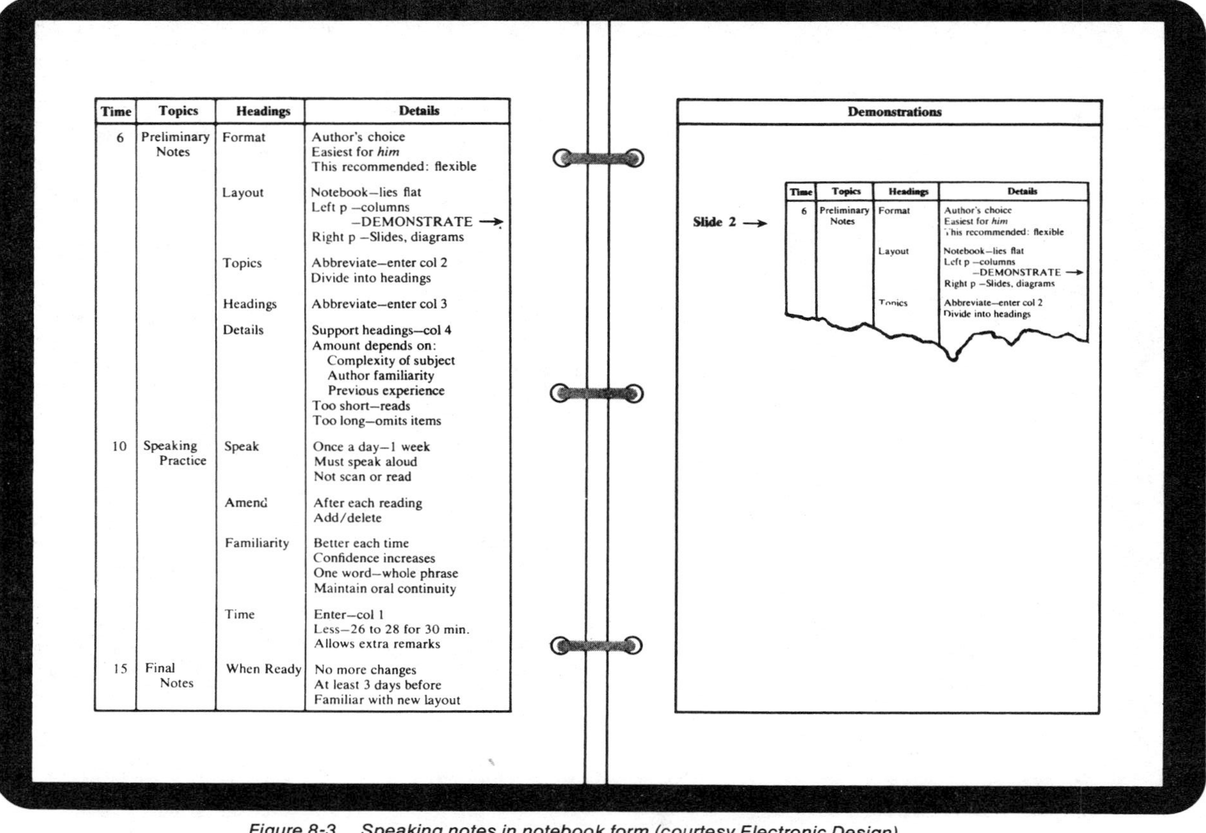

Time	Topics	Headings	Details
6	Preliminary Notes	Format	Author's choice Easiest for *him* This recommended: flexible
		Layout	Notebook–lies flat Left p –columns –DEMONSTRATE → Right p –Slides, diagrams
		Topics	Abbreviate–enter col 2 Divide into headings
		Headings	Abbreviate–enter col 3
		Details	Support headings–col 4 Amount depends on: Complexity of subject Author familiarity Previous experience Too short–reads Too long–omits items
10	Speaking Practice	Speak	Once a day–1 week Must speak aloud Not scan or read
		Amend	After each reading Add/delete
		Familiarity	Better each time Confidence increases One word–whole phrase Maintain oral continuity
		Time	Enter–col 1 Less–26 to 28 for 30 min. Allows extra remarks
15	Final Notes	When Ready	No more changes At least 3 days before Familiar with new layout

Demonstrations

Slide 2 →

Time	Topics	Headings	Details
6	Preliminary Notes	Format	Author's choice Easiest for *him* This recommended: flexible
		Layout	Notebook–lies flat Left p –columns –DEMONSTRATE → Right p –Slides, diagrams
		Topics	Abbreviate–enter col 2 Divide into headings

Figure 8-3. Speaking notes in notebook form (courtesy Electronic Design).

his initial speaking notes and for speaking practice, the other, which he will prepare just before the presentation, for his final speaking notes. He should take his list of topics, abbreviate them as much as possible, and enter them in the "Topics" column of the initial notebook. He should then expand each topic into a series of brief general headings, and enter these in the "Headings" column. Finally, he should support these two columns of cryptic notes with details (information he may need to refresh his memory) which he should enter in the "Details" column. On the right-hand page he should enter comments on the visual aids he plans to use, and key them into the "Details" column at the proper points.

The amount of information he provides in these columns will depend on the complexity of the subject, on Mickey's familiarity with it, and on his previous speaking experience. An experienced speaker familiar with the subject needs less information than an inexperienced speaker such as Mickey. As a general rule, the notes should not be so lengthy that Mickey cannot extract pertinent points at a glance, for then there will again be a tendency to read. Neither should they be so brief that he has to rely too much on his memory, which could cause him to stumble haltingly through his talk.

PRACTICE

Mickey's next task is to practice speaking from the notes at least once a day for several days. It will not be enough simply to scan or read the notes and assume that he is thus becoming familiar with them. He must speak from them aloud, as though presenting the paper to an audience.

Using Preliminary Notes. After each reading he should modify the notes, including more information or deleting unnecessary words. As he grows familiar with the notes his confidence will increase, and certain sentences and apt phrases will spring readily to mind at the sight of a single word or short heading. Thus he will soon find that he can easily maintain oral continuity between headings.

As he rehearses his paper, Mickey should time himself and enter time marks at suitable intervals in the left-hand column. He should aim to speak for less time than the Papers Committee allows; the ideal is to allow 26 to 28 minutes speaking time for a paper scheduled for a maximum of 30 minutes. Then, if he wishes to make some previously unanticipated remarks when he presents his paper (possibly referring to statements in papers presented prior to his), he will have the time.

If he is using visual aids—and he should whenever possible, to insert variety into his presentation—he should practice using them, first on their own and then as part of the whole paper. This will give him a chance to check whether he has keyed them in at the right places, and whether the entries are sufficiently clear to permit him to adjust from speech to visual aid and back again without losing continuity.

Using Final Notes. When he is satisfied that his preliminary notes are satisfactory and will not need any major changes, he can prepare his final speaking notes. These should be typed with a large typeface, or hand-lettered (in ink) in clearly legible capital letters.

Mickey should plan to have these final notes ready at least three days before leaving for the conference. To a certain extent the headings in the first set of notes have helped to trigger familiar phrases and sentences. Now he has to familiarize himself with new pages. During these last practice sessions, he should attempt a full dress rehearsal by presenting the paper before some of his colleagues, among whom there may be someone qualified to comment on his platform techniques. If this is not possible, he should at least try speaking the paper alone, standing at a rostrum or desk to simulate actual conditions. This "dry run" will also give him the opportunity to check his speaking time.

PRESENTATION

Overcoming Nervousness. There are very few speakers who are not at least a little nervous when the time comes to present their paper. Some nervous tension is perfectly normal, and can even help Mickey give a good performance. He will find that once he starts speaking most of this nervousness disappears. A lot will depend on his speaking notes. If he has done his job well, knows that they are reliable, and is confident using them, he will find the familiar phrases and sentences forming easily. Then he will begin to relax, and speak with even greater confidence.

Improving Platform Manner. Mickey was lucky; he was able to learn some elementary platform techniques from Ron Brophy, a member of the Winman electrical engineering staff who has been doing some research into educational training methods (his technical brief on audiovisual programmed learning appears in Chapter 5 (Figure 5-3). Ron was present at Mickey's "dry run" and spent some time afterwards telling him how he could improve his presentation. Here are some of Ron's suggestions, which apply to any speaking situation:

Speak at a moderate rate—120 to 140 words per minute is recommended.

Speak up. If possible, try speaking without a microphone, since this gives you much greater freedom of movement and tonal flexibility. If a microphone has to be used, try to maintain a moderately constant speaking level.

Pause occasionally to study your speaker's notes. Never be afraid to stop speaking for a few moments while consolidating your position and establishing that every major topic has been covered. This also gives you an opportunity to check elapsed time.

Look at your audience. Try to speak to individuals in turn, rather than the group as a whole, picking them out in different parts of the room so that every listener will feel he or she is being addressed personally.

Use humor, but only if it fits naturally into the paper and you are adept at speaking humorously. Make sure that your audience laughs *with* you and not *at* you.

Avoid distracting habits that tend to divert audience attention. Examples are pacing back and forth or balancing precariously on the edge of the platform (the audience will be far more interested in seeing whether you fall off than in following your paper). Nervous afflictions, such as jingling keys or coins in your pocket (put them in a back pocket, out of reach), playing with objects on the speaker's table (remove them before you start speaking), or cracking your knuckles, should also be avoided.

Ron Brophy also pointed out to Mickey that although adequate pre-platform preparation and knowledge of platform techniques can give an author confidence, they are not sufficient in themselves to break down the initial barrier between speaker and audience. Successful speakers and instructors develop a well-rounded personality which they use continuously and unconsciously to establish a sound speaker-audience relationship. For the author of a technical paper who faces an audience once and then only briefly, the most important personality attribute is enthusiasm.

Enthusiasm is contagious. If a speaker likes his subject, really enjoys describing it, his enthusiasm will be demonstrated in his presentation by the vigorous manner in which he tackles his material. If he is also businesslike and cheerful he will quickly reach his audience, who will respond to his approach by listening attentively.

In summary, a technical person who wants to make a good presentation before an audience should:

1. Prepare the material thoroughly.
2. Practice speaking.
3. Learn a few platform techniques.

If you are willing to devote the time and have the interest to try these methods, you will discover that you are speaking *to* your listeners; not *at* them, as you would have done had you simply read your original paper. Even more important, you will win the support and respect of your audience, who will applaud your attempts to speak extemporaneously.

Taking Part in Meetings

We all have occasion to attend meetings. In industry you may be asked to sit on a committee set up for a multitude of reasons, from resolving technical problems that are tying up production to organizing the company's annual picnic. The effectiveness of such meetings is controlled entirely by those taking part. Meetings attended by persons *aware of their role* as participants can move quickly and achieve good results; those attended by individuals who seize the opportunity to air personal complaints can be deadly dull and cripple action. Unfortunately, cumbersome, long-winded meetings are much more common than short, efficient ones.

Meetings can be either structured or unstructured, depending on their purpose. A structured meeting follows a predetermined pattern: Its chairperson prepares an agenda that defines the purpose and objectives of the meeting and the topics to be covered. The meeting then proceeds logically to each point. An unstructured meeting uses a conceptual approach to derive new ideas. Only its purpose is defined, since its participants are expected to introduce suggestions and comments which may generate new concepts (this approach is sometimes known as "brainstorming"). It is the structured type of meeting that you are most likely to encounter in industry, and which I discuss here.

THE PARTICIPANT'S ROLE

You can contribute most to a meeting by arriving prepared, stating clearly your facts, ideas, and opinions when called on, and keeping quiet the remainder of the time. If you observe these three basic rules you will do much to speed up affairs. Let's examine them more closely.

Come Prepared. If a meeting is scheduled to start at 3:00 P.M. do not wait until 2:30 to gather the information you need. Arriving with a sheaf of papers in hand and shuffling through them for the first 25 minutes creates a disturbance and makes you miss much of what is being said. Start gathering information as soon as you know what is required of you, sort it out to identify the items you need, then jot down topic headings and specific data you will have to quote. Take into the meeting only those papers you will need.

Be Brief. In your opening remarks summarize what you have to say, then follow with facts and details. Present only those items your listeners need to know. If you have statistical data to offer, print copies and distribute them when you begin to speak. If you have a lot of information to distribute, ask the chairperson to distribute copies with the agenda so that everyone can look it over before coming to the meeting. Be ready to answer questions and analyze your facts in greater depth, but be sure to keep to the main topic. Finally, address your remarks to the chairperson.

Keep Quiet. There are many parts of a meeting when your role is to be only an interested observer. At these times you should keep quiet unless you have a relevant question, an additional piece of evidence, or an educated opinion. Avoid the annoying habit of always having something to add to the information others are presenting (recognize what others already know: you are not an authority on everything). At the same time, do not withhold information if it would be a genuine contribution. Be ready to present an opinion when the chairperson indicates that a topic should be discussed, but only if you have thought it out and are sure of its validity. Recognize, too, that a discussion should be a one-to-one conversation between you and the chairperson, or sometimes between you and the topic specialist. It should never become a free-for-all with each person arguing a point with his or her neighbor.

THE CHAIRPERSON'S ROLE

Good chairpersons are difficult to find. A good chairperson controls the direction of a meeting with a firm hand, yet leaves ample room for the participants to feel they are making the major contribution. He must be a good organizer, an effective administrator, and a diplomat (to smooth ruffled feathers if opinions differ too widely). Much of the success of a meeting will result from the chairperson's preparation before the meeting starts, and his ability to maintain control as it proceeds.

Prepare an Agenda. Some time before the meeting starts, the chairperson should prepare an agenda of topics to be discussed and circulate it to all committee members. If certain members have specific contributions to make, the agenda should indicate by name who will be presenting the information. A typical agenda might follow the pattern in Figure 8-4, which also reminds committee members of the time and date of the meeting.

Run the Meeting. The chairperson's first responsibility is to start the meeting on time: a person with a reputation for being slow in getting meetings started will encourage latecomers. The second responsibility is to keep the meeting as short as possible without seeming to "railroad" decisions. The third responsibility is to maintain adequate control.

The meeting should be run roughly according to the rules of parliamentary procedure. (Since most inplant meetings are relatively informal, full parliamentary procedure would be too cumbersome.) The chairperson should introduce each topic on the agenda in turn, invite the person specializing in the topic to present a report, then open the topic for discussion. The discussion offers the greatest challenge, for the chairperson must permit a good debate to generate among the members, yet be able to steer a member who digresses back to the main topic. He must be able to sense when a discussion on a subject has gone on too long, and be ready to break in and ask for a decision. Similarly, he must know when strong opinions are likely to block resolution of a knotty problem, and assign a person or subcommittee to investigate further.

The best way to learn to be a good chairperson is to watch others undertake the role. Study those who seem to get a lot of business done without appearing to intrude too much in the decision-making. Learn what you should not do from those whose meetings seem to wander from topic to topic before a decision is made, have many "contributors" all speaking at the same time, and last far too long.

THE SECRETARY'S ROLE

Sometimes a stenographer is brought in to act as secretary and record the minutes of a meeting, but more often the chairperson appoints one of the participants to take minutes. If you happen to be selected, you should know how to go about it.

Recording minutes does not mean writing down everything that is said. Minutes should be brief (otherwise they will not be read), so there is room only to mention the highlights of each topic discussed. Items that must be recorded are: (1) main conclusions reached; (2) decisions made (with, if necessary, the name[s] of the person[s] who made them, or the results of a vote); and (3) what is to be done next and who is to do it. The best way to get this information quickly is to write the agenda topics on a lined sheet of paper, spacing them about two inches vertically. In these spaces jot down the highlights in note form,

H L Winman and Associates

INTER - OFFICE MEMORANDUM

From: R Davis

Date: November 5, 19__

To: A Rittman
B Chansois
G Hyl
M Kevin
D Calaban
D Smithson
B Brewster
A King

Subject: Notice of Meeting -- Children's Christmas Party Committee

The third meeting of the Children's Christmas Party Committee will be held in the large conference room at 15:00 hr on Monday, November 8. The agenda will be:

1. Unfinished business from October 22 meeting.

2. Selection of hall (D Calaban).

3. Purchasing of gifts (A King, A Rittman).

4. Catering arrangements (B Chansois).

5. Entertainment (B Brewster).

6. Other business.

R. Davis

Rick Davis, Chairperson
Children's Christmas Party Committee

Figure 8-4. A typical agenda for a meeting.

leaving room to write in more information from memory immediately after the meeting.

The completed minutes should be distributed to everyone present, preferably within 24 hours. They should be a permanent record on which the chairperson can base the agenda for the next meeting (if there is to be one), and participants can depend for a reminder of what they are supposed to do. I like the format shown in Figure 8-5, which provides an "action" column to draw participants' attention to their particular responsibilities.

Attending an Interview

Chapter 3 discussed how you should prepare a letter of application and a resume or personal biography. Here are a few pointers on how to present yourself during an employment interview.

PREPARING FOR THE INTERVIEW

The key to a good interview is preparation. If you have prepared yourself well, the interview will most likely run smoothly and you will present yourself confidently.

As soon as you are invited to attend an interview—or, better still, before you are called—start researching facts about the company (or organization, if it is a government establishment). Presumably, you will have done some research before submitting your letter of application. Now you need to identify additional information, such as the number of persons the company employs, specific fields in which it is involved, work for which it is particularly well known, its major products and services, important contracts it has received (news of which has been released to the media), locations of branch offices, and the company's involvement in community activities. Such knowledge can be extremely useful during the interview, because it permits you to ask intelligent questions at appropriate places—questions which indicate to the interviewer that you have done your homework.

You also need to prepare for difficult questions an interviewer may pose to test your readiness for the interview and the sincerity of your application. You may be asked:

Why do you want to join our organization?

Why do you want to leave your present employer? (Asked only of persons who are already employed.)

Why did you leave such-and-such a company on such-and-such date? (Asked of persons whose resumes show no explanation for a previous employment termination.)

What salary do you expect?

If lack of preparation for such questions causes you to hesitate too long before answering, the interviewer may interpret your hesitation to mean you find the questions difficult to answer or that there are factors you would rather

H L Winman and Associates

PROFESSIONAL CONSULTING ENGINEERS

CHILDREN'S CHRISTMAS PARTY COMMITTEE

Minutes of Meeting

Main Conference Room, 15:00 hr, November 8, 19__

In attendance:

R Davis (Chairperson)	A Rittman
B Chansois (Secretary)	D Calaban
G Hyl	D Smithson
M Kevin	B Brewster
A King	

MINUTES	ACTION
1. R Davis informed the committee that he had obtained $1200 to spend on the party ($500 from the company and $700 from the social club). It will be divided as follows:	
Children's gifts — $600	A Rittman
Rental and decoration of hall — $175	D Calaban
Entertainment — $200	B Brewster
Catering — $225	B Chansois
2. The hall used last year cannot be used because of recent fire damage. Discussion of an alternative resulted in three choices: Viceroy Lodge, Eldwood Club, and Ramona Room. D Calaban will investigate these and report his findings at the next meeting.	D Calaban
3. Now that funds are available, the gift selection subcommittee will purchase gifts. Lists of employees' children have been updated. A King will be responsible for girls' gifts; A Rittman for boys' gifts. Each will enlist two staff members to help in gift selection, wrapping and labelling.	A King A Rittman
4. Last year's caterers have been contacted and are available on the proposed date. B Chansois will obtain firm price quotations and place the catering order.	B Chansois
5. There will be one hour of entertainment. B Brewster has engaged "Maurice and Mickey" (a ventriloquist/magic act) and will rent a film projector and cartoons.	B Brewster
6. G Hyl volunteered to be Santa Claus.	
7. The meeting adjourned at 16:05. Next meeting: 15:00, Wednesday November 17.	

B Chansois

B Chansois, Secretary

Figure 8-5. Minutes of a meeting.

conceal. In either case, you may inadvertently be providing an entirely misleading impression of yourself.

An interviewer who asks what salary you expect is partly testing your preparation for the interview and partly assessing how accurately you value yourself. For an undergraduate at a university or college, the question is largely academic: undergraduates compare notes and quickly learn what starting salaries are being offered. But for a person who recently has been or currently is employed, the question is important and must be anticipated. Always know the salary you would like to receive and think you are worth. Avoid quoting a salary range, such as "between 15 and 18 thousand dollars," because it seems to indicate unsureness. Quote a definite figure, such as $16,500, and you will sound much more confident.

You should be ready to ask questions during the interview. Just as the interviewer wants to acquire information about you, so should you want to learn things about the company and the opportunities it can offer. Consider what questions you would like answered, and jot them onto a small card to be stored in a convenient pocket or your purse. Then when the interviewer asks, "Now, do you have any questions?" you can pull out the card.

Make the entries on your card brief and clearly legible; ideally, use capital letters. Your list should be short, because you need to scan it quickly and the interviewer won't have time to answer a lot of questions. So limit your choice to the really important topics. Remember, too, that the quality of your questions will demonstrate how carefully you have given thought to the interview.

THE INTERVIEW'S PARTS

When you enter the interview room, the interviewer will have read the documents (letter of application, resume, application form) you submitted previously, and so will have some knowledge of you. You may be asked, however, to fill in some details orally, partly to set the interview in motion, and partly to refresh the interviewer's memory. This occurs particularly when you are being interviewed by several persons, who jointly are called an interview board. Usually, one member of the board has read your information thoroughly, but the others may have had time for only a quick glance at it just before you enter the room. For them, some repetition of your background can be useful.

An interview normally falls into three fairly easy-to-distinguish parts. The initial part is an exchange of pleasantries between yourself and the interviewer, who wants you to be at ease. To help you adjust to the interview environment, he or she may ask questions on topics you can answer confidently, such as a major news item or something from the hobbies and interests section in your resume. This initial part of the interview normally is short.

In comparison, the middle part of the interview is quite long. During this part the interviewer tries to find out as much as possible about you. He or she will want to hear your opinions and have you demonstrate your knowledge on certain topics. The interviewer will want to control the direction the interview takes but will expect you to develop your answers and to comment on each topic in sufficient depth to establish that you have real knowledge and experience, backed up by well-thought-out opinions.

The closing portion of the interview also is short. The interviewer will ask if you have questions to ask and will discuss details about the company and employment with it. By this stage the interviewer should have a pretty good impression of you, and you should know whether you want to be employed by the company he or she represents.

PRESENTING YOURSELF WELL

An interview should be a reasonably informal experience during which interviewer and applicant exchange views, talk about the position, and discuss the applicant's background. Both parties should be reasonably relaxed so that a maximum of good information is exchanged. Most often, however, an applicant's nervousness may inhibit the effectiveness of his or her presentation.

Here are some pointers to help you overcome that initial nervousness, and so present yourself well during an interview:

1. Try to appear comfortable and confident, even though the opposite may be true. Enter the interview room purposefully, and shake hands firmly. A firm handshake shows that you are a confident, definite person; a limp, weak, wishy-washy handshake seems to say you are a nervous individual rather unsure of yourself. Which impression do you want to convey?
2. When invited to sit, make yourself comfortable. Sit well back in the chair rather than on the edge. Let your hands relax comfortably in your lap. Try to relax.
3. As in a more formal speaking situation, concentrate on eye contact. Look at and talk directly to the interviewer. If you are being interviewed by more than one person, address each person in turn. If a particular person asks a question, direct your reply to and look at that person as you reply. But when you have questions to ask, address them to the board chairperson.
4. Control your voice carefully, and make sure everyone can hear you. Speak at a moderate speed, carefully thinking out your answers. Don't be afraid to let your enthusiasm for a topic show.
5. Give well-developed answers. Even though a question may seem to require only a yes or no answer, try to include additional factors to give your answers greater depth. On the other hand, don't try to run away with the conversation and steer it only in a direction that suits you.
6. Don't be afraid to ask questions, but have a clear idea of what you want to ask before you pose them. The interviewer will recognize a good question and the clarity of thought behind it.

And now for the answers to four questions job applicants frequently ask about interviews:

Should I smoke during the interview?
Yes, if invited to do so. Take your cue from the interviewer: if he or she smokes, then you certainly should be able to do so. Look for an ashtray on the desk as a clue to whether the interviewer really accepts smoking.

Should I bring in extra materials to show to the interviewer?
Generally no, unless you have a specific piece of information which you want to add to your job application or resume. To bring in samples of reports and technical papers you have written, with the intention of showing them as examples of your work, is almost futile because the interviewer will be unable to give them more than a cursory glance. If you have documentation of important work that the interviewer might ask about, then carry it with you in a briefcase or envelope, ready to show if asked for it.

Should I use humor?
If a humorous answer fits exactly into a particular situation, then by all means use it. But be sure that it does suit the situation and that you will not be judged as being sarcastic rather than humorous. Humor used in the right place at the right time can be useful; humor used to "warm up an audience" can create the wrong impression.

How should I answer questions when I'm not sure of the answer?
Don't bluff. If you don't know the answer to a question, say so. If you are not sure what the interviewer is trying to ask, ask the interviewer to rephrase the question. Or even rephrase the question yourself, and then ask if you have interpreted it correctly.

Finally, try to be yourself. Remember that interviewers want to see the kind of person you really are. If you relax, and answer questions comfortably and purposefully, they will gain a good impression of you. If you try too hard to be the kind of person you think the interviewers want you to be, or to give the kind of answers you think they want rather than the answers you really believe in, they may detect it and judge you accordingly.

Assignments

Speaking situations you are likely to encounter in industry will develop from projects on which you are working. Hence assignments for this chapter are assumed to grow naturally out of the major writing assignments presented in other chapters.

TECHNICAL BRIEFINGS

Many of the projects in Chapters 4 to 7 offer opportunities for you to brief a client, management, or other members of your department on the results of a technical investigation. Projects that particularly suit oral reporting are:

Chapter	*Project*	
4	11	Coping with a Noisy Neighbor
5	1	Evaluating Microfilm Readers

5	2	Choosing a Site for an Auto-Mart
5	3	Testing Highway Marking Paint
5	5	To Buy or to Lease?
6	3	Individual versus Bulk Metering of Apartment Blocks
6	5	Stormwater Disposal, Cayman Flats
6	6	Examining the Effects of Aircraft Noise
6	7	Installing Dial-a-Wash in the United States and Canada
6	10	A Minicomputer for Wesman Distributors?
7	11	Researching a New Manufacturing Material or Process

MEETINGS

Some of the projects in Chapters 4, 5, and 6 also offer excellent opportunities for group participation. If you are assigned one of these projects, try tackling it this way:

1. Install a class manager to coordinate the project (he or she should be voted into office).
2. Hold a meeting to discuss the project, and assign different individuals or teams to undertake specific aspects. Appoint a secretary to keep minutes, and set up a schedule for the project.
3. Let each team research its part of the project separately.
4. During the research or investigation phase, hold a meeting for the teams to report progress and compare ideas.
5. At the end of the research phase, hold a meeting for the teams to report their initial results.
6. Give the teams time to write their final reports and to prepare technical briefings.
7. Let the class manager and secretary write a management report that outlines the project as a whole and summarizes the teams' general findings.
8. Hold a briefing for the teams to present their findings orally (to individuals defined by your instructor). Let the class manager act as chairperson and introduce the speakers.

Projects that particularly suit group participation are:

Choosing a Site for an Auto-Mart (Chapter 5, Project 2), and **Testing Highway Marking Paint** (Chapter 5, Project 3). For these two assignments the participants will have been manipulating the same information. After individual choices have been made, hold a meeting to discuss the individual choices and to arrive at a group decision for the best choice.

Building a Toothpick Tower (Chapter 6, Project 1). Here, you can hold two meetings:

1. After construction but before testing, when each participant can describe his or her design, discuss why it was chosen, and predict the maximum load it will sustain.
2. After testing, when each participant can describe how his or her tower failed and discuss what should have been done to prevent premature failure.

Individual versus Bulk Metering of Apartment Blocks (Chapter 6, Project

3). After initial evaluation of the data has been made by each participant, hold a meeting to discuss:

1. Which is the better metering method.
2. The effectiveness of energy awareness programs, and whether they influence the results.
3. The abnormalities and what causes them.

Decide what factors should be discussed in the report.

Installing Dial-a-Wash in the United States and Canada (Chapter 6, Project 7). For this project, small teams can be dispatched to different parts of the city to research suitable sites. When their research is complete and reports have been written, hold a meeting to discuss the relative suitability of the various sites:

1. Have each team describe its selected site in turn and tell why it has selected the site.
2. Discuss the merits of all sites presented, and try to reach a consensus as to which site offers the best location to recommend to the Dial-a-Wash company.

A Minicomputer for Wesman Distributors?(Chapter 6, Project 10). When individual or group research is complete, have each individual (or group) present his/her findings at a meeting. Open the meeting to discussion. Try to decide which is the best minicomputer/microprocessor and peripherals to recommend to Wesman Distributors, or whether it is better to recommend the company does not invest in a minicomputer.

INTERVIEWS

Assume that you have applied for an advertised position (it may be one of the advertisements at the end of Chapter 3, or an advertisement you have seen locally), and that you have been asked to attend an interview. Write down four pertinent questions you would want to ask the interviewer. Explain *why* you would ask each question, and tell the kind of answer you hope it will elicit.

Be sure to indicate which position you have applied for, if it has been drawn from Chapter 3, or attach a copy of the job advertisement if you obtained it locally.

9

ILLUSTRATING TECHNICAL DOCUMENTS

If you open any well-known technical magazine you will immediately notice that illustrations are an integral part of most articles. Some are photographs that display a new product, a process, or the result of some action; others are line drawings that illustrate a new concept; some demonstrate how a test or an experiment was tackled; another group may consist of charts and graphs that illustrate progress or show technical data in an easy-to-visualize form.

Good illustrations are not limited solely to magazine articles. They serve an equally useful purpose in technical reports, where their primary role is to help readers understand the topic. Interesting illustrations attract a reader's eye and encourage him or her to read a report. They can also break up dull-looking pages of narrative that lack eye appeal.

This chapter discusses the types of illustrations seen most often in technical reports, indicates the overlaps that exist between types, and suggests occasions when they can be used most beneficially.

Graphs

Graphs are a simple means for showing a change in one function in relation to a change in another. A function used frequently in such comparisons is time. The other function may be temperature, erosion, wear, speed, strength, or any of many factors that vary as time passes.

Suppose, for instance, that technician John Greene wants to find how long it takes a newly painted manometer case to cool down after it comes out of the drying oven. He goes to the paint shop armed with a stopwatch and a special thermometer. When the next manometer case comes out of the oven he starts taking readings at half-minute intervals, and records the results as in Table 9-1.

These readings are part of a study he is undertaking into the cooling rate of

Table 9-1 Cooling Rate, Manometer Case MM–7

Time Elapsed (min:sec)	Temperature (deg C)	Time Elapsed (min:sec)	Temperature (deg C)
:30	152.9	5:30	43.9
1:00	123.4	6:00	41.1
1:30	106.7	6:30	38.9
2:00	91.2	7:00	37.2
2:30	77.8	7:30	35.6
3:00	69.5	8:00	34.5
3:30	61.7	8:30	33.4
4:00	55.6	9:00	32.8
4:30	51.5	9:30	31.7
5:00	47.3	10:00	31.1

Ambient temperature 22.8°C Oven temperature 180°C

different components manufactured by Robertson Engineering Company. The information will also be used by the production department to establish how long manometer cases must cool before assemblers can start working on them with bare hands (the maximum bare-hand temperature has been established by management/union negotiation to be 38°C).

A quick inspection of this table shows that the temperature of the case drops continuously, is within 8.3 degrees of the ambient temperature after ten minutes (*ambient* means "surrounding environment"), and is down to the bare-hand temperature after seven minutes. A much closer examination determines that the temperature drops rapidly at first, then progressively more slowly as time passes.

Single Curve. John Greene can make the data he has recorded in Table 9-1 much more readily understood if he converts it into the graph in Figure 9-1. Now it is immediately evident that the temperature drops very rapidly at first, then slows down until the rate of change is almost negligible. There is no point in measuring or showing further drops in temperature unless John wants to demonstrate how long it takes the component to cool right down to the ambient temperature (probably 30 minutes or more).

Multiple Curves. In Table 9-2 John compares the temperature readings he has recorded for the cooling manometer case with measurements he has taken under similar conditions for a cover plate and a panel board. This time, however, he simplifies the table slightly by showing the temperatures at one-minute intervals.

What can we assess from this table? The most obvious conclusion is that in ten minutes the cover plate has cooled down less than the manometer case, and even less than the panel board. We can also see that the initial rate at which the components cooled varied considerably (the panel board, very quickly; the manometer case, fairly quickly; the cover plate, seemingly quite slowly), but it is difficult to assess whether there were any *changes in rate of cooling* as time progressed.

This can be shown more effectively in a graph, which John Greene has

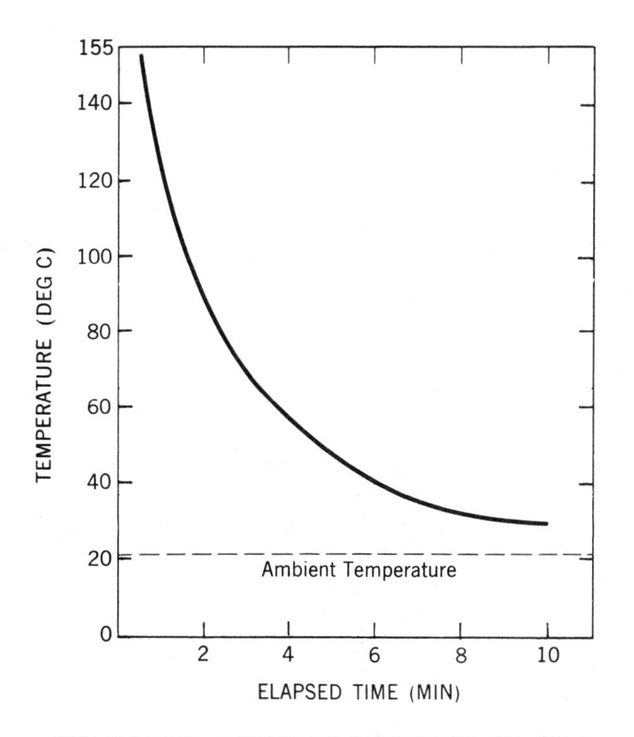

COOLING RATE – MANOMETER CASE MODEL NO. MM-7

Figure 9-1. Graph with a single curve.

plotted in Figure 9-2. The rapid initial drop in temperature is evident from the initial steepness of the three curves, with each curve flattening out to a slower rate of cooling after two to four minutes. The difference in cooling rates for the three components is much more obvious than in the table.

Graphs will be essential when John Greene presents this data in a future report. If his intent is to present a general description of temperature trends, his narrative can be accompanied only by graphs like these, or by charts. But if he

Table 9-2 Cooling Rates for Three Components

Time Elapsed (minutes)	*Temperature (deg C)*		
	Cover Plate	*Panel Board*	*Manometer Case*
0:30	154.5	145.1	152.9
1	136.7	97.3	123.4
2	112.3	67.8	91.2
3	95.1	51.7	69.5
4	82.3	42.3	55.6
5	71.7	36.1	47.3
6	63.9	32.2	41.1
7	55.6	29.5	37.2
8	49.5	27.2	34.5
9	43.4	26.1	32.8
10	38.9	25.0	31.1

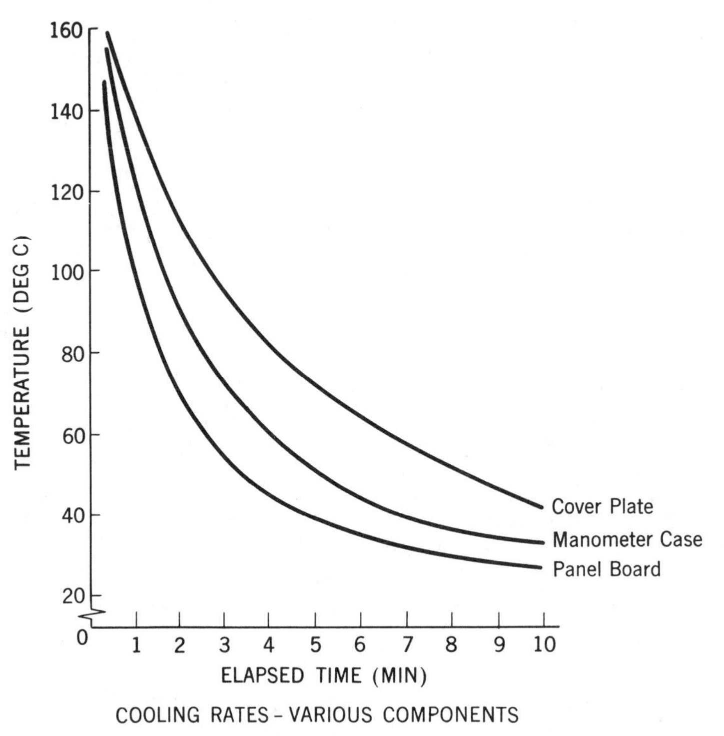

Figure 9-2. Graph with multiple curves.

also wants to discuss exact temperature at specific times for each material, then the narrative and graphs will have to be supported by figures similar to those in Table 9-2. Factors that both you and John should consider when preparing graphs, charts, and tables are outlined below.

Constructing a graph usually offers no problems to technical people because they recognize a graph as a logical means for conveying statistical information. But constructing a graph that also tells a story is an aspect they may easily overlook. In order to illustrate your reports with graphs designed to emphasize the right information, you must first know the tools you have to work with.

Scales. The two functions to be compared in the graph are entered on two scales: a horizontal scale along the bottom and a vertical scale along the left-hand side. (On large graphs the vertical scale is sometimes repeated on the right-hand side to simplify interpretation.) The scales meet at the bottom left-hand corner, which normally—but not always—is designated as the zero point for both. The curved lines in Figures 9-1 and 9-2 are the actual graphs; they are known as curves even though they may be straight lines or a series of short straight lines joining points plotted on the graph.

The functions represented by the two scales are commonly known as the dependent and independent variables, so named because a change in the dependent variable *depends* on a change in the independent variable. This can be demonstrated best by an example. If I wanted to show how the fuel consumption of my car increases with speed, I would enter speed as the independent variable along the bottom scale, and fuel consumption as the dependent variable along the left-hand side (see Figure 9-3). Fuel consumption *depends* on speed (or, if you prefer, speed *influences* fuel consumption); the speed does not depend on the fuel consumption. The same applies to John Greene's temperature measurement graphs: temperature is the dependent variable because it depends on the *time that has elapsed* since the components came out of the oven (the independent variable).

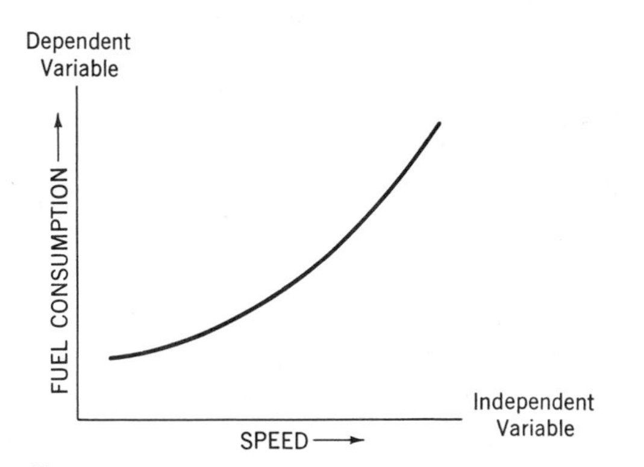

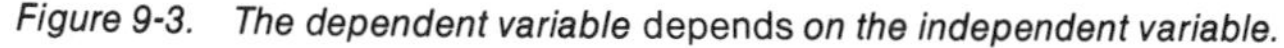

Figure 9-3. The dependent variable depends *on the independent variable.*

When you construct a graph, the first step is to identify which function should form the horizontal scale and which the vertical scale. Table 9-3 lists some typical situations which show that the same function (e.g. temperature) can be an independent variable in one situation and dependent in another. Selection of the independent variable depends on which function can be more readily identified as influencing the other function in the comparison.

The second factor to consider is scale interval. Poorly selected scale intervals, particularly scale intervals that are not balanced between the two variables, can defeat the purpose of a graph by distorting the story it conveys. Suppose John Greene had made the verticle scale interval of his time vs. temperature graph in Figure 9-1 much more compact, but had retained the same spacing for the horizontal scale. The result is shown in Figure 9-4(a). Now the rapid initial decrease in temperature is no longer evident; indeed, the impression conveyed by the curve is that temperature dropped only moderately at first, remained almost constant for the last three minutes, and will never drop to the ambient temperature. The reverse occurs in Figure 9-4(b), which shows the effect of compressing the horizontal scale: now the curve seems to say that temperature plummets downward and it will be only a minute or two until the ambient temperature is reached. Neither curve creates the correct impression, although technically the graphs are accurate.

I said earlier that both scales normally start at zero, which would be the case when the resulting curve is comfortably balanced in the graph area. If it is crowded against the top or right-hand side, then a zero starting point is

Table 9-3 Identifying Dependent and Independent Variables

Graph Illustrates	*Dependent Variable (vertical scale)*	*Independent Variable (horizontal scale)*
1. The effect that frequency has on the gain of a transducer	Gain	Frequency
2. How attendance at a ball game varies with temperature	Attendance	Temperature
3. How much a motor's speed affects the noise it produces	Noise	Speed
4. The changes in temperature brought about by changes in pressure	Temperature	Pressure
5. How much an increase in payload reduces an aircraft's range by limiting the amount of fuel it can carry	Aircraft range (or fuel load)	Payload
6. How much increasing the fuel load of an aircraft to achieve greater range reduces its effective payload	Payload	Fuel load (or aircraft range)

Note: A function can be either dependent or independent, depending on its role in the comparison (see temperature in examples 2 and 4, and both functions in examples 5 and 6).

unrealistic. In Figure 9-1 the curve occupies the top 75% of the graph area. Since it is obvious that no points will ever be plotted below the ambient temperature (which will hover around 23°C), the bottom portion of the vertical scale is unnecessary. This can be corrected by starting that scale at a higher value (say 20°, as in Figure 9-5), or by breaking the scale to indicate that some scale values have been omitted (Figures 9-2 and 9-6).

A graph should be easy to read. Multiple-curve graphs should not have too many curves or they will be hard to interpret. If a graph contains more than three curves, it is probably getting too crowded, particularly if the curves cross each other. You can help a reader identify the most important curve by making it heavier than the others (Figure 9-6), and can differentiate among curves that cross by using different symbols for each (Figure 9-7):

Most important curve	—	bold line
Next most important curve	—	light line
Third curve	—	dashes
Least important curve	—	dots

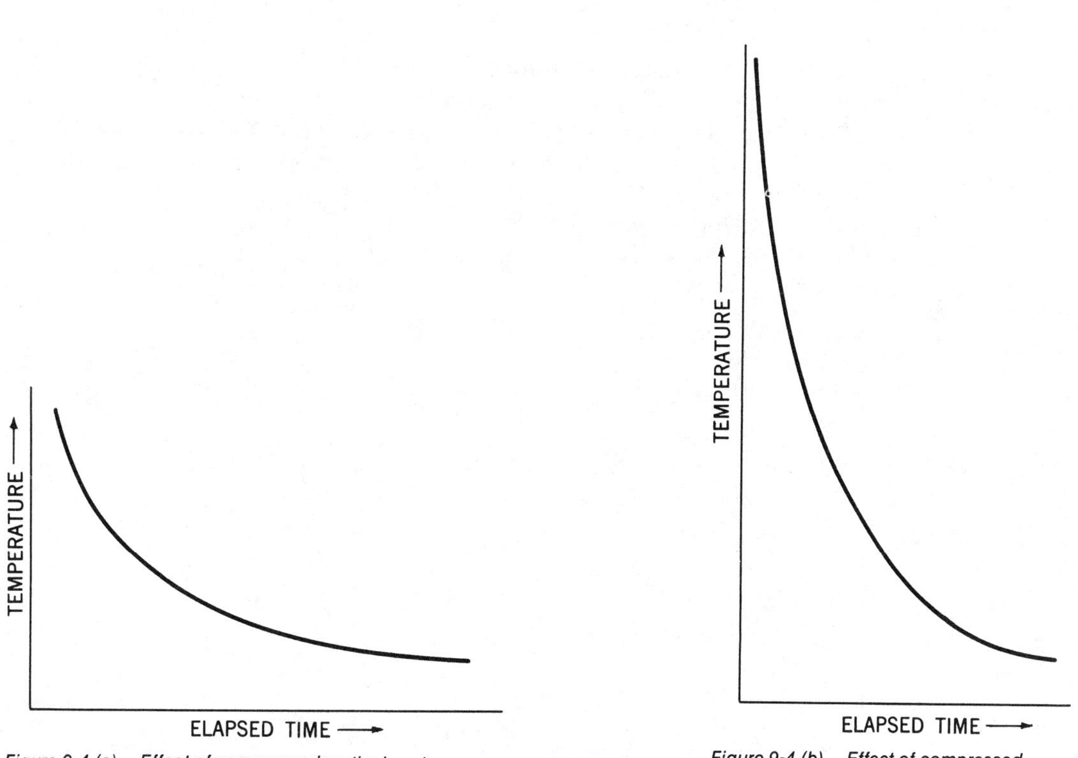

Figure 9-4 (a). Effect of compressed vertical scale.

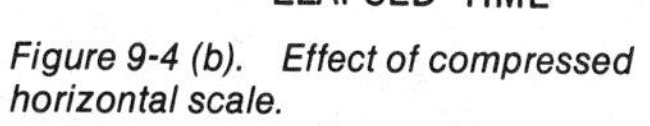

Figure 9-4 (b). Effect of compressed horizontal scale.

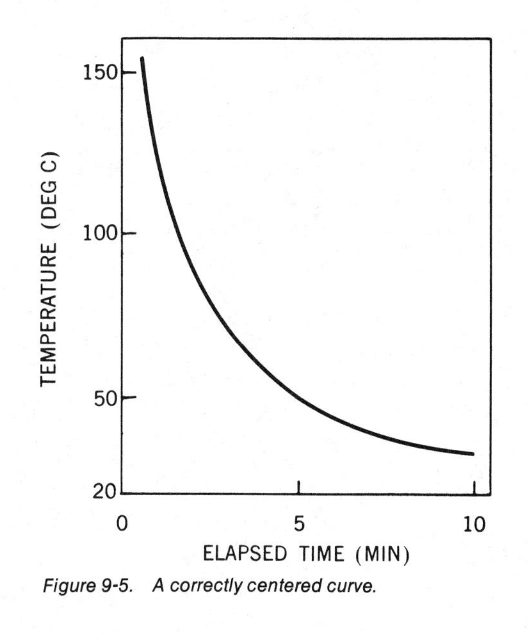

Figure 9-5. A correctly centered curve.

Simplicity. Simplicity is important in graph construction. If a graph illustrates only trends or comparisons, and the reader is not expected to extract specific data from it, then a grid is not necessary. But if the reader will want to extrapolate quantities, a grid must be included. In the graphs we have examined, Figure 9-5 has no grid, Figure 9-1 has an implied grid that only suggests the grid pattern, and Figures 9-6 and 9-7 have full grids so that quantities can be extracted from them. Graphs without grids do not need top and right-hand borders (Figure 9-3).

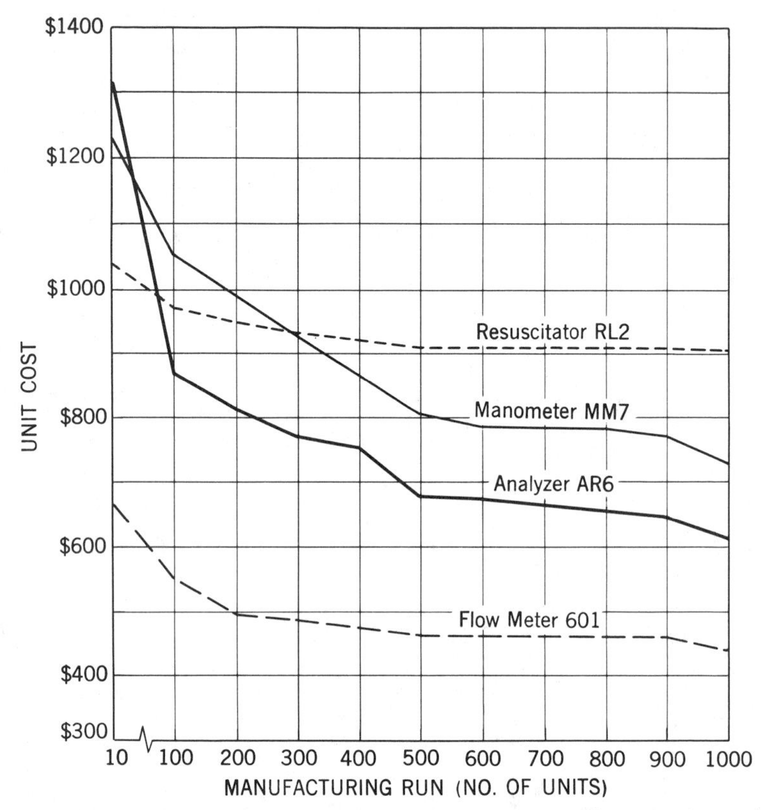

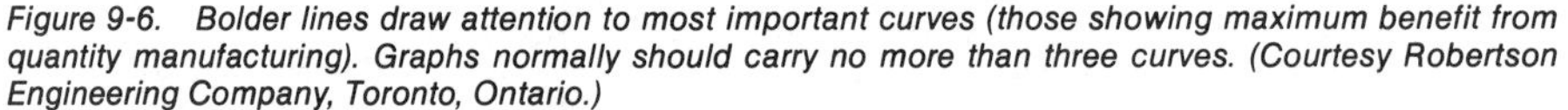

Figure 9-6. Bolder lines draw attention to most important curves (those showing maximum benefit from quantity manufacturing). Graphs normally should carry no more than three curves. (Courtesy Robertson Engineering Company, Toronto, Ontario.)

Plot points should be omitted and all captions should be horizontal. Captions for the curves should appear at the end of the curve whenever possible (Figure 9-2), or, alternatively, above or below the curve (Figures 9-6 and 9-7). They should never be written along the slope of the curve. The only caption that may be entered vertically is the identification for the vertical scale function.

Charts

Most charts show trends or compare only general quantities. Although it can be said that many graphs fall within this definition, I choose to separate them from

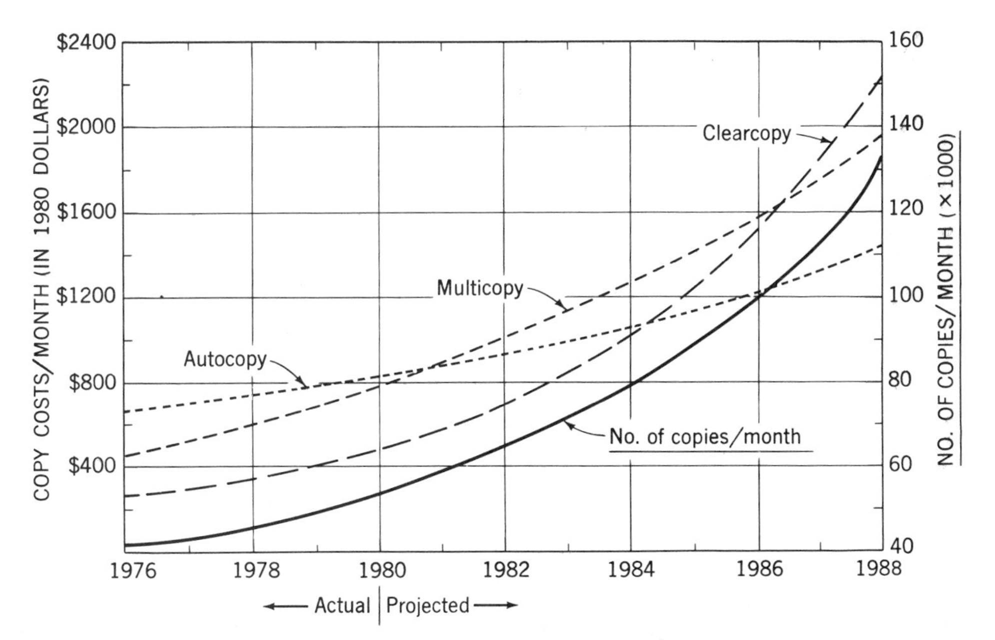

Figure 9-7. Different symbols distinguish between curves showing current and projected copying costs for three copiers. Note the two vertical scales, which permit three functions to be shown on one graph. (Courtesy H. L. Winman and Associates, Cleveland, Ohio.)

charts because graphs have the ability to provide accurate interpretation, whereas charts generally do not. Hence, charts are more often seen in reports, technical articles, and papers intended for a general readership.

BAR CHARTS

You can use bar charts to compare functions that do not necessarily vary continuously. In the graph in Figure 9-1, John Greene plotted a curve to show how temperature decreased continuously with time. He could do this because both functions were varying continuously (time was passing and temperature was decreasing). For the production department, however, he has to prepare a report on how long it takes various components coming from the oven to cool to a safe temperature for bare-hand work. He prepares a bar chart to depict this because he knows the report will be read by both management and union representatives, and some of the readers may need easy-to-interpret data. He also has only one continuous variable to plot: elapsed time. The other variable is noncontinuous because it comprises the various components he has tested. In this case elapsed time is the dependent variable, and the components the independent. The bar chart John constructs is shown in Figure 9-8.

Scales for a bar chart can be made up of such diverse functions as time, age groups, heat resistance, employment categories, percentages of population, types of soil, and quantities (of products manufactured, components sold, oscilloscopes in use, and so on). Charts can be arranged with either vertical or horizontal bars depending on the type of information they portray. The bars

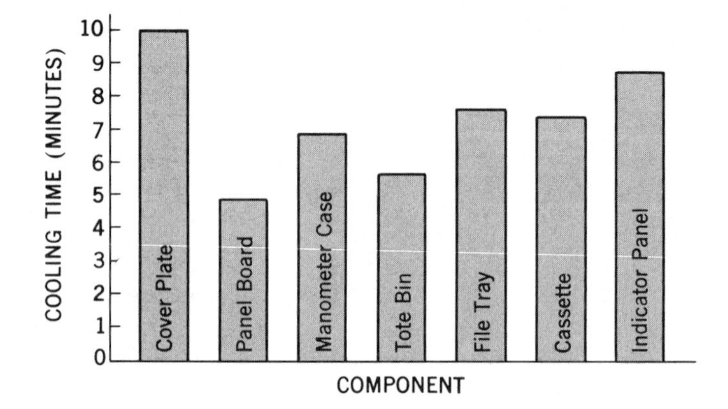

Figure 9-8. Vertical bar chart with one continuous variable (cooling time).

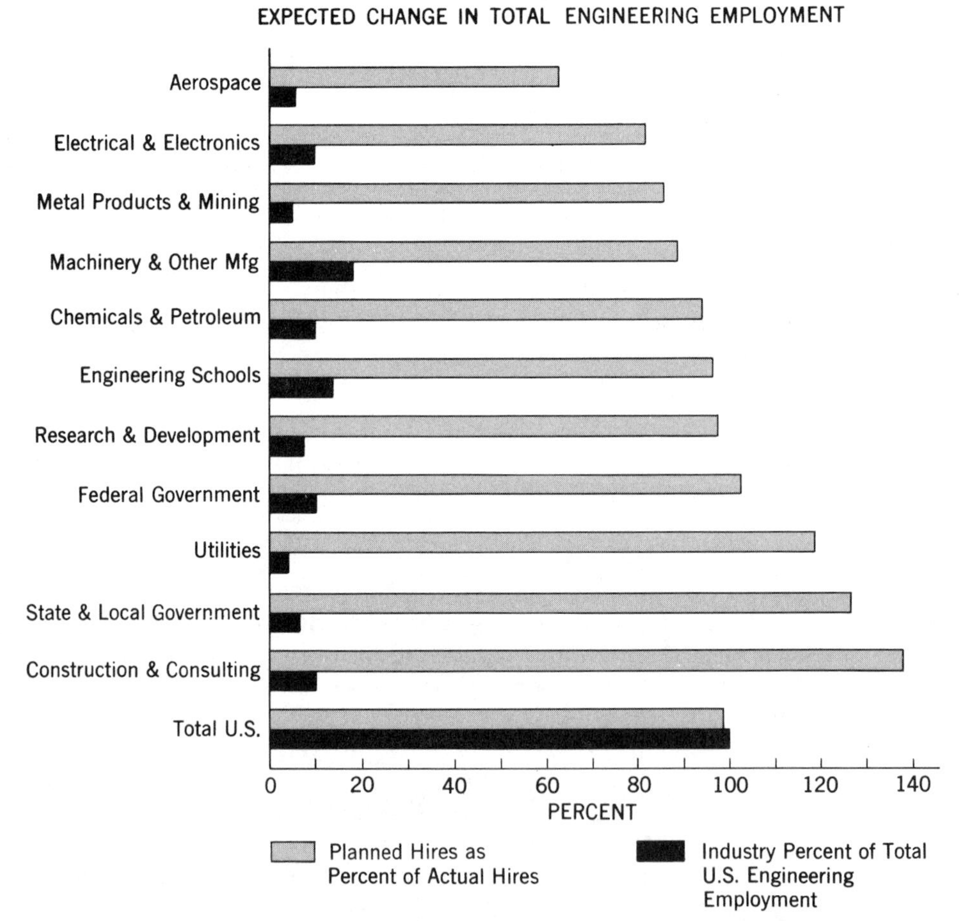

Figure 9-9. This horizontal bar chart does three things: it shows the percentage of engineers employed in major industries, it predicts hirings, and it demonstrates the reduction in employment for engineers in the aerospace industry. (Courtesy Machine Design.*)*

normally are separated by spaces the same width as each of the bars.

In a complex bar chart, the bars may be shaded or colored to indicate comparisons within each factor being considered. The horizontal bar chart in Figure 9-9 uses two shades to describe two factors on the one chart, and has a legend to help the reader identify what each shade represents. Sometimes individual bars can be shaded to show proportional content, like those shown in Figure 9-10.

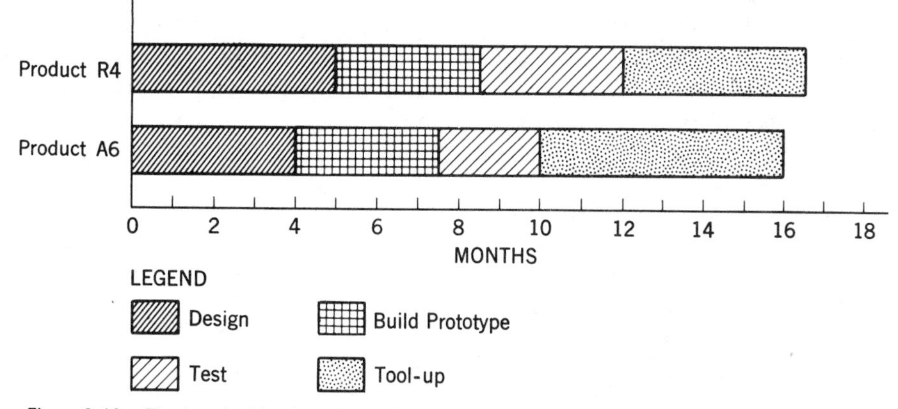

Figure 9-10. The bars in this chart show development times for proposed new products. The legend is included with the chart.

Horizontal bar charts can be used in an unconventional way when information can be arranged naturally on either side of a zero point, as when comparing negative and positive quantities, satisfactory and defective products, or passed and failed students. The chart in Figure 9-11 divides products returned for repair into two groups: those that are covered by warranty, and

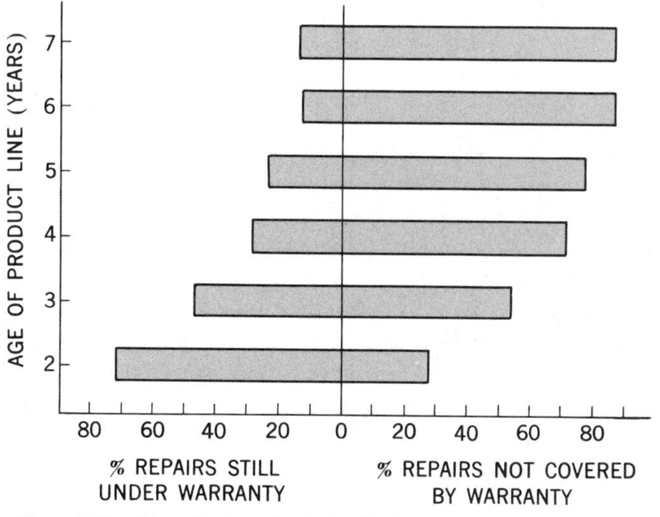

Figure 9-11. A bar chart constructed on both sides of a zero point.

those that are not. Each bar represents 100% of the total number of items repaired in a particular product age group and is positioned about the zero line depending on the percentage of warranty and nonwarranty repairs.

HISTOGRAMS

A histogram looks like a bar chart, but functionally it is similar to a graph because it deals with two continuous variables (functions that can be shown on a scale to be increasing or decreasing). It is usually plotted like a bar chart because it does not have enough data on which to plot a continuous curve (see Figure 9-12). The chief visible difference between a histogram and a bar chart is that there are no spaces between the bars of a histogram.

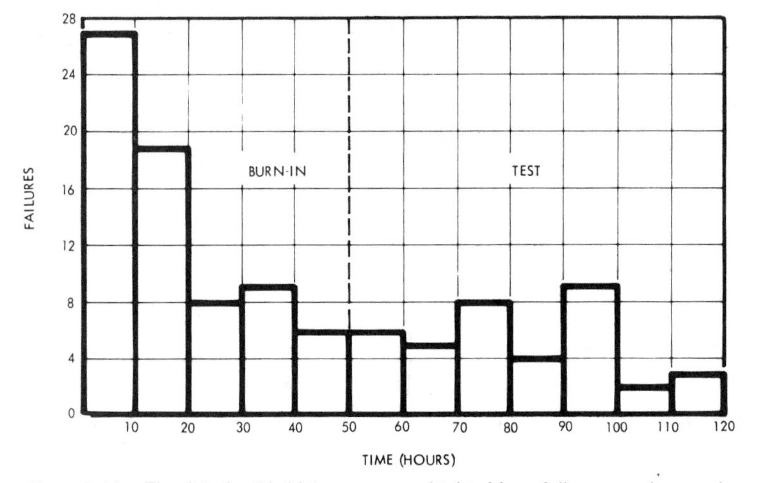

Figure 9-12. The data for this histogram was obtained from failure records on only 14 units. Considerably more data would have been required to construct a curve. Courtesy IEEE [Proceedings, 1968 Symposium on Reliability] and the authors [H. S. Minner and H. A. Romero], General Dynamics, Fort Worth, Texas.)

SURFACE CHARTS

A surface chart (Figure 9-13) may look like a graph, but it is not. Its construction may seem so awkward that a technical person might wonder when it would be necessary to use one. Yet as a means for conveying information pictorially to nontechnical readers, it can serve a very useful purpose.

Like a graph, a surface chart has two continuous variables that form the scales against which the curves are plotted. But unlike a graph, individual curves cannot be read directly from the scales. The uppermost curve on a surface chart shows the *total* of the data being presented. This curve is achieved as follows:

1. The curve containing the most important or largest quantity of data is drawn in first, in the normal way. This is the Hydro curve in Figure 9-13.
2. The next curve is drawn in above the first curve, using the first curve as a base (i.e. "zero") and adding the second set of data to it. For example, the energy resources shown as being variable in 1980 are:

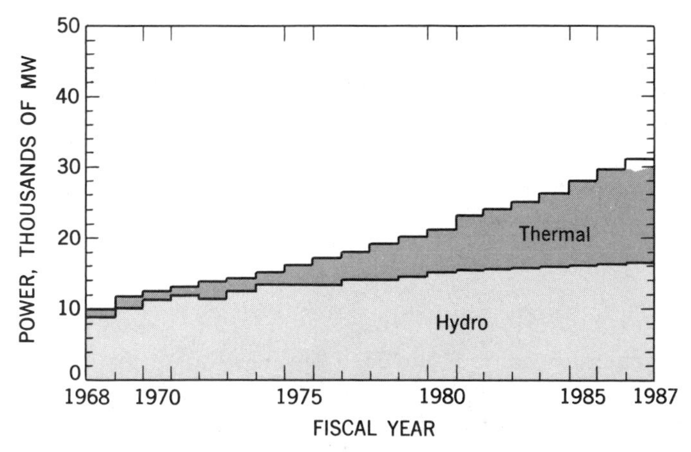

Figure 9-13. Surface chart adds Thermal data to Hydro data to show total firm energy resources in the Pacific Northwest area of U.S. (Courtesy IEEE Spectrum.*)*

Hydro: 15,000MW
Thermal: 7,000MW

In Figure 9-13, the lower curve for 1980 is plotted at 15,000MW. The 1980 data for the next curve is 7,000MW, which is added to the first set of data so that the second curve indicates a *total* of 22,000MW. (If there is a third set of data, it is added on in the same way.)

The area between curves is shaded to indicate that the curves represent the boundaries of a cumulative set of data. Normally, the lowest set of data has the darkest shade, and each set above it is progressively lighter.

PIE CHARTS

A pie chart is aptly named, because it looks like a whole pie viewed from above with cuts in it ready for people of varying appetites. It is a pictorial device for showing approximate divisions of a whole unit. The pie chart in Figure 9-14

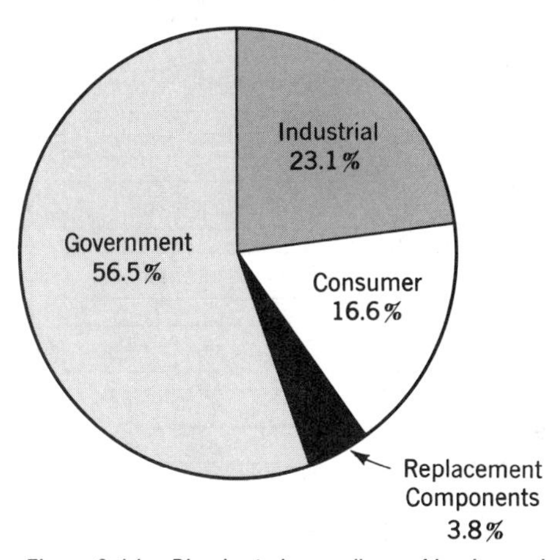

Figure 9-14. Pie chart shows slices of business done in major product categories. (Courtesy IEEE Spectrum.*)*

depicts the percentage of electronic equipment manufactured in four major product categories.

If a pie chart has a lot of tiny wedges which would be difficult to draw and hard to read, some of them are combined into a larger single wedge and given a general heading, such as "miscellaneous expenses," "other uses," or "minor effects." All the wedges must add up to a whole unit, such as 100%, $1.00, or 1 (unity).

Diagrams

Under this general heading is included any illustration that helps the reader understand the narrative yet does not fall within the category of graph, chart, or table. It can range from a schematic drawing of a complex circuit to a simple plan of an intersection. I must add one restriction: If included in the narrative part of a report, it must be clear enough to read easily. This means that complex drawings should be placed in an appendix and treated as supporting data.

Diagrams should be simple, be easy to follow, and contribute to the story. They can comprise organization charts (see Chapter 2), flow diagrams (Figure 9-15, and Figures 4-10 and 5-1), site plans (Figure 9-16), and sketches.

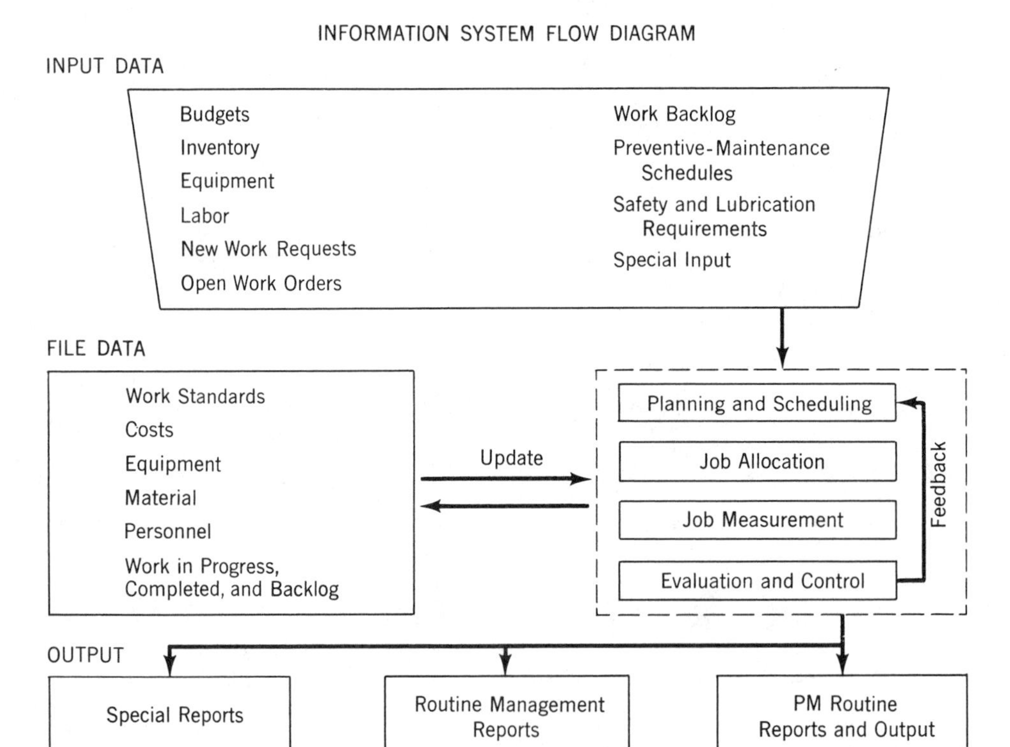

Figure 9-15. Flow diagram of a Maintenance Information System. (Courtesy Mechanical Engineering.*)*

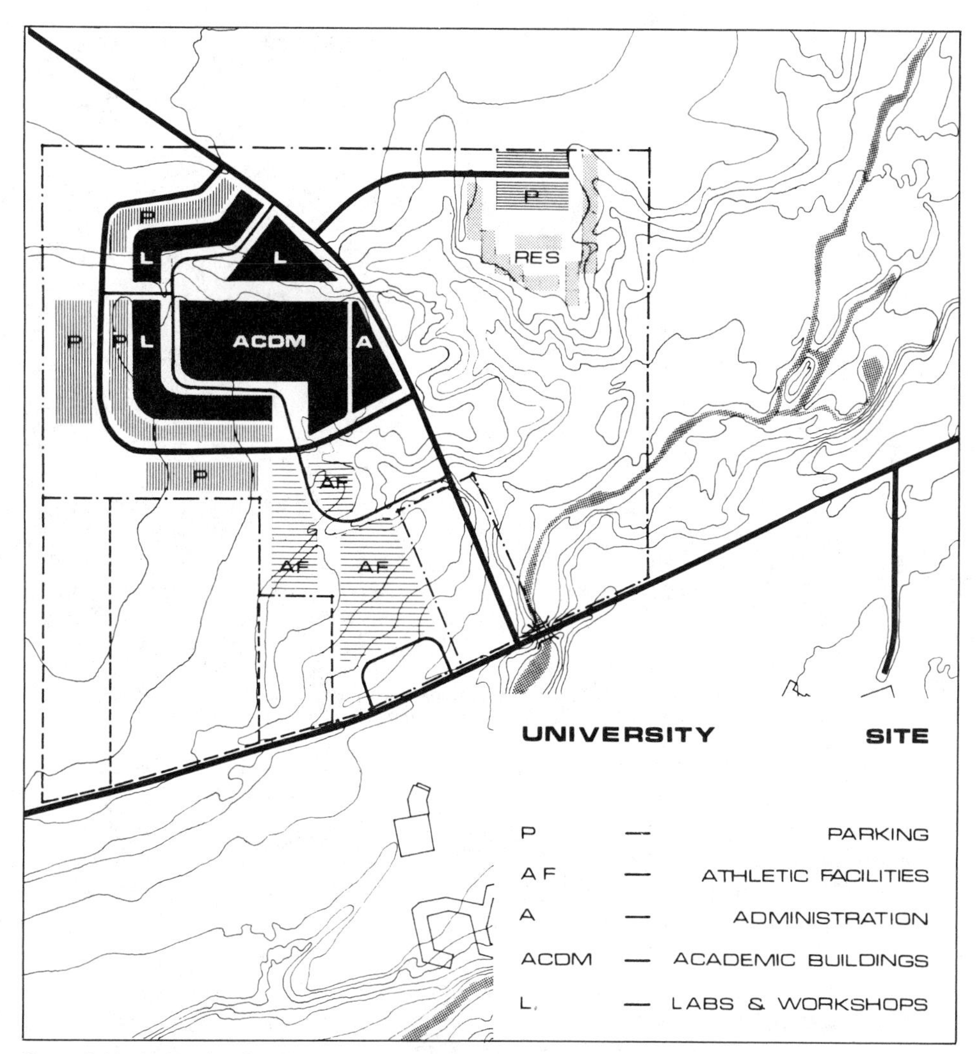

Figure 9-16. A site plan that illustrates where college facilities are to be located. (Courtesy Smith, Carter, Searle—W. L. Wardrop & Associates Ltd., Winnipeg, Canada.)

Photographs

A photograph can do much to help a reader visualize shape, appearance, complexity, or size. For example, the photographs of Harvey Winman, Wayne Robertson and Anna King add depth to the narrative description of the two engineering companies in Chapter 2.

The criterion when selecting a photograph is that it be clear and contain no extraneous information that might distract the reader's attention. The photographs in Figure 9-17 show two views of the same equipment. The right-hand view offers only a general impression of the whole tape recording unit. In the

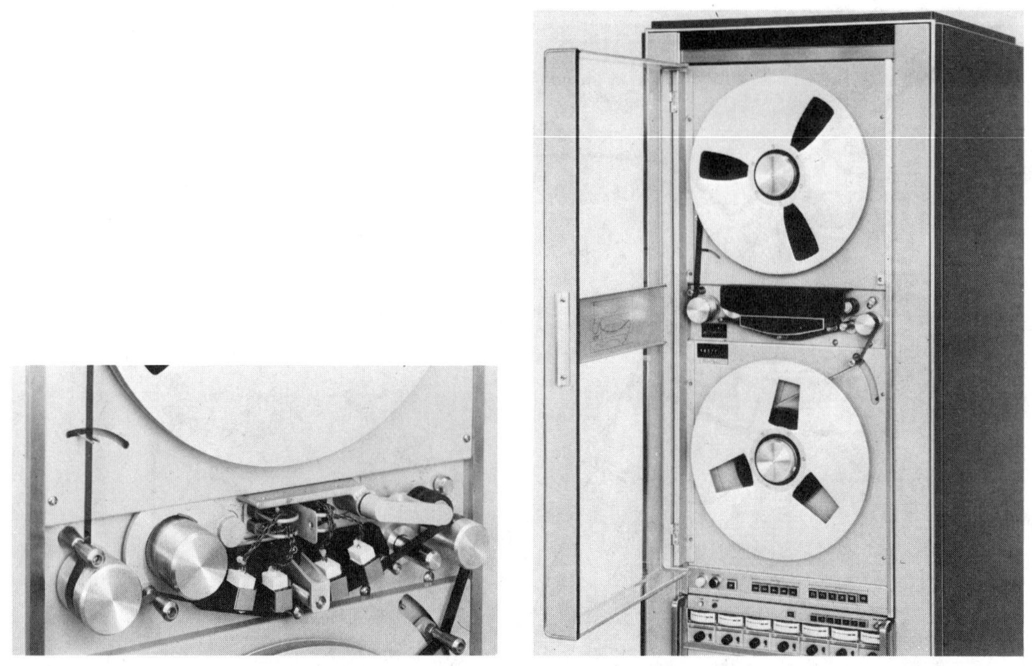

Figure 9-17. Close-up photograph (left) shows details of tape drive; more distant view (right) shows general appearance of HP-3950 Tape Recorder. (Courtesy Hewlett-Packard Co., Palo Alto, California.)

left-hand view, the cover over the tape drive system has been removed and the photographer has zoomed right in to show details of the open-loop tape drive on the transport. The latter photograph is much more useful to users of the equipment.

Photographs are more difficult to copy and print than drawings. For really clear reproduction, they need professional services often beyond the capability of an inplant printing department. Sometimes a photograph will lose so much essential detail that it fails to support a report effectively. Before deciding to use photographs you should find out if your printer can reproduce them clearly and economically. If the printer cannot handle them easily, you may have to glue individual copies of the photographs into your report (an awkward process), have copies printed professionally, or replace the photographs with good sketches.

Tables

The criterion for inserting a table into a report is whether the reader will need to refer to it. If it is going to be used as the report is read, then it should be included in the discussion. If the reader will be able to understand the report without referring to it, but may want to consult it later, it should be included as an appendix. If the information in the table can be expressed more simply by words, a graph, or a chart, then the table should be omitted.

A table inserted as an illustration should be as short as possible so that it can be read easily. If the information compiled during a series of tests is lengthy, the essentials should be summarized and built into a short table, as has been done for Table 1 on page 201, with a reference there to the main body of information, which is in the appendix. Similarly, it should have as few columns as possible, and each column should contain only data that the reader will need.

From the design viewpoint, I prefer a table that is "open," that is, without lines between the vertical and horizontal columns. (For comparison, Tables 9-1 and 9-2 are open, while Table 9-3 is closed.) The captions at the head of each column should be clear and specific and should include units of measurement (e.g. decibels, volts, seconds) so that the units need not be repeated throughout the table. This has been done in Tables 9-1 and 9-2.

It is not enough simply to insert a table and assume that the reader will know what to infer from it. The narrative should refer to the table and comment on its reason for being there. This entry draws the reader's attention to a specific area of the table:

> The voltage fluctuations were recorded at ten-minute intervals and entered in column 3 of Table 7, which shows that fluctuations were most marked between 8:15 and 11:20 A.M.

Alternatively, a similar comment can be inserted as a note beneath the table.

So far we have assumed that tables contain only technical information, such as the results of tests. As Table 9-3 shows, this is not necessarily true. Frequently, the best way to show comparative data or summaries of analyses is to insert informative abstracts or comments in a table, as has been done at the end of Fred Stokes's evaluation report (Chapter 5) and in Appendix A of the Montrose Residential Teachers' College site evaluation (Report No. 1 in Chapter 6).

Mechanical Considerations

POSITIONING

Not only must the narrative refer to every illustration in a report, but each illustration should be on the same page as or facing the narrative it supports. A reader who has to keep flipping pages back and forth between narrative and illustrations will soon tire, and your reasons for including the illustrations will be defeated.

When reports are printed on only one side of the paper, full-page illustrations can become an embarrassment. The only feasible way to place them conveniently near the narrative is to print them on the back of the preceding page, facing the words they support. But this in turn may pose a printing problem. A more logical solution is to limit the size of illustrations so that they can be placed beside, above, or below the words, and then to make sure that the stenographer who is typing the report keys them in at the right points.

Horizontal full-page illustrations may be inserted sideways on a page but

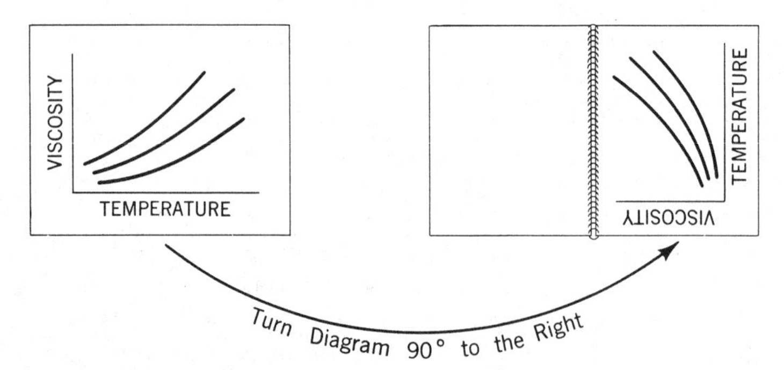

Figure 9-18. Page-size horizontal drawings should be positioned so they can be read from the right.

must always be positioned so that they are read from the right (see Figure 9-18). This holds true whether they are placed on a left- or right-hand page.

When an illustration is too large to fit on a normal page, or is going to be referred to frequently, you should consider printing it on a foldout sheet and inserting it at the back of the report. If the illustration is printed only on the extension panels of the foldout, the page can be left opened out for continual reference while the report is being read. This technique is particularly suitable for circuit diagrams and flow charts.

If the equipment used to print your reports cannot reproduce large fold-out sheets, you can prepare the illustrations on tracing paper and make blueline prints of them. An ideal size is 11 × 22 (28 × 56 cm), which folds conveniently into a standard 8½ × 11 inch (21.5 × 28 cm) report, as shown in Figure 9-19. Such a fold-out can be cut from a standard size C (17 × 22 inches; 43 × 56 cm) sheet of blueprint paper. Even larger fold-outs can be made in this way, but they tend to be unwieldy.

PRINTING

Always discuss printing methods with the person who will be making copies of your report *before* you start making reproduction copy. Certain reproduction equipment cannot handle some sizes, materials, and colors, and few can reproduce photographs clearly.

ILLUSTRATING A TALK

An illustration for a talk must be utterly simple. Its message must be so clear that the audience can grasp it in seconds. An illustration that forces an audience to read and puzzle out a curve detracts from a talk rather than complements it.

If you want to convert any illustration to a large visual aid or slide, you will have to observe the following limitations:

1. Make each illustration *tell only one story.* Avoid the temptation to save preparation time by inserting too much data on one chart. Be prepared to make two or three simple charts in place of a single complex one.

(a) Fold-out sheet opened out for reading

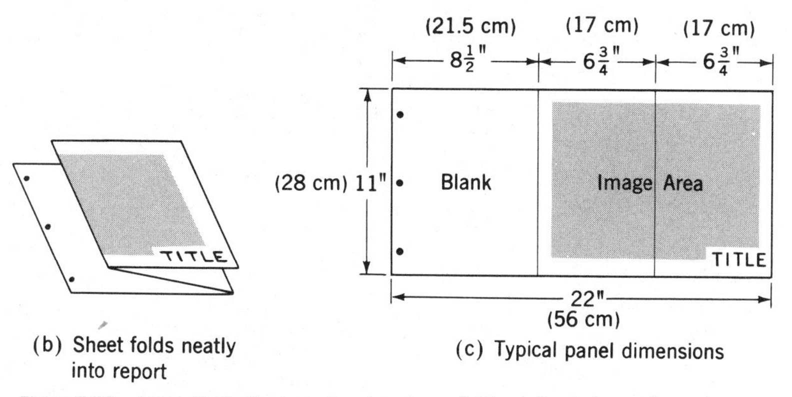

Figure 9-19. Large illustrations can be placed on a fold-out sheet at rear of report.
(a) Fold-out sheet opened out for reading.
(b) Sheet folds neatly into report.
(c) Typical panel dimensions.

2. Use bold letters large enough to be read easily by the back row of your audience.
3. Use very few words, and separate them with plenty of white space.
4. Give the illustration a short title.
5. Insert only the essential points on a graph. Let the curves tell the story, rather than be buried in construction detail.
6. Accentuate key figures and curves with a bold or colored pen (but remember that from a distance some colors look very similar).
7. Avoid clutter—a simple illustration will draw attention to important facts, whereas a busy one will hide them.

Diagrams that are to illustrate an inplant briefing may be hand-lettered with a felt pen on large sheets. But those to be used for a technical paper presented before a large audience should be prepared more professionally. In the first instance you are expected to do a workmanlike job at no great expense. In the second, you are conveying an image of yourself and the company you represent; in effect, your diagrams are demonstrating the technical quality of your company's products.

Assignments

PROJECT NO. 1: WATER CONSUMPTION IN MONTROSE

The table below is a record of the average daily water consumption for the City of Montrose, Ohio, over the past calendar year. The City Engineer has to prepare two reports in which he will identify how much water has been consumed by different segments of the community, and analyze when and why variations in consumption occurred. He asks you to prepare a graph, chart, or diagram to accompany each of these reports. They are to comprise:

1. An illustration for a technical report that will be read by engineers and technicians involved in water supply and distribution.
2. An illustration for a water consumption analysis that will be sent to all consumers. (This part of the project should be done in conjunction with the writing assignment outlined in Project No. 10 of Chapter 7.)

City of Montrose, Ohio
Average Daily Water Consumption
(assume gallons or liters)

Month	*Business & Industry*	*Private Homes*	*Schools & Colleges*
January	4,256,000	3,608,000	1,910,000
February	4,310,000	3,673,000	2,296,000
March	4,318,000	4,127,000	2,501,000
April	4,325,000	4,980,000	2,507,000
May	4,331,000	5,641,000	2,610,000
June	4,417,000	6,775,000	2,192,000
July	4,484,000	8,926,000	807,000
August	4,491,000	9,681,000	762,000
September	4,369,000	6,810,000	1,418,000
October	4,323,000	5,604,000	2,298,000
November	4,298,000	4,254,000	2,304,000
December	4,243,000	3,672,000	1,641,000

PROJECT NO. 2: NOISE LEVELS AT PROPOSED COLLEGE SITES

Prepare an illustration comparing the variations in noise levels at Bluff Heights, Varsity Valley, and Hartland Point (the three sites considered in the formal report, "Evaluation of Sites for Montrose Residential Teachers' College," in Chapter 6). Base your illustration on the data recorded in Appendix B to the report.

Describe why you selected a particular type of illustration (graph, chart, and so on) to depict this data.

PROJECT NO. 3: NOISE LEVELS AT MANSASK AND ADANAC

Prepare graphs or charts to show how noise levels differ under the four sets of conditions at Mansask Insurance Corporation:

1. When only Mansask is working.
2. When both Mansask and Adanac Novelties (the noisy neighbor) are working.
3. When only Adanac is working.
4. When neither office is working.

Include a "normal" office working noise level. Remember that your illustrations should *compare* the noise levels, not simply show the levels recorded for the four sets of conditions. (For details, refer to Project 11 of Chapter 4.)

PROJECT NO. 4: COST OF AUTOMOBILES USED BY STAFF

Design an illustration to compare the costs of buying, leasing, or reimbursing employees for using their own cars, for the situation described in Project 5 of Chapter 5.

PROJECT NO. 5: ELECTRICITY CONSUMPTION IN APARTMENTS

Prepare illustrations to compare the electricity consumption figures in the tables attached to Project 3 of Chapter 6.

PROJECT NO. 6: EMPLOYMENT PATTERNS

Prepare graphs or charts to compare:

1. Salaries
2. Employment potential

for the different employer categories described in Project 4 of Chapter 6.

PROJECT NO. 7: AIRCRAFT NOISE

Prepare graphs or bar charts comparing the sound levels recorded for the different types of aircraft in Project 6 of Chapter 6.

PROJECTS COMBINED WITH OTHER ASSIGNMENTS

Some of the other major report writing assignments in Chapters 5 and 6 also need good graphic aids if they are to convey their information effectively. Specific projects that can combine illustrating with report writing are:

Testing Highway Marking Paint (Chapter 5, Project 3)
Designing an Electronic Meeting Timer (Chapter 5, Project 6)
Building a Toothpick Tower (Chapter 6, Project 1)
Installing Dial-a-Wash in the United States and Canada (Chapter 6, Project 7)
Building and Testing a Meeting Timer (Chapter 6, Project 9)
A Minicomputer for Wesman Distributors? (Chapter 6, Project 10)

10

THE TECHNIQUE OF TECHNICAL WRITING

This chapter concentrates on a few writing techniques that will enable you to convey information both quickly and efficiently. I assume you are already proficient in grammar or "English," and can recognize and correct basic writing problems. If you need practice in basic writing, I suggest you refer to a textbook such as the *Prentice-Hall Handbook for Writers.* You can also refer to the glossary of terms in Chapter 11 for information on how to resolve many of the problems that crop up in technical writing (how to form abbreviations and compound adjectives, how to spell problem words, when to use numerals or spell out numbers in narrative, and so on).

The Whole Document

Throughout this book I have stressed the need to tell readers immediately what they most want to know. This means structuring your writing so that the very first paragraph (if possible, the first sentence) satisfies their curiosity. Most executives and many technical readers are busy people who have time to read only essential information. By presenting the most important items first, you can help them decide whether they need to read the whole document or whether they should hand it to someone more familiar with the topic.

THE PYRAMID TECHNIQUE

The pyramid technique is so named because it starts at the top with a small morsel of essential information and then substantiates it with a broad base of details, facts, and evidence. In most letters and short reports it has only two stages: a brief *summary* followed by the *full development,* as shown in Figure 10-1(a). In long reports an additional stage, the *essential details* (see Figure 10-1[b]), is inserted between the summary and the full development.

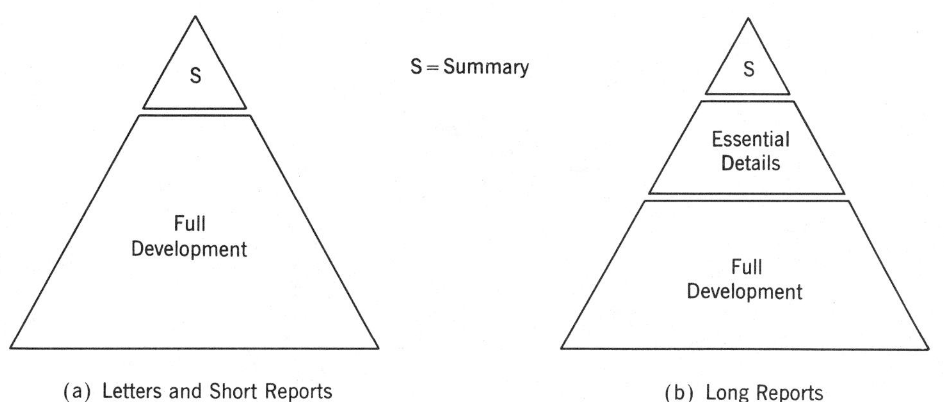

Figure 10-1. The pyramid technique.

Readers normally are not consciously aware of the pyramid technique. They simply find that a document in which it is employed is very easy to follow. In the letter report in Figure 10-2, Stanley Roning summarizes in the first paragraph the two things Wayne Robertson wants to know right away: whether the training course was a success and what results were achieved. He uses the remainder of the letter to fill in background details, to state briefly how the course was run, to report on student participation, and to comment on student reaction.

The pyramid technique requires that the reader know the main elements of the story by the end of the first paragraph. He can then read the full development more intelligently. If he is very busy and the report is long, he may choose to stop reading at the end of the summary and continue with the full development later, or he may scan the full development quickly and pass the report on to someone else for action.

In the full development you tell the whole story. You amplify what you have already stated in the summary by inserting all the technical details readers need to understand the subject fully. You may develop the subject in whatever order you like, although much of the time you will find that a past-present-future order is the simplest and most effective. This in turn may present a problem, since you will have to maintain continuity between the summary (which frequently ends with a comment on future action) and the first paragraph of the full development (which normally starts by saying how the project began).

To make a good transition from the summary *back* to the background information requires skill. It has been done well in Figure 10-2 because the writer begins paragraph two by referring to the training course mentioned in the previous paragraph (he starts the full development by saying: "This was a pilot course. . . ."). Further examples of good transitions can be found in the letter reports in Figures 4-3 and 4-9 (Chapter 4).

There is no problem in making this transition in the long formal report because the summary and the introduction (which contains the background

THE RONING GROUP
COMMUNICATION CONSULTANTS
Box 181 Postal Station "C" Winnipeg, Manitoba R3M 3S7
Tel. 452-6480

14 August 19__

Mr Wayne D Robertson, President
Robertson Engineering Company
600 Deepdale Drive
Toronto, ON
M5W 4R9

Dear Mr Robertson

Results of Pilot Report Writing Course

The report writing course we conducted for members of your engineering staff was completed successfully by 14 of the 16 participants. The average mark obtained was 63%.

This was a pilot course set up in response to a 13 August 19__ enquiry from Mr F Stokes. At his request, emphasis was placed on giving students practical experience in writing business letters and technical reports. Attendance was voluntary, the 16 students being selected at random from 29 applicants.

Best results were achieved by students who recognized their writing problems before they started the course, and willingly became very actively involved in the practical work. A few presumably had expected it to be an "information" type of course, and hence were less willing to take part in the heavy writing program. Our comments on the work done by individual students are attached.

Course critiques completed by students indicate that the course met their needs from a letter and report writing viewpoint, but that they felt more emphasis could have been placed on technical proposals and oral reporting. Perhaps such topics should be covered in a short follow-up course.

We enjoyed developing and teaching this pilot course for your staff, and particularly appreciated their enthusiastic participation.

Sincerely

Stanley G. Roning

Stanley G Roning
President

SGR:ib
enc

Figure 10-2. A letter report written using the pyramid technique.

information) are on on separate pages and are often separated by the table of contents. Thus a technical writer can develop a brief but complete story in the summary without having to consider how to build a transition to the next paragraph. Indeed, the organization of many modern formal reports will assist writers to adapt their writing to the pyramid technique.

In Figure 10-3 the three main sections of the alternative format for the formal report discussed in Chapter 6 are paralleled with the three stages of development demanded by the pyramid technique. The summary of the formal report is the initial stage of the pyramid; the introduction, conclusions, and recommendations are the essential details; and the discussion is the full development. In effect, the formal report written in this way becomes three reports: the brief but informative *summary* written in general terms that can be understood by any reader; the *essential details* of the introduction, conclusions, and recommendations, written for the knowledgeable but not necessarily technical reader; and the *full development* of the discussion, written for the reader who has the interest and technical capability to understand all the evidence the writer has to present.

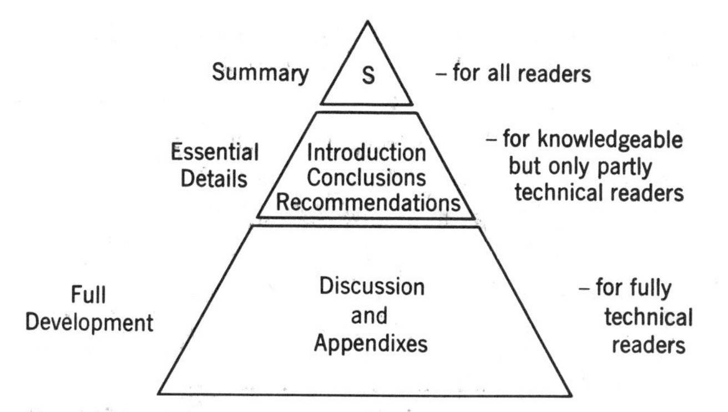

Figure 10-3. Pyramid technique applied to the formal report.

What does this technique mean to the writer of scientific and engineering documents? Principally, it provides the opportunity to cater to more than one level of reader within the same document. This is particularly true of long formal reports, less so of short business letters. But simply adopting the two- or three-stage approach of the pyramid technique is insufficient in itself to ensure good writing. It is a mechanical means to effective writing that must be supported by two less tangible but equally important factors: tone and writing style.

TONE

The fact that it is possible to write for several levels of reader in the same document does not imply that you can omit identifying a specific reader (see Chapter 1). Without this important step you will not be able to set the right tone. The reader you must identify is always the person who will be reading the full development. Thus, you should write primarily for your most technically knowledgeable reader, while directing the essential details and the summary to progressively less technically knowledgeable persons.

Whether your writing should be formal or informal will depend on the situation and your familiarity with the reader. Formal reports should adopt a formal tone. (Note, however, that a formal tone is not stiff or pompous; there is no room for writing that makes readers feel uncomfortable because they are not as knowledgeable as you are.) A formal report conveys an image of your company's reputation even before the first page is opened; the words between the covers must maintain that image. Business letters are generally less formal, depending on their importance. For example, a management-level letter proposing a joint venture on a major defense project would be formal, whereas letters between engineers discussing mutual technical problems would be informal. A memorandum report can be informal, since normally it is an inplant document most often written between persons who know each other.

Varying levels of tone are evident in the following extracts from three separate H. L. Winman and Associates' documents, all written on the same subject.

1. *Extract from a Memorandum.* John Wood's Material Testing Laboratory has compression-tested samples of concrete for Karen Woodford of the Civil Engineering Department. In his memorandum reporting the test results, John writes:

   Informal Tone — I have tested the samples of concrete you took from the sixth floor of Tarryton House and none of them meets the 4800 psi you specified. The first failed at 4070 psi; the second at 3890 psi; and the third at 4050 psi. Do you want me to send these figures over to the architect, or will you be doing it?

2. *Extract from a Letter Report.* Karen Woodford conveys this information to the architect in a brief letter report:

   Semiformal Tone — Our tests of three samples taken from the sixth floor of Tarryton House show that the concrete at 52 days still was 800 psi below your specification of 4800 psi. We doubt whether further curing will increase the strength of this concrete more than another 150 psi. We suggest, however, that you examine the design specifications before embarking on an expensive and time-consuming remedy.

3. *Extract from a Formal Report.* The architect rechecked the design specifications and decided that 4200 to 4300 psi still would not satisfy the design requirements. He then requested H. L. Winman and Associates to prepare a formal report that he could present to the general contractor and the concrete supplier. Karen Woodford's report said, in part:

   Formal Tone — At the request of the architect we cut three 30 × 15 cm diameter cores from the sixth floor of Tarryton House 52 days after the floor had been poured. These cores were subjected to a standard compression test with the following results (detailed calculations are attached at Appendix A):

Core No.	*Location*	*Failed at:*
1	0.46 m W of col 18S	4070 psi
2	0.84 m N of col 22E	3890 psi
3	1.42 m N of col 46E	4050 psi

The average of 4003 psi for the three cores is 797 psi below the design specification of 4800 psi. Since further curing will increase the strength of the concrete by no more than 150 psi, we recommend that this concrete pour be rejected.

Although the information conveyed by these three examples is similar, the tone the writer adopts varies sufficiently to help set the scene for each situation.

STYLE

The pyramid technique has no effect on writing style, other than consideration of the reader at each stage. Complexity of subject and technical level of reader have some effect on style, in that you should use careful, simple language when describing a very complex topic to a moderately knowledgeable reader. As the reader's technical level increases, or the complexity of the subject decreases, you can use longer sentences and words, and more complex sentences. Here are some suggestions:

1. When presenting low-complexity background information, and descriptions of nontechnical or easy-to-understand processes, write in an easygoing style that tells readers they are encountering information that does not require total concentration. Use slightly longer paragraphs and sentences, and insert a few adjectives and adverbs to color the description and make it more interesting.
2. For important or complex data, use short paragraphs and sentences. Present one item of information at a time. Develop it carefully to make sure it will be fully understood before proceeding to the next item. Use simple words. This punchy style will warn readers that the information demands their full attention.
3. When writing instructions or step-by-step descriptions, start with a narrative-type opening paragraph that introduces the topic and presents any information that the readers should know or would find interesting; then follow it with a series of subparagraphs each describing a separate step. These subparagraphs should conform to the following general rules:
   - 3.1 Each should develop only one item of the process.
   - 3.2 They should be short; even single sentence subparagraphs are acceptable, particularly for technical instructions.
   - 3.3 They should be parallel in construction, that is, they should generally conform to the same "shape." The importance of parallelism is discussed later in this chapter.
   - 3.4 Each step of an instruction should start with an active verb (see p. 237 for suggestions).

WRITING SEQUENCE

If you are to set the right tone throughout, you must write in reverse order, starting with the full development. Writing a report in the order in which it will be read is difficult, if not impossible. An engineering technician who writes the summary before the full development will use too many adjectives and adverbs, big words when shorter words would be more effective, and dull opening statements such as "This report has been written to describe the investigation into defective MN-1 compasses carried out by H. L. Winman and Associates." He will be writing without having established exactly what he will say in the full development.

The best writing sequence therefore should be the reverse of that shown in the pyramid in Figure 10-3:

Step 1 Write the full development. Direct it to the type of technical reader who will use or analyze your report in depth.

Step 2 Write the essential details (omitted in short informal reports). Make them brief and direct them to a semitechnical reader or person in a supervisory or managerial position. They should comprise:

a. Introductory information to help the reader understand why the project (and hence the report) was undertaken and what the terms of reference were.
b. The results of the project, or the main conclusions that can be drawn from the full development; this should follow naturally from the introductory information and satisfy the terms of reference.
c. A recommendation (if one is to be made) showing what needs to be done next.

Step 3 Write the summary. Direct it to a nontechnical reader who has absolutely no knowledge of the project or the contents of the report. Make it very brief and make sure it tells:

a. Why the report was written.
b. What was found out.
c. What are the main conclusions and recommendations.

PARAGRAPH NUMBERING

The types of industrial documents that most often carry paragraph numbers are military reports, specifications, and technical instructions. Some companies stipulate that all their reports bear paragraph numbers, but they are in the minority. Their reasons for doing so, however, are valid: paragraph numbers are an excellent aid to organization, and they provide a useful means for cross-referencing (both within the report itself and from one report to another).

A paragraph numbering system must be obvious to the reader, simple for you (the writer) to manipulate, and easy for the typist to arrange on the page. The simplest paragraph numbering system starts at 1 and numbers paragraphs consecutively to the end of the document. More complex systems combine numbers, letters, and decimals to allow for subparagraphing. Three possible arrangements appear on the next page.

Method (A) is a numerals-only decimal system, simple and unambiguous; it is often used by the military services and by specification writers. However, the subparagraphing does become difficult to follow in long documents,

TYPICAL PARAGRAPH NUMBERING SYSTEMS

METHOD (A)	METHOD (B)	METHOD (C)
1.	A.	1.
2.	1.	2.
3.	2.	3.
3.1	(a)	3.1
3.1.1	(b)	3.2
3.1.2	(1)	a)
3.1.2.1	(2)	b)
	B.	(1)
etc., up to	1.	(2)
6 digits	2.	etc.
	etc.	

particularly when indentation cannot be used. (Indentation of multiple-digit subparagraphs can quickly result in a very narrow band of information down the right-hand side of the page.)

Method (B) permits indentation and uses a combination of letters and numerals. Extremely simple, it is popular with many report writers even though it can be ambiguous (there can be several paragraph 1's, 2's, etc., one set under A, another under B, and so on).

The third method, (C), combines the decimal and letter-number arrangements for a system with no ambiguity between main paragraphs and subparagraphs. It is the method chosen by Anna King for H. L. Winman and Associates' reports that carry paragraph numbers. She suggests that engineers use normal-size paragraphs for the first two levels, and short paragraphs for the lower-level subparagraphs:

1. 2.	main paragraphs
2.1 2.2	full subparagraphs
a) b)	short subparagraphs
(1) (2)	very short subparagraphs (i.e. one sentence only)

This paragraph numbering system combines well with the heading arrangements illustrated in Figure 10-4. For a partial example of this system used in an informal report, turn to Figure 5-2 in Chapter 5.

ARRANGEMENT OF HEADINGS

There are three main types of headings you are likely to encounter: center headings, side headings, and paragraph headings. Their use is demonstrated in Figure 10-4.

MAIN CENTER HEADING

SUBSIDIARY CENTER HEADING

The Main Center Heading is always capitalized, and may be underlined. Subsidiary Center Headings (if used) may be capitalized but not underlined, as shown above, or may be in lower case letters and underlined, as shown immediately below.

Subparagraphing Without Paragraph Numbering

Side Headings

Side headings, as the one immediately preceding this paragraph shows, are usually typed in lower case letters (except for the first letter of each major word), and are underlined.

Each paragraph is typed level with the same margin as the side heading. In technical writing, the first line of each paragraph is seldom indented.

Subparagraph Headings and Subparagraphing

Subparagraph headings are indented about five typewriter spaces.

Each subparagraph is typed as a solid indented block. It should not run back to the left-hand margin because that would make the visible evidence of the subparagraphing much less obvious.

Secondary Subparagraphing

If further subparagraphing is necessary, the headings and subparagraphs are indented a further five typewriter spaces.

Heading Built into the Paragraph. In this lesser-used arrangement, the text continues immediately after the heading. Paragraph headings like these can be used for main paragraphs, subparagraphs, and secondary subparagraphs.

Subparagraphing Combined with Paragraph Numbering

1. Side Headings

1.1 Normally, when paragraph numbers are used, side headings are assigned simple paragraph numbers, almost like section numbers, as has been done here.

1.2 Where only a single paragraph follows a side heading or subparagraph heading, it is not assigned a separate paragraph number. But it is typed with its left-hand margin indented slightly, level with the start of the heading above it (see the paragraph immediately following heading 1.3).

1.3 Subparagraph Headings and Subparagraphing

If a single paragraph follows the heading, no number is assigned to the paragraph. If more than one paragraph follows the heading, each is assigned an identification number or letter:

a) This would be the first subparagraph.

b) This would be the second subparagraph.

c) Each subparagraph can be further subdivided into a series of very short subparagraphs:

(1) Here is such a subparagraph.

(2) Preferably, each such subparagraph should contain no more than one sentence.

d) Lower level subparagraphs can also be assigned subparagraph headings.

Figure 10-4. Combined system of headings, paragraphs, and paragraph numbers.

Paragraphs

The role of the paragraph is complex. It should be able to stand alone but normally is not expected to. It must contribute to the document of which it is a part, yet it must not be obtrusive except when called on to emphasize a specific point. It should convey only one idea, although made up of several sentences each containing a separate thought.

Experienced writers construct effective paragraphs almost subconsciously. They adjust length, tone, and emphasis to suit their topic and the atmosphere they want to create. Literary writers, in particular, knowingly stretch and bend the rules of good paragraph construction to obtain exactly the right impact. But even they once had to study and master the techniques of good paragraph writing, although now they let the rhythm of the words guide them far more than the rules.

We are not so fortunate. We have to learn the rules and apply them consciously. But in doing so we must take care not to let our approach become too pedantic, or become so bound by the rules that we write in a stilted manner that is uninteresting and unrhythmic. So when considering the factors designed to help you construct good paragraphs, remember that they are blocks on which to build your writing, not bars to imprison your creativity.

The three elements essential to good paragraph writing are:

Unity
Coherence
Adequate Development

These elements cannot stand alone. All three must be present if a paragraph is to be useful to its reader.

UNITY

For a paragraph to have unity, it must be built entirely around a central idea. This idea is expressed in a topic sentence and developed in supporting sentences. In effect, this permits us to construct paragraphs using the pyramid technique, with the topic sentence taking the place of the summary and the supporting sentences representing the full development (see Figure 10-5).

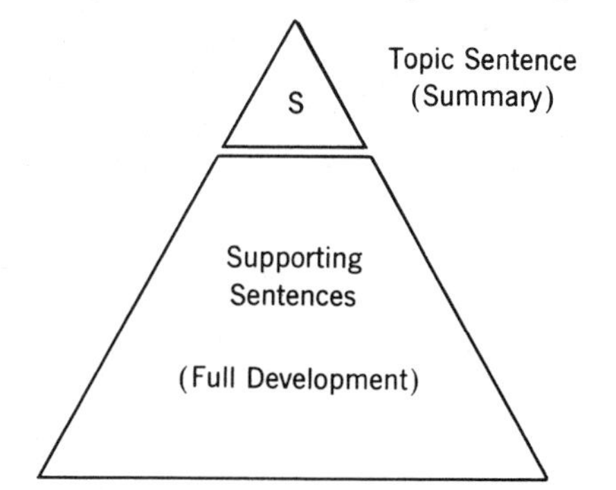

Figure 10-5. Pyramid technique applied to the paragraph.

The paragraph below is strongly unified because its topic is clearly expressed and the supporting sentences develop it fully (the numbers that precede each sentence are for reference):

> (1) *Joey, when we began our work with him, was a mechanical boy.* (2) He functioned as if by remote control, run by machines of his own powerfully creative fantasy. (3) Not only did he himself believe that he was a machine but, more remarkably, he created this impression in others. (4) Even while he performed actions that are intrinsically human, they never appeared to be other than machine-started and executed. (5) On the other hand, when the machine was not working we had to concentrate on recollecting his presence, for he seemed not to exist. (6) A human body that functions as if it were a machine and a machine that duplicates human functions are equally fascinating and frightening. (7) Perhaps they are so uncanny because they remind us that the human body can operate without a human spirit, that body can exist without soul. (8) And Joey was a child who had been robbed of his humanity.[1]

This opening paragraph of a technical article demonstrates that technical writing does not have to be dull. Let's analyze it, identify why it has unity, and discover why it holds our interest and makes us want to read on.

The paragraph is summarized in sentence (1), the topic sentence, which sets the scene in nontechnical terms that create interest. Sentence (2) amplifies the topic sentence by using slightly more technical terms and introducing the fact that Joey's mechanical aspects are self-created. Then sentence (3) underscores the strength of Joey's imagination by telling us that he caused adult onlookers to feel that he really was a machine. This is carried further in sentence (4), which relates his machinelike actions to normal activities, and sentence (5), which shows how very real his machinelike qualities were, even in moments of inactivity. Then sentences (6) and (7) introduce the fascination (and an element of apprehension) that Joey's condition caused in persons observing him. Sentence (8) restates the idea expressed in sentence (1), but this time in human as opposed to mechanical terms.

The topic sentence does not always have to be the first sentence in a paragraph; there are even occasions when it does not appear at all and its presence is only implied. But until you are a proficient writer you would be wise to place your topic sentences right up front, where both you and your readers can see them. Then, when you have experience to support your actions, you can experiment and try placing topic sentences in alternative positions.

The two examples that follow are paragraphs that do not start with a topic sentence. In this descriptive passage on baseball the topic sentence is at the end of the paragraph (compare it with the similarly structured technical paragraph describing how exploration crews affected permafrost, p. 236).

> The reason, obviously, is that baseball came up from the sand lots—the small town, the city slum, and the like. It had a rowdy air because rowdies played it. One of the stock tableaux in American sports history is the aggrieved baseball player jawing with the umpire. In all our games, this tableau is unique; it belongs to baseball, from

[1]From Bruno Bettelheim, "Joey, A 'Mechanical Boy.'"

> the earliest days it has been an integral part of the game, and even in the carefully policed major leagues today it remains unchanged. *Baseball never developed any of the social niceties.*[2]

In this continuing description of Joey, the "Mechanical Boy," the topic sentence is only implied. If it had been stated it would have read something like this: "Joey's machinelike actions were realistic."

> For long periods of time, when his "machinery" was idle, he would sit so quietly that he would disappear from the focus of the most conscientious observation. Yet in the next moment he might be "working" and the center of our captivated attention. Many times a day he would turn himself on and shift noisily through a sequence of higher and higher gears until he "exploded," screaming "Crash, crash!" and hurling items from his ever present apparatus—radio tubes, light bulbs, even motors or, lacking these, any handy breakable object. (Joey had an astonishing knack for snatching bulbs and tubes unobserved.) As soon as the object thrown had shattered, he would cease his screaming and wild jumping and retire to mute, motionless nonexistence.[3]

COHERENCE

Coherence is the ability of a paragraph to hold together as a solid, logical, well-organized block of information. A coherent paragraph is abundantly clear to its readers; they can easily follow the writer's line of reasoning and have no problem in progressing from one sentence to the next.

Most technical people are logical thinkers and should be able to write logical, well-organized paragraphs. But the organization must not be kept a secret; it must be apparent to every reader who encounters the work. Simply summarizing a paragraph in the topic sentence and then following it with a series of supporting sentences does not make a coherent paragraph. The sentences must be arranged in an identifiable order, following a pattern that assists the reader to understand what is being said.

This pattern will depend on the topic and the type of document. Paragraphs describing an event or a process most likely will adopt a sequential pattern; those describing a piece of equipment will probably be patterned on the shape of the equipment or the arrangement of its features. Patterns that can be used for typical writing situations are illustrated in Table 10-1.

Narrative Patterns. You can write narrative-type paragraphs whenever you want to describe a sequence of steps or events. The past-present-future pattern of a progress report or occurrence report, such as Bob Walton's accident report in Figure 4-3 is a typical example. The pattern should be clearly evident, as in two of the following three paragraphs:

A coherent paragraph (in chronological order)

> The accident occurred when D. Friesen was checking in at the Remick Airlines counter. He placed the company Polaroid camera on the counter while he completed flight boarding procedure. When the passenger ahead of him lifted a carry-on bag from the

[2]From Bruce Catton, "The Great American Game," *American Heritage,* April 1959. Copyright © 1959 by American Heritage Publishing Co., Inc. Reprinted by permission.

[3]Bettleheim, "Joey, 'A Mechanical Boy.' "

Table 10-1 Paragraph Patterns Used in Technical Writing

Topics or Subjects			Writing Patterns	Definitions and Examples
Event Occurrence Situation Process Procedure Method	Generally Narrative		Chronological order	The sequence in which events occurred; e.g. the steps taken to control flooding, how a new product was developed, how an accident happened
			Logical order	Used when chronological order would be too confusing (when describing a multiple-activity process, or several events that occurred concurrently); requires careful analysis to achieve clear narrative
			Cause to effect	The factors (causes) leading up to the result (effect) they produce; e.g., dirty air vents and failure to lubricate bearings cause overheating and eventual failure of equipment
			Evidence to conclusion	The factors, data, or information (evidence) from which a conclusion can be drawn; e.g. how results obtained from a series of tests help identify the source of a problem
			Comparison	A comparison of factors (price, size, weight, convenience) to show the differences between products, processes, or methods
Equipment Scene Appearance Building Location Features	Generally Descriptive	By Shape	Vertical	From top to bottom* of a tall, narrow subject
			Horizontal	From left to right* of a shallow, wide subject
			Diagonal	From corner to corner (for items arranged across a subject)
			Circular	From center to circumference* (for items arranged concentrically) Clockwise* (for items arranged around a subject)
		By Features	In order of: Size	From smallest item to largest item*
			Importance	From most important to least important item*
			Operation	The sequence in which the items are used when the equipment is operated

*Or vice versa.

counter, its shoulder strap tangled with the carrying strap of the camera and pulled the camera to the floor. D. Friesen examined the camera and discovered a 1½-inch crack across its back. Remick Airlines' representative K. Trane took details of the incident and will be calling you to discuss compensation.

A much less coherent paragraph (containing the same information but not presented in an identifiable pattern)

The accident occurred when D. Friesen was checking in at the Remick Airlines counter. K. Trane, a Remick Airlines representative, took details of the incident and will be calling you to discuss compensation. The damaged camera received a 1½-inch crack across the back. When the passenger ahead of D. Friesen removed a carry-on bag from the counter, its shoulder strap tangled with the carrying strap of the company Polaroid camera and pulled it to the floor. D. Friesen had placed the camera on the counter while he completed flight boarding procedure.

A coherent paragraph (tracing events from evidence to conclusion)

A mild shimmy at speeds above 52 mph (84 km) was noticed about 10 days after the new tires had been installed. A visual check of all four wheels revealed no obvious defects. Rotation of the four wheels to different positions on the vehicle did not eliminate the shimmy but seemed to change its point of origin. To pin down the cause I replaced each wheel in turn with the spare wheel, and found that the shimmy disappeared when the spare was in the left front position. The wheel removed from that position was tested and found to have been incorrectly balanced.

Descriptive Patterns. You can write descriptive paragraphs to describe scenes, buildings, equipment, and any subject having physical features. The shape of the subject, or the order in which parts are operated, their order of importance, or their arrangement from largest to smallest, will define the pattern. For example:

Excerpt from paragraph (features in order of importance)

The most important control on the bombardier's panel is the firing button, which when not in use is held in the black retaining clip at the bottom left-hand corner. Next in importance is the fusing switch at the top right of the panel; when in the "OFF" position it prevents the bombs from being dropped live. Two safety switches, one immediately above the firing button retaining clip and the other to the right of the bank of selector switches, prevent the firing button from being withdrawn from its clip unless both are in the "UP" position.

Continuity. A fully coherent paragraph must also have smooth transitions between the sentences. Smooth transitions give a sense of continuity that makes readers feel comfortable. As they finish one sentence, there is a logical bridge to the next. This can be accomplished by using linking words and by referring back to what has already been said. In the example just quoted there is a natural flow from "The most important . . ." in the first sentence to "Next in importance . . ." in the second; the third sentence then refers back to the firing button to relate the newly introduced safety switches to what has gone before. The transitions are equally good in the first paragraph describing damage to a Polaroid camera, each sentence containing a component that is a development from one of the previous sentences. This is not true of the second paragraph, in which each new sentence introduces a new subject with no reference to what has already been said.

ADEQUATE DEVELOPMENT

The element of adequate development demands good judgment. You must identify your readers clearly enough so that you can look at each paragraph from their point of view. Only then can you establish whether your supporting sentences amplify the topic sentence in enough detail to satisfy their interest.

Simple insertion of additional supporting sentences does not necessarily meet the requirements for adequate development. The supporting sentences must contain just the right amount of pertinent information, all directed to a particular reader. There must never be too little or too much information. Too little results in fragmented paragraphs that offer snippets of information which arouse readers' interest but do not satisfy their needs. On the other hand, too much information can lead to long, repetitious paragraphs that annoy readers. Compare the following paragraphs, all describing the result of exploration crews' first venture with machinery across the Peel Plateau east of Alaska, intended for a reader who is interested in the problems of working in the north but who has never seen what the terrain is like.

Inadequate development Trails left by tractors look like long narrow scars cut in the plateau. Many of them have been there for years. All have been caused by permafrost melting. They will stay like this until the vegetation grows in again.

Adequate development To the visitor viewing this far northern terrain from the air, the trails left by tractors clearing undergrowth for roads across the plateau look like long, narrow scars. Even those that have been there for as long as 28 years are still clearly defined. All have been caused by melting of the permafrost, which started when the surface moss and vegetation were removed and will continue until the vegetation grows in again—perhaps in another 30 years.

Over-development To the conservationist, whose main interest is the protection of the environment, viewing this far northern terrain from the air is a heartrending sight. To him, the trails left by tractors clearing undergrowth for roads across the plateau look like long, narrow scars. The tractors were making way for the first roads to be built by man over an area that until now had been trodden only by Indians indigenous to the area, and the occasional trapper. Some of these trappers had journeyed from Quebec to seek new sources of revenue for their trade. But now, in the very short time span of 28 years, man has defiled the terrain. With his machines he has cut and gouged his way, thoughtlessly creating havoc that will be visible to those that follow for many decades. Those that preceded him for centuries had trodden carefully on the permafrost, leaving no trace of their presence. The new trails, even those that have been there for 28 years, are visible almost as though they had been cut yesterday. And all were caused by melting of the permafrost. . . .

In these examples the descriptive pendulum has swung from one extreme to the other. The first paragraph leaves the reader with questions: What were the tractors doing? How many years? How soon will the vegetation grow in again? The second paragraph develops the topic sentence in just enough detail; it explains why the tractors left semipermanent scars and predicts how long they will remain. The third paragraph, though interesting, is filled with irrelevant information (e.g. where the trappers came from) and repetitive statements that detract from the main theme. It might be suitable for a novel but not for a technical report or description.

The first paragraph of the article on Joey, the "Mechanical Boy" also was adequately developed because it provides just sufficient information to support the author's original intent: to introduce the boy. It confines itself to a brief word picture of Joey's problem (but not of Joey himself) and its effect on both the boy and the adults who worked with him. Because its readers would be technical men and women who specialize in other technical fields, the paragraph introduces its subject in general terms that will be understood by and provoke interest in a wide range of readers. The author has not described who Joey is, where he is, and what his surroundings are like. Instead, he has started with a thought-provoking topic sentence and used the whole paragraph to develop the image of the "mechanical" child. To have attempted to introduce names, locations, and similar information would have destroyed the impact of this very effective introductory paragraph.

CORRECT LENGTH

How long should your paragraphs be? The answer is simple: no longer than you need to cover a particular topic for a particular reader. But also keep these points in mind:

1. Variety in paragraph length has a lively visual effect. A series of equal-length paragraphs creates the impression of dullness. If the paragraphs are consistently short, readers will feel cheated because they are not presented with enough information. (Too many short paragraphs create a jackrabbit effect—all starts and stops.) Conversely, a succession of very long paragraphs will make readers feel they are swimming in a sea of information in which they cannot readily identify the main topic.
2. Readers attach importance to a paragraph that is clearly longer or shorter than those surrounding it. A very short paragraph among a series of generally longer paragraphs attracts attention, just as a longer paragraph among a series of short paragraphs implies that the information it contains is particularly important.
3. Paragraph length needs to be adjusted to suit the complexity of your topic and the technical level of your readers. Generally, complex topics demand short paragraphs containing small portions of information. But their length should be conditioned by your knowledge of specific readers. Even a complex topic can be covered in long paragraphs for persons who are technically knowledgeable.

Sentences

Although sentences normally form an integral part of a larger unit—the paragraph—they still must be able to stand alone. While helping to develop the whole paragraph, each has to develop a separate thought. In doing so they play an important part in placing emphasis—in stressing points that are important and playing down those that are not.

Just as experienced writers first had to learn the elements of good paragraph writing, so they also had to learn the elements of good sentence construction. These are:

Unity
Coherence
Emphasis

Unity and coherence perform a similar function in the sentence to their role in a well-written paragraph.

UNITY

Although the comparison is not quite so clearly defined, we can still apply the pyramid technique to the sentence in the same way it is applied to the paragraph and the whole document. In this case the summary is replaced by a primary clause that presents only *one thought,* and the full development by subsidiary clauses and phrases that develop or condition that thought (see Figure 10-6). Thus a unified sentence, as its name implies, presents and develops only a single thought.

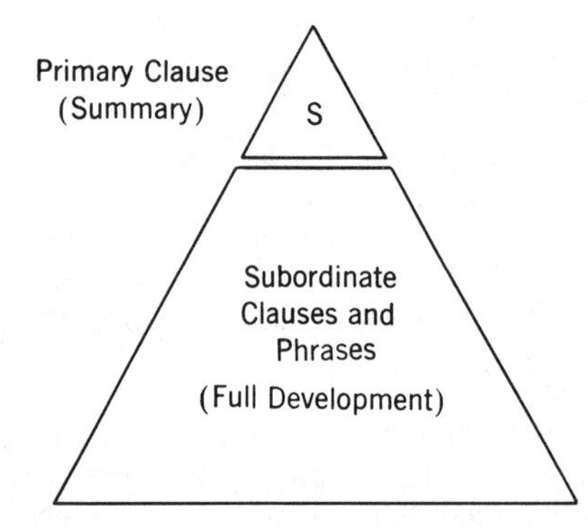

Figure 10-6. Pyramid technique applied to the sentence.

Compare these two sentences:

A unified sentence that expresses one main thought	The Amron Building will make an ideal manufacturing plant because of its convenient location, single-level floor, good access roads, and low rent.
A complicated sentence that tries to express two thoughts	The copier should never nave been placed in the general office, where those using it interrupt the work being done by the stenographic pool, which has been consistently understaffed since the beginning of the year.

The first sentence has unity because everything it says relates to only one topic: that the Amron Building will make a good manufacturing plant. The second sentence fails to have unity because readers cannot tell whether they are supposed to be agreeing that the copier should have been placed elsewhere, or sympathizing that there has been a shortage of stenographers. No matter how complex a sentence may be, or how many subordinate clauses and phrases it may have, every clause must either be a development of or actively support only one thought, expressed in the primary clause.

COHERENCE

Coherence in the sentence is very similar to coherence in the paragraph. Coherent sentences are continuously clear, even though they may have numerous subordinate clauses, so that the message they convey is apparent throughout. Like the paragraph, they need good continuity, which can be obtained by arranging the clauses in logical sequence, by linking them through direct or indirect reference to the primary clause, and by writing them in the same grammatical form. Parallelism, discussed at the end of this chapter, plays an important role here.

The unified sentence discussing the Amron Building has good coherence because its purpose is continuously clear and each of its subordinate clauses links comfortably back to the lead-in statement (through the phrase "because of its . . ."). The clauses are written in parallel form (i.e. they have the same grammatical pattern), which is the best way to carry the reader smoothly from point to point. The first sentence below lacks coherence because there is no logic to the arrangement or form of the subordinate clauses. Compare it with the second sentence, which despite its length is still coherent because it continuously develops the thought of "late" and "damaged" expressed in the primary clause.

An incoherent sentence The Amron Building will make an ideal manufacturing plant because of its convenient location, which also should have good access roads, the advantage of its one-level floor, and it commands a low rent. (34 incoherent words)

A coherent sentence Many of the 61 samples shipped in December either arrived late or were damaged in transit, even though they were shipped one week earlier than usual to avoid the Christmas mail tie-up, and were packed in polyurethane as an extra precaution against rough handling. (45 coherent words)

EMPHASIS

Whereas unity and coherence ensure that a sentence is clear and uncluttered, properly placed emphasis helps the reader identify its important parts. By arranging a sentence effectively you can help the reader attach importance to the whole sentence, to a clause or phrase, or even to a single word.

Emphasis on the Whole Sentence. You can give a sentence more emphasis than sentences that precede and follow it by manipulating its length or by stressing or repeating certain words. You can also imply to the reader that all the clauses within a sentence are equally important (without affecting its relationship with surrounding sentences) by balancing its parts.

Although you should aim for variety in sentence length, there may be occasions when you will want to adjust the length of a particular sentence to give it greater emphasis. Readers will attach importance to a short sentence placed among several long sentences, or to a long sentence among predominantly short ones. They will also detect a sense of urgency in a series of short sentences that carry them quickly from point to point. This technique is used effectively by story tellers:

> The prisoner huddled against the wall, alone in the dark. He listened intently. He could hear the guards, stomping and muttering. Cursing the cold, probably.

In technical writing we have little occasion to write very short sentences, except perhaps to impart urgency to a warning of a potentially dangerous situation:

> Dangerously high voltages are present on exposed terminals. Before opening the doors:
> 1. Set the master control switch to "OFF."
> 2. Hang the red "NO!" flag on the operator's panel.
>
> Never cheat the interlocks.

Neither must we err in the opposite direction and write overly long sentences that are confusing. The rule that applies to paragraph writing—adjust the length to suit complexity of topic and technical level of reader—applies equally to sentence writing.

Repetition of key words is another effective technique for making a sentence strong and emphatic, but it must be used sparingly. Its sudden use will catch a reader's attention; repeated use will seem awkward and contrived.

Similarity of shape can signify that all parts of a sentence are of equal importance. Clauses separated by a coordinate conjunction (mainly *and, or, but,* and sometimes a comma) tell a reader that they have equal emphasis. The following sentences are "balanced" in this way:

> Eight test instruments were used for the rehabilitation project, *and* were supplied free of charge by the Dere Instrument Company.
>
> The gas pipeline will be 300 miles shorter than the oil pipeline, *but* will have to cross much more difficult terrain.
>
> The upper knob adjusts the instrument in the vertical plane, *and* the lower knob adjusts it in the horizontal plane.

Emphasis on Part of a Sentence. While coordination means giving equal weight to all parts of a sentence, subordination means emphasizing a specific part and deemphasizing all other parts. It is effected by placing the most important information in the primary clause and placing less important information in subordinate clauses. In each of the following sentences the main thought is italicized to identify the primary clause:

> When the technician momentarily released his grip, *the control slipped out of reach.*

The bridge over the underpass was built on a compacted gravel base, partly to save time, partly to save money, and partly because materials were available on site.

He lost control of the vehicle when the wasp stung him.

When the wasp stung him, *he lost control of the vehicle.*

Emphasis on Specific Words. Where we place words in a sentence has a direct bearing on their emphasis. The first and last words in a sentence are automatically emphasized. But if we place unimportant words in one of these impact bearing positions, they can rob a sentence of its emphasis:

Emphasis misplaced	Such matters as equipment calibration will be handled by the standards laboratory however.
Emphasis restored	Equipment calibration will, however, be handled by the standards laboratory.
Emphasis misplaced	Without exception change the oil every six days at least.
Emphasis restored	Change the oil at least every six days.

The verbs we use have a powerful influence on emphasis. Strong verbs attract the reader's attention, whereas weak verbs tend to divert it. Active verbs are strong because they tell *who did what.* Passive verbs are weak because they merely pass along information; they describe *what was done by whom.* The sentences below are written using both active and passive verbs. Note that the versions written in the active voice are consistently shorter:

Passive Verbs	*Active Verbs*
Elapsed time is indicated by a pointer.	A pointer indicates elapsed time.
The project was completed by the installation crew on May 2.	The installation crew completed the project on May 2.
It is suggested that meter readings be recorded hourly.	I suggest you record meter readings hourly.
The samples were passed first to quality control for inspection, and then to the shipping department where they were packed in polyurethane.	Quality control inspected the samples and then the shipping department packed them in polyurethane.

There are occasions when you will have to use the passive voice because you are reporting an event without knowing who took the action, or prefer not to name a person. For instance you may prefer to write:

The strain gauge was read at ten-minute intervals.

rather than:

Kevin McCaughan read the strain gauge at ten-minute intervals.

Or, if you want to deemphasize Kevin's role:

A technician read the strain gauge at ten-minute intervals.

Unfortunately, many scientists and engineers in industry still believe that everything should be written in the passive voice. They shun the active voice because they feel it is too strong and does not fit their professional character. With the passage of time, this outdated belief is slowly being eroded.

Words

The right word in the right place at the right moment can greatly influence your readers. A heavy, ponderous word will slow them down; an overused expression will make them doubt your sincerity; a complex word that they do not recognize will annoy them; and a weak or vague word will make them think of you as indefinite. But the right word—short, clear, specific, and necessary—will help them understand your message quickly and easily.

WORDS THAT TELL A STORY

Words should convey images. We have many strong, descriptive words in our individual vocabularies, but most of the time we are too lazy to employ them because the same old routine words spring easily to mind. We write "put" when we would do better to write "place," "position," "insert," "drop," "slide," or any one of numerous descriptive verbs that describe the action better. Compare these examples of vague and descriptive words:

Vague Words	*Descriptive Words*
He *got it out* using the MX extraction tool.	He *pulled* it out He *extracted* it He *unlocked* it He *unscrewed* it
While the crew was in town they *picked up* spare parts.	they *bought* they *purchased* they *ordered* they *borrowed* they *stole*
The project will *take a long time.*	*require 300 work hours* *employ two persons for two weeks* *last four months*

Story writers use descriptive words to convey active images to their readers. When Bruce Catton writes (on page 310): "the aggrieved baseball player *jawing* with the umpire," the reader can see the jaw going up and down much more clearly than if Catton had written "arguing" in its place. Because we are concerned with *technical* writing does not mean we should avoid seeking colorful words. One descriptive word that defines size, shape, color, smell, taste, and so on is much more valuable than a dozen words that only generalize. (But a word of warning: reserve most of these colorful, descriptive words for your verbs and nouns rather than for adverbs and adjectives.)

Analogies can offer a useful means for describing an unfamiliar item in terms a nontechnical reader will recognize:

> A resistor is a piece of ceramic-covered carbon about the size of a cribbage peg, with a two-inch length of wire protruding from each end.

Specific words tell the reader that you are a definite, purposeful individual. Vague generalities imply that you are unsure of yourself. If you write:

> It is considered that a fair percentage of the samples received from one of our suppliers during the preceding month contained a contaminant.

you are giving the reader four opportunities to wonder whether you really know much about the topic:

1. "It is considered"	Who has voiced this opinion?
2. "a fair percentage"	How many?
3. "one of our suppliers"	Who? One in how many?
4. "contained a contaminant"	What contaminant? In how strong a concentration?

All these generalities can be avoided in a shorter, more specific sentence:

> We estimate that 60% of the samples received from RamSort Chemicals last June were contaminated with 0.5% to 0.8% mercuric chloride.

This statement tells the reader that you know exactly what you are talking about. As a technical person, you should never create any other impression.

LONG VERSUS SHORT WORDS

Big words create a barrier between writer and reader. Some writers use big words to hide their lack of knowledge; others, because they start writing without first defining clearly what they want to say. There are many long scientific words that we have to use in technical writing—we should surround them with short words whenever possible so our writing will not become ponderous and overly complex.

Sometimes we use a certain word or expression because we see another writer use it. Vance Richards, H. L. Winman and Associates' top architect, writes as flamboyantly as he dresses, and his draftspersons tend to copy him. Jim Martin models his writing on Vance's, and has replaced the simple, direct style he brought with him as a new employee with wordy writing like this:

Vague and Wordy

> Pursuant to the client's original suggestion, Mr. Richards was of the opinion that the structure planned for the client would be most suitable for immediate erection on the site until recently occupied by the old established costume manufacturer known as Garrick Garments. In accordance with the client's anticipated approval of this site, Mr. Richards has taken great pains to design a multi-level building that can be considered to use the property to an optimum extent.

"Waffle" is the word technical editor Anna King uses to describe such cumbersome writing. By eliminating unnecessary expressions (*pursuant to; of the opinion that; for immediate erection on; in accordance with; can be considered to use; an optimum extent*), Anna was able to cut Jim's 75 words to a much more effective 42 words:

Clear and Direct Mr. Richards believes the type of building the client wants should be erected on the site previously occupied by Garrick Garments. He has assumed that the client will approve this site, and has designed a multilevel building which fully develops the property.

LOW INFORMATION CONTENT WORDS

Words of low information content (LIC) contribute little or nothing to the facts conveyed by a sentence. They are untidy lodgers. Remove them and the sentence appears neater and says just as much. The problem is that practically everyone inserts LIC words into sentences, and we become so accustomed to them that we do not notice how they fill up space without adding any information.

Table 10-2 contains some of the words and phrases we must try to delete from our writing. They are difficult to identify because they often sound like good but rather vague prose. In the following sentences the LIC words have been italicized; they should be either deleted or replaced, as indicated by the notes in parentheses.

The control is actuated by *means of* No. 3 valve. (delete)
Adjust the control *as necessary* to obtain maximum deflection. (delete)
Tests were run for *a period of* three weeks. (delete)
If the project drops behind schedule *it will be necessary to* bring in extra help. (*I, you, he, we,* or *they will* bring in extra help)
By Wednesday we had a backlog of 632 units, *and for this reason* we adopted a two-shift operation. (*so*)
A new store will be opened *in an area* where market research has *given an indication that there actually is a* need for more retail outlets. (delete; *indicated the*)

Clichés and hackneyed expressions are similar to LIC words and phrases, except that their presence is more obvious and their effect can be more damaging. Whereas LIC words are intangible in that they impart a sense of vagueness, hackneyed expressions can make a writer sound pompous, garrulous, insincere, or artificial. Referring to oneself as "the writer," starting and ending letters with such overworked phrases as "We are in acknowledgement of" and "please feel free to call me," and using semilegal jargon ("the aforementioned correspondent," "in the matter of," "having reference to") are all typical hackneyed expressions. Further examples are listed in Table 10-3.

Presenting the right word in the correct form is an important aspect of technical writing. Comments on how to abbreviate words, how to form compound adjectives, whether or not you should split an infinitive (generally, you may), and when you should spell out numbers or present them as numerals are contained in a glossary of usage in Chapter 11.

Table 10-2
Examples of Low Information Content (LIC) Words and Phrases

The LIC words and phrases in this partial list are followed by an expression in parenthesis (to illustrate a better way to write the phrase) or by an (X), which means that it should be dropped entirely.

actually (X)
a majority of (most)
a number of (many; several)
as a means of (for; to)
as a result (so)
as necessary (X)
at the rate of (at)
at the same time as (while)
bring to a conclusion (conclude)
by means of (by)
by use of (by)
communicate with (talk to; telephone; write to)
connected together (connected)
contact (talk to; telephone; write to)
due to the fact that (because)
during the time that (while)
end result (result)
exhibit a tendency (tend)
for a period of (for)
for the purpose of (for; to)
for this reason (because)
in an area where (where)
in an effort to (to)
in close proximity to (close to; near)
in connection with (about)
in fact (X)
in order to (to)
in such a manner as to (to)
in terms of (in; for)
in the course of (during)
in the direction of (toward)
in the event that (if)
in the form of (as)
in the neighborhood of; in the vicinity of (about; approximately; near)
involves the use of (employs; uses)
involves the necessity of (demands; requires)
is designed to be (is)
it can be seen that (thus; so)
it is considered desirable (I or we want to)
it will be necessary to (I, you, or we must)
of considerable magnitude (large)
on account of (because)
prior to (before)
subsequent to (after)
with the aid of (with)

Note: Many of these phrases start and end with words such as *as, at, by, for, in, is, it, of, to,* and *with.* This knowledge can help you to identify LIC words and phrases in your writing.

Table 10-3
Typical Clichés and Hackneyed Expressions

and/or
as per
attached hereto
at this point in time
enclosed herewith
for your information (as an introductory phrase)
if and when
in our opinion
in reference to
in short supply
in the foreseeable future
in the matter of
last but not least
please feel free to
pursuant to your request
regarding the matter of
this will acknowledge
we are pleased to advise
we wish to state
with reference to
you are hereby advised

Parallelism

Parallelism in writing means "similarity of shape." It is applied loosely to whole documents, more firmly to paragraphs, sentences, and lists, and tightly to grammatical form. It is important in all forms of writing because it creates a sense of balance. Readers generally do not notice that parallelism is present in a good piece of writing—they only know that it reads smoothly. But they are certainly aware that something is wrong when parallelism is lacking. Good parallelism makes readers feel comfortable, so that even in long or complex sentences they never lose their way. Its ability to help readers through difficult passages makes parallelism particularly applicable to technical writing.

THE GRAMMATICAL ASPECTS

If you keep your verb forms similar throughout a sentence in which all parts are of equal importance (i.e. in which the sentence has coordination, or is balanced), you will have taken a major step toward preserving parallelism. For example, if I write "unable to predict" in the early part of a balanced sentence, I should write "able to convince" rather than "successful in convincing" in a latter part:

Parallelism violated Mr. Johnson was *unable to predict* the job completion date, but was *successful in convincing* management that the job was under control.

Parallelism maintained (A) Mr. Johnson was *unable to predict* the job completion date, but was *able to convince* management that the job was under control.

More direct alternative (B) (parallelism maintained) Mr. Johnson *predicted* no completion date, but *convinced* management that the job was under control.

The rhythm of the words is much more evident in the latter two sentences, particularly the alternative sentence, in which the parallelism has been stressed by converting the still awkward ". . . unable . . . able . . ." of sentence (A) to the positively parallel "predicted . . . convinced . . ." of sentence (B). Two other examples follow:

Parallelism violated Pete Hansk likes surveying airports and to study new construction techniques.

Parallelism maintained Pete Hansk likes to survey airports and to study new construction techniques.

or

Pete Hansk likes surveying airports and studying new construction techniques.

Parallelism violated His hobbies are designing solid-state circuits and hi-fi component construction.

Parallelism maintained His hobbies are designing solid-state circuits and constructing hi-fi components.

or

His hobbies are solid-state circuit design and hi-fi component construction.

Parallelism is particularly important when you are joining sentence parts with the coordinating conjunctions *and, or,* and *but* (as in the examples above), sometimes with a comma, or with correlatives, which are:

either . . . or
neither . . . nor
not only . . . but also

Each part of a correlative must be followed by an expression with the same grammatical form. That is, if *either* is followed by a verb, then *or* must also be followed by the same form of verb:

Parallelism violated You may either repair the test set or it may be replaced under the warranty agreement.

Parallelism maintained You may either repair the test set or replace it under the warranty agreement.

Similarly, if one part of a correlative is followed by a phrase, then the second part must be followed by the same form of phrase:

Parallelism violated The instrument not only requires mechanical repair, but also it will have to be realigned electrically.

Parallelism maintained The instrument requires not only mechanical repair but also electrical realignment.

APPLICATION TO TECHNICAL WRITING

Although parallelism is a useful means for maintaining continuity in general writing, it is in technical writing that it seems to have special application. Parallelism can clarify difficult passages and give rhythm to what otherwise might be dull material. When building sentences that have a series of clauses, you can help the reader see the connection between elements by molding them in the same shape throughout the sentence. There is complete loss of continuity in this sentence because it lacks parallelism:

Parallelism violated In our first list we inadvertently omitted the 7 lathes in room B101, 5 milling machines in room B117, and from the next room, B118, we also forgot to include 16 shapers.

When the sentence is written so that all the items have a similar shape, the clarity is restored:

Parallelism restored In our first list we inadvertently omitted 7 lathes in room B101, 5 milling machines in room B117, and 16 shapers in room B118.

Emphasis can also be achieved by presenting clauses in logical order. If they are arranged climactically (in ascending order of importance), or in order of increasing length, the reader's interest will gain momentum right through the sentence:

> No one ever offered to help the navigator struggle out to the aircraft, a parachute hunched over his shoulder, a roll of plotting charts tucked under his arm, a sextant and an astrocompass dangling from one hand, and his precious navigation bag heavy with instruments, log tables, radar charts, and the route plan for the ten-hour flight clenched firmly in the other.

Within the paragraph parallelism has to be applied more subtly. If it is too obvious, the similarity in construction can be dull and repetitive. The verbs often are the key: keep them generally in the same mood and they will help bind the paragraph into one cohesive unit (this is closely tied in with coherence, discussed under "Paragraphs"). This paragraph has good parallelism:

> One of these structures is the student union, a rallying point for snacking, dalliance and amusement. From morning until night it resounds to the blare of the jukebox, the clink of coffee cups, the clatter of bowling pins, the click of billiard balls, the slap of playing cards, the gentle creak of lounge chairs and, in the plushier ones, the splash of languid bodies in tepid swimming pools. There's likely to be an informal dance here every Friday and Saturday night. They have a ball—banquet or name-band dance—about every weekend in the ballroom.[4]

Similarity of shape is most obvious in the second sentence, with its rhythmic:

> *blare* of the jukebox
> *clink* of coffee cups
> *clatter* of bowling pins
> *click* of billiard balls
> *slap* of playing cards
> *creak* of lounge chairs
> *splash* of languid bodies in tepid swimming pools

Here we have words that paint strong images. They not only have the same grammatical form but also are bound together because they relate to the same sense: hearing. The longest clause is at the end of the sentence, where the reader can relax a moment with the "languid bodies in tepid swimming pools."

Of course, parallelism can be of particular effect in descriptive writing like this, where the author wants to convey exciting images. In technical writing the subjects are more mundane, giving us less opportunity for imaginative imagery. Nevertheless, you should try to use parallelism to carry your reader smoothly through your paragraphs, as in this description of part of a surveyor's transit:

> Two sets of clamps and tangent screws are used to adjust the leveling head. The upper clamp fastens the upper and lower plates together, while the upper tangent screw permits a small differential movement between them. The lower clamp fastens the lower plate to the socket, while the lower tangent screw turns the plate

[4]From Jerome Ellison, "Are we Making a Playground out of College?" *The Saturday Evening Post,* March 7, 1959. Reprinted by permission.

through a small angle. When the upper and lower plates are clamped together they can be moved freely as a unit; but when both the upper and lower clamps are tightened the plates cannot be moved in any plane.

APPLICATION TO SUBPARAGRAPHING

Subparagraphing seldom occurs in literature, but is used frequently in technical writing to separate events or steps, describe an operation or procedure, or list parts or components. Subparagraphing always demands good parallelism.

In the following paragraph, a series of tests has been divided into subparagraphs (for clarity, only the initial words of each test are shown here):

> Three tests were conducted to isolate the fault:
>
> In the first test a matrix was imposed upon the face of the cathode ray tube and. . . .
>
> The second test consisted of voltage measurements taken at. . . .
>
> A continuity tester was connected to the unit for test 3 and. . . .

The last subparagraph is not parallel with the first two. To be parallel it must adopt the same approach as the others (i.e. first mention the test number and then say what was done):

> For the third test, a continuity tester was connected to the unit and. . . .

To be truly parallel the subparagraphs should start in *exactly* the same way:

> In the first test a matrix was. . . .
>
> In the second test a series of. . . .
>
> In the third test a continuity tester was. . . .

But this would be too repetitive. The slight variations in the original version make the parallelism more palatable.

If more than three tests have to be described, a different approach is necessary. To continue with "A fourth test showed. . .," "For the fifth test. . .," and so on, would be dull and unimaginative. A better method is to insert a number in front of each paragraph:

> Seven tests were conducted to isolate the fault:
>
> 1. A matrix was imposed upon the face of the cathode ray tube and. . . .
> 2. Voltage measurements were taken at. . . .
> 3. A continuity tester was connected to the unit and. . . .

Parallelism has been retained, and we now have a more emphatic tone that lends itself to technical reporting.

A third version employs active verbs together with parallelism to build a strong, emphatic description that can be written in either the first or third person:

> In laboratory tests conducted to isolate the fault we:
>
> 1. *Imposed* a matrix upon the face of the cathode ray tube and. . . .
> 2. *Measured* voltage at. . . .
> 3. *Connected* a continuity tester to the unit and. . . .

To change from the first person to the third person, only the lead-in sentence has to be rewritten:

In tests conducted to isolate the fault, the laboratory:

1. *Imposed* . . . .
2. *Measured* . . . .
3. *Connected* . . . .

Parallelism also plays an important role in parts listing. This very specific application is discussed in Chapter 7.

Assignments

Exercise 1. Identify words and phrases of low information content in the following sentences. In some cases you will be able to delete the unnecessary words; in others you will have to revise a phrase or rewrite the sentence.

1. Measure the current with the aid of an ammeter connected to test points 9 and 14.
2. The flywheel turns at the rate of 1200 revolutions per minute.
3. Lake Wabagoon is located in close proximity to Montrose, Ohio.
4. We have examined your concrete batcher and consider it will be necessary to effect repairs that will cost in the region of $647.49.
5. The new output phase involves the use of cathode followers in cascade.
6. We plan tentatively to conduct a preliminary heat run on March 16.
7. An automatic switch-off valve is a feature that prevents damage to the equipment should the temperature rise above 180°C.
8. The end result was in the form of a mixture that required further analysis.
9. We are pleased to be able to report that your receiver has been tested and aligned and found to meet all the manufacturer's specifications.
10. We wish to state that if you have any further queries regarding this matter, please feel free to contact Mr. A. B. Jones at any time.
11. The training program has been planned in such a manner as to reduce waiting time to a minimum.
12. The installation of an additional machine would result in an increase in operating expenditure in the approximate amount of $1800.00.
13. For your information, the new laboratory is now ready to receive the first shipment of test samples.
14. If the needle drops below 70 on the scale, it will be necessary to switch over to the standby unit.
15. A check in the amount of $237.42 is attached, in respect of laboratory services performed during the preceding month.
16. In an effort to increase production, it was found necessary by the plant superintendent to hire four additional assemblers.
17. Since all the spare parts needed for the overhaul have been received, work will be able to start immediately. Orders for additional spare parts are neither envisaged nor necessary.
18. A stability test did in fact prove that damaged instruments were a contributing factor in reducing measurement accuracy.

19. The angle is measured in terms of degrees and minutes of arc, and is recorded on a roll chart by means of a special type of pen.
20. We are in agreement with the committee's decision to make an effort to encourage greater student participation in activities conducted by the community.

Exercise 2. The following short passages lack compactness, simplicity, or clearness. Improve them by deleting unnecessary words, or by partial or complete rewriting.

1. Mr. Wasalusky has worked very efficiently as company janitor for seven years and for this reason it is recommended that he be considered for an increase in remuneration.
2. Personnel desirous of availing themselves of the opportunity of joining the bowling league are requested to contact J. Soames.
3. Connect the signal generator to the microphone terminals on the junction box with shielded wire.
4. The technicians entertained a feeling of distrust toward government regulations.
5. It is entirely within the realms of possibility that the inspector is not cognizant of the specification.
6. The radioactive source must be maintained in an area where unauthorized personnel cannot gain access.
7. We apologize for the prolonged delay in settling your account, and wish to advise that your request for payment has again been forwarded to the Accounts Department with a request that a check be issued to you at the earliest possible opportunity.
8. Our project has now dropped rather far behind schedule, although we plan to make up as much as possible of this loss within a reasonable length of time.
9. On completion of the inspection, we found no evidence to support the view that negligence had occurred.
10. In an effort to consolidate purchases of consumable goods by different departments, the chief buyer will screen all purchase requisitions for the same with a view to reducing the quantity of orders for similar items.
11. Pursuant to your request, the instruments you ordered (P.O. 2610) will be shipped to the Gatsano site. In the event that you have not received them by the end of next month, please forward the form attached hereto.
12. All the components supplied for the installation were examined and many were found to be defective or missing.
13. This instrument is relatively inexpensive, although it could be improved by the addition of a muting control, and will undoubtedly appeal to the majority of "ham" operators.
14. A replacement klystron was installed in the transmitter and appeared to operate satisfactorily for two to three hours. During the lunch hour there was a distinct flash, followed by what seemed like intermittent arcing, which signaled its early demise.
15. With regard to your valued inquiry of January 21, we can supply four type J locks at $12.75 and eight type MR control units at $16.24, and will be glad to ship same immediately on receipt of your valued order.
16. Integration occurred well below the minimum time stipulated by the specification (18 minutes).

17. If it is intended that the instruments are to be monitored hourly, then it will be necessary to arrange for the maintenance technician to record the meter readings on form 168.
18. As well as presenting a monthly outage analysis, each field technician also gave a brief report on the serviceability status of the monitoring equipment at his pumping station.
19. In an effort to reduce the number of failures of operational equipment, it was decided to perform mechanical tests of all components before assembly. This was done for a period of one month to assess whether the failure rate would in fact decrease. At the end of the month a 43% decrease in failures had been recorded.
20. Although the new contract was authorized by the government prior to the end of the old contract, it was not received until subsequent to a layoff of 320 men.

Exercise 3. The following sentences offer choices between words that sound similar or are frequently misused. Select the correct word in each case.

1. Vic Braun's panel truck was (stationary/stationery) 47 yards north of the intersection.
2. From Dr. Hilbury's remarks, the staff (implied/inferred) that they were expected to work harder.
3. The BRILLIANCE control (affects/effects) the intensity of the image on the screen.
4. Ground work could not be started this week (as/because/for) the soil was still frozen.
5. The (amount/number) of books on numerical control held by the library has increased significantly since last year's count.
6. This year's consumption was slightly (fewer/less) than in previous years, probably because the maintenance crew conducted (fewer/less) inspections.
7. It is a lot (farther/further) from Cleveland to Montrose, Ohio, than from Duluth to Reece, Minnesota.
8. No outside work was done last week because of the inclement (climate/weather).
9. Please (accept/except) my apologies for the delay in replying to your request for technical information.
10. Mr. Winman's suggestion that the company purchase an executive jet (appears/seems) to be feasible.
11. During May the Laboratory processed 67 of the 103 samples received from the Roper Corporation. The (balance/remainder) will be processed in June.
12. Instrument serviceability increased by 20% two months after the preventative/preventive) maintenance program was started.
13. To test fuel consumption, we started the engine on February 8 at 9:06 A.M. and allowed it to run (continually/continuously) until it ran out of fuel, which occurred on February 13 at 3:23 P.M.
14. Andy Rittman reported from the construction (cite/sight/site) that progress was being hampered by (defective/deficient) excavation equipment that had been subject to many breakdowns.
15. Before continuing with step 23, check that the sheet is free (from/of) surface defects.

16. We concluded that there was little difference (among/between) the four pantographs evaluated.
17. The frequency and loudness of the noise decreased as the flywheel (deac-celerated/decelerated).
18. The project group cannot write (its/their) report until all the data (has/have) been examined.
19. To obtain an objective appraisal of his draft report, Hank Williams had it read by (a disinterested/an uninterested) technician who was not familiar with the topic.
20. The (principal/principle) reason I have included this exercise is that it will introduce you to the glossary of technical usage.

Exercise 4. Rewrite the following sentences to make them more emphatic (i.e. change from passive voice to active voice).

1. When a check was made of the aircraft's compasses, it was discovered by the calibration crew that a +3° installation error existed.
2. It was recommended by the architect that a sulfate-resistant concrete should be used by the installation contractor.
3. At a project group meeting held on September 28, it was unanimously agreed that no further action should be taken until after Mr. Dawes had returned from New York.
4. Just before fracture of the specimen occurred, the needle on the gauge was noted to fluctuate wildly.
5. Because of low cloud and drizzle, Remick Airlines Flight 176 was instructed by the duty Air Traffic Controller to overfly Montrose Municipal Airport and to make a landing at Syracuse.
6. Accommodation for the night was arranged by the airline at its own expense for passengers who were stranded.
7. It was noted in the report that the cause of the accident was due to a bearing that was defective.
8. Although there had been a decrease in noise level of 12.7 dB for three weeks, complaints of "uncomfortable" noise were still being received by the Industrial Relations Department.
9. A new book on construction materials was received by Mr. Winman in the morning's mail, which the Materials Testing Laboratory was asked to evaluate.
10. It is our considered opinion that 167 yards of pavement are in need of rebuilding. Further, it is recommended that the remaining 421 yards of pavement be resurfaced.

Exercise 5. Improve the parallelism in the following sentences.

1. I recommend that we purchase the Smoothset blueprint machine because it is fast, quiet, and is not expensive.
2. The technicians were given training in organizing technical data and in how to present their conclusions.
3. Three of the applicants were given promotions, and transfers were arranged for the other four applicants.
4. We have found that the new system has four disadvantages:

   Too costly to operate.
   It causes delays.

Fails to use any of the existing equipment.
It permits only one in-process examination.

5. If you purchase only one set the price will be $350.00, whereas the price will be $275.00 per set if you purchase six or more sets.
6. Although relatively inaccessible, the Kettle Generating Station is frequently visited by engineers interested in the technicalities of the project and artists.
7. A night crew was started to accelerate construction, rather than as a hindrance to progress by causing further problems.
8. The analysis was needed not only to determine permafrost regression, but also as a help in planning the proposed pipeline route.

Exercise 6. Check that the numbers in the following sentences have been expressed in the proper form (refer to Article 4 of Chapter 11 for rules for writing numbers in narrative). Rewrite those that have been written incorrectly; in some examples, also check for correct abbreviations and parallelism.

1. The run-up will start at 10 o'clock on Tuesday morning.
2. As stated in chapter four, page 127, para 14.2, a machining tolerance of .003 in. is specified.
3. The survey covered 10857 males and 9881 females.
4. The heat tests must be conducted during week 3 and the results must be tabulated by the end of the 4th week.
5. 100 of the units are to be manufactured with a 1¾ cm diameter. 287 units are to be manufactured with a 3.25 cm diameter.
6. Twenty-six students wrote test number eight, with approximately twenty percent obtaining a mark of better than eighty percent.
7. For the first phase of test 14 2 spring balances will be required, plus 7 10 kg weights. In phase 2, ½ kilogram of mercury and 7½ liters of oil will be needed. These items will be used for ½ a day.
8. In reply to your letter of October 21st, and your subsequent Telex of third November, our price quotation for four (4) vertical drafting tables is six hundred and forty-seven dollars each. If you decide to purchase six (6) or more units, the price will be twenty percent less.

Exercise 7. Abbreviate the terms shown in *italics* in the following sentences. In some cases you will also have to express numerals in the proper form. (Rules for forming abbreviations appear in Article 2 of Chapter 11.)

1. Please supply *twenty-four kilograms* of bitumastic compound and *three* rolls of *decimal ninety-two meter wide* roofing material.
2. By driving at a constant 86 *kilometers an hour* we achieved a fuel consumption of 25.4 *kiloliters per kilometer.*
3. A noise level of 52.6 *decibels* was recorded at 825 *hertz.*
4. Rotate the control *clockwise* until a *continuous wave* signal is heard in the headset.
5. The transformer is rated for operation up to *eighty-five degrees Celsius.*
6. The municipality's 60 *centimeter inside diameter* storm sewers are fed into a 2.8 *meter diameter* culvert.
7. A meeting has been called for 10:30 *in the morning on Tuesday the fifth of September 1972.*

8. The heater produces *fifty-six thousand British thermal units* of heat per hour.
9. The base of the antenna is 726.3 *meters above mean sea level.*
10. Lake Wabagoon is 3.8 *kilometers northeast* of Montrose, Ohio.
11. The sun's elevation at *zero-eight hours, seventeen minutes, and twenty-six seconds Greenwich mean time* was calculated to be *twenty-three degrees, four minutes, and fourteen seconds* of arc.
12. We are shipping our voltmeter *serial number* 4257 to you for recalibration.
13. Television station DMON of Montrose, Ohio has been allocated a transmitting frequency in the *ultrahigh frequency* band (*three hundred megahertz to three gigahertz*).
14. The metric liter is equivalent to 0.264 *United States gallons* or 0.220 *Imperial gallons.*
15. Production from *number three* well averages *four hundred barrels per day.*
16. Strip *twenty-two millimeters* of insulation from both ends of *sixty-eight centimeters* of *number twenty American Wire Gauge* hookup wire.

Exercise 8. Select the correctly spelled words among the choices offered below.

1. Ian Bailey (recommended/reccomended/reccommended) we rent first class (accomodation/acommodation/accommodation) while staying at Gatsano.
2. The project team was commissioned to (develop/develope) a lubricant which would (supercede/supersede) the product we currently use.
3. A (seperate/separate) bag of (desiccant/dessicant) was packed with each module in the shipment.
4. Each member of the field team (received/recieved) three different (innoculations/inocculations/inoculations) before leaving for the project in East Africa.
5. For the Varlon installation (its/it's) better to use heavy (gauge/guage) cable.
6. We noted barely (discernable/discernible) meter fluctuations and (similarly/similarily) mild wave distortion.
7. The (auxiliary/auxilliary) units were (paraleled/paralleled/paralelled/parallelled) for maximum circuit protection.
8. In the (forward/forword/foreword) to his report, Mr. Winman (complemented/complimented) his engineers on their comprehensive analysis of the problem.

Exercise 9. Use the following information to write a single paragraph. Underline the topic sentence (summary statement).

a. A battery is composed of one or more cells connected in series.
b. Within a battery each cell is a source of chemical energy. This chemical energy is converted into electrical energy—but only when needed.
c. There are two types of cells.
d. One type of cell is the primary cell, which can supply electricity until its chemical energy is exhausted; then you have to throw it away.
e. Some cells can be recharged after use. In this type, an electrical charge can be used to return the chemicals within the cell to their original state (i.e. prior to discharge). This type of cell is known as the secondary cell.

Exercise 10. Rewrite this one-paragraph notice to make it clearer and more personal.

PROCEDURE RE EXPENSE CLAIMS

Expense claims must be handed to the Accounts Section before 10:30 A.M. on Wednesday for payment on Friday. Personnel failing to hand in their forms at the proper time, may do so at any time until 4:30 P.M. on Thursday but must wait until Monday for payment. Under no account will a late claim be paid in the same week that it was filed. Claims handed in after Thursday will be processed with the following week's claims and will be paid on the next Friday.

Exercise 11. Write a single paragraph describing the following situation. Underline your topic sentence (summary statement).

a. Your company has a lot of telephone switching equipment to maintain. Most is old, but some is new.
b. New telephone switching equipment is under warranty for one year.
c. The manufacturer's warranty specifies new equipment must be lubricated with Vaprol.
d. You currently use Vaprol for all telephone switching equipment.
e. Tests by the Engineering Department show Gra-Lub to be a better lubricant.
f. Gra-Lub has a graphite base; Vaprol has not.
g. You want to adopt Gra-Lub for all telephone switching equipment.
h. You don't want to stock and use two types of lubricant for the same equipment.
i. You have written to the manufacturer to request authorization to use Gra-Lub on equipment still under warranty.

11

GLOSSARY OF TECHNICAL USAGE

A standard glossary of usage contains rules for combining words into compound terms, for forming abbreviations, for capitalizing, and for spelling unusual or difficult words. This glossary also offers suggestions for handling many of the technical terms peculiar to industry. Hence, it is oriented toward the technical rather than the literary writer.

The information, arranged alphabetically, is preceded by six brief supplementary articles on specific aspects that would be too lengthy to include in the glossary. These are:

Article 1: Combining Words into Compound Terms
Article 2: Abbreviating Technical and Nontechnical Terms
Article 3: Capitalization and Punctuation
Article 4: Writing Numbers in Narrative
Article 5: Summary of Numerical Prefixes and Symbols
Article 6: Introduction to Metric Units

The glossary also contains a selection of words most likely to be misused or misspelled. These include:

Words that are similar and frequently confused with one another; e.g. *imply* and *infer, diplex* and *duplex, principal* and *principle.*

Common minor errors of grammar, such as *comprised of* (should be *comprises*), *most unique* (*unique* should not be compared), *liaise* (an unnatural verb formed from *liaison*).

Words that are particularly prone to misspelling; e.g. *desiccant, oriented, immitance.*

Words for which there may be more than one "correct" spelling; e.g. *programer* and *programmer.*

Where two spellings of a word are in general use (e.g. *symposiums* and *symposia*), both are entered in the glossary and a preference shown for one of them.

Definitions have been included when they will help you select the correct word for a given purpose, or to differentiate between similar words having different meanings. These definitions are intentionally brief and intended to act only as a guide; for more comprehensive definitions, consult an authoritative dictionary (I recommend *Webster's New Collegiate Dictionary*).

All entries in the glossary are in lower case letters. Capital letters are used where capitals are recommended for that specific word, phrase, or abbreviation. Similarly, periods have been eliminated except where they form part of a specific entry. For example, the abbreviation for "inch" is *in.*, and the period that follows it has been inserted intentionally to distinguish it from the word "in."

Finally, think of this glossary as a guide rather than a collection of hard and fast rules. Our language is continually changing, and what was fashionable yesterday may seem pedantic today and a cliché tomorrow. I expect that in some cases your views will differ from mine. Where they do, I hope that the comments and suggestions I offer will help you to choose the right expression, word, or abbreviation, and that you will be able to do so both consistently and logically.

Articles

ARTICLE 1: COMBINING WORDS INTO COMPOUND TERMS

One of the biggest problems facing technical writers (and particularly the typists who have to record their work) is to know whether multiword expressions should be compounded fully, joined by hyphens, or allowed to stand as two or more separate words. For example, should you write:

cross check, cross-check, or crosscheck?
counter clockwise, counter-clockwise, or counterclockwise?
change over, change-over, or changeover?

The tendency today is to compound a multiword expression into a single term. But this bare statement cannot be applied as a general rule because there are too many variations, some of which appear in the glossary.

Most multiword expressions are compound adjectives. When two words combine to form an adjective they are either joined by a hyphen or compounded to form one word. They are usually joined by a hyphen if they are formed from a noun-adjective expression:

Noun-Adjective	*As a Compound Adjective*
vacuum tube	vacuum-tube voltmeter
cathode ray	cathode-ray tube
high frequency	high-frequency oscillator

But when one of the combining words is a verb, they often combine into a one-word adjective. Under these conditions they normally will compound into a single-word noun:

Two Words	*As a Noun*	*As an Adjective*
lock out	lockout	lockout voltage
shake down	shakedown	shakedown test
cross over	crossover	crossover network

Three or more words that combine to form an adjective in most cases are joined by hyphens. For example, *lock test pulse* becomes *lock-test-pulse generator.* Occasionally, however, they are compounded into a single term, as in *counter-electromotive force.* Specific examples are in the glossary.

Obviously, these "rules" cannot be taken at full face value because there are occasions when they do not apply. Useful guides for doubtful combinations are the *Government Printing Office Style Manual*[1] which contains a list of over 19,000 compound terms, and Rudolph Flesch's admirable handbook *Look It Up.*[2]

ARTICLE 2: ABBREVIATING TECHNICAL AND NONTECHNICAL TERMS

You may abbreviate any term you like, and in any form you like, providing you indicate clearly to the reader how you intend to abbreviate it. This can be done by stating the term in full, then showing the abbreviation in parentheses to indicate that from now on you intend to use the abbreviation. Here is an example:

> Always spell out single digit numbers (sdn). The only time sdn are not spelled out is when they are being inserted in a column of figures.

When forming abbreviations of your own, take care not to form a new abbreviation when a standard one already exists. For example, if you did not know that there is a commonly accepted abbreviation for pound (weight), you might be tempted to use *pd*. This would not sit well with your readers, who might resent replacement of their old friend *lb.*

There are three basic rules that you should observe when forming abbreviations:

1. *Use lower case letters,* unless the abbreviation is formed from a person's name:
   centimeter — cm
   kilogram — kg
   approximately — approx
   decibel — dB (the *B* represents *Bell* (Alexander Graham Bell)

[1]*United States Government Printing Office Style Manual,* Superintendent of Documents, Washington, D.C., 20402.

[2]Rudolph Flesch, *Look It Up: A Deskbook of American Spelling and Style* (New York: Harper and Row, 1977).

2. *Omit all periods,* unless the abbreviation forms another word:

horsepower	— hp
cubic centimeter	— cm^3
cathode ray tube	— crt
foot/feet	— ft
inch	— in.
singular	— sing.

3. *Write plural abbreviations in the same form as the singular abbreviation:*

inches	— in.
pounds	— lb
kilograms	— kg
hours	— h *or* hr

Of course there are exceptions which have grown as part of the language. Through continued use these unnatural abbreviations have been generally accepted as the correct form and now cannot easily be displaced. A few examples follow:

for example *(exempli gratia)*	— e.g.	*(There is a slowly growing trend to write these as* eg *and* ie*)*
that is *(id est)*	— i.e.	
morning *(ante meridiem)*	— a.m.	
inside diameter	— ID	
number(s)	— No.	

Other unusual abbreviations exist in specific technical disciplines. The glossary offers a partial list of standard abbreviations and some technical abbreviations. Consult a good dictionary for general guidance; for specific technical terms refer to the list of standard abbreviations compiled by one of the technical societies in your discipline.

ARTICLE 3: CAPITALIZATION AND PUNCTUATION

Lower case letters (that is, not capitalized) should be used as much as possible in technical writing. Too many capital letters cause an untidy appearance and reduce the effectiveness of capital letters inserted for emphasis. The glossary therefore recommends lower case letters except (1) where usage has resulted in general adoption of a capitalized form (as in *No., GCA,* and the Brinell and Rockwell hardness tests), (2) for proper nouns, and (3) for expressions coined from proper nouns which have not yet been accepted as common terms. (Note also that although we still use capital letters when referring to Doppler's principle, we write of doppler radar in lower case letters.)

The tendency today is to underpunctuate technical writing. This does not mean you may omit punctuation with a blasé "when in doubt, leave out" attitude. But you need not punctuate every clause and subclause. The intent is to obtain smooth reading; the criterion is to make the message clear. Hence, you should insert punctuation where it is needed rather than rigorously divide the information into tight little formal compartments.

ARTICLE 4: WRITING NUMBERS IN NARRATIVE

The conventions that dictate whether a number should be written out or expressed in figures differ between ordinary writing and technical writing. In technical writing you are much more likely to express numbers in figures.

The rules listed below are intended mainly as a guide. There will be many times when you have to make a decision between two rules that conflict. Your decision should then be based on three criteria:

Which method will be most readable?
Which method will be simplest to type?
Which method did I use previously, under similar circumstances?

Good judgment and a desire to be consistent will help you to select the best method each time.

The basic rule for writing numbers in technical narrative is:

Spell out single-digit numbers (one to nine inclusive).
Use figures for multiple-digit numbers (10 and above).

Exceptions to this rule are:

Always Use Figures:

1. When writing specific technical information, such as test results, dimensions, tolerances, temperatures, statistics, and quotations from tabular data.
2. When writing any number that precedes a unit of measurement: 3 mm; 7 lb; 121.5 MHz.
3. When writing a series of both large and small numbers in one passage: During the week ending 27 May we tested 7 transmitters, 49 receivers, and 38 power supplies.
4. When referring to section, chapter, page, figure (illustration), and table numbers: Chapter 7; Figure 4.
5. For numbers that contain fractions or decimals: 7¼; 7.25.
6. For percentages: 3% gain; 11% sales tax.
7. For years, dates, and times: At 3 P.M. on January 9, 1982; 08:17, 20 Feb 82.
8. For sums of money: $2000; $28.50; 20 dollars; $0.27 (preferred) or 27 cents.
9. For ages of persons.

Always Spell Out:

1. Round numbers that are generalizations: about five hundred; approximately forty thousand.
2. Fractions that stand alone: repairs were made in less than three-quarters of an hour.
3. Numbers that start a sentence. (Better still, rewrite the sentence so that the number is not at the beginning.)

Additional rules are:

Spell out one of the numbers when two numbers are written consecutively and are not separated by punctuation: 36 fifty-watt amplifiers or thirty-six 50-watt amplifiers. (Generally, spell out whichever number will result in the simplest or shortest expression.)

Insert a zero before the decimal point of numbers less than one: 0.75; 0.0037.

Use decimals rather than fractions (they are easier to type), except when writing numbers that are customarily written as fractions.

Insert commas in large numbers containing five or more digits: 1,275,000; 27,291; 4056. (Insert a comma in four-digit numbers only when they appear as part of a column of numbers.)

Write numbers that denote position in a sequence as 1st, 2nd, 3rd, 4th . . . 31st . . . 42nd . . . 103rd . . . 124th. . . .

ARTICLE 5: SUMMARY OF NUMERICAL PREFIXES AND SYMBOLS

The table below summarizes the numerical prefixes and abbreviations used in the glossary, and conforms to the requirements for writing SI (metric) units.

Multiple/ Submultiple	*Prefix*	*Symbol*	*Multiple/ Submultiple*	*Prefix*	*Symbol*
10^{18}	exa	E	10^{-1}	deci	d
10^{15}	peta	P	10^{-2}	centi	c
10^{12}	tera	T	10^{-3}	milli	m
10^{9}	giga	G	10^{-6}	micro	μ
10^{6}	mega	M	10^{-9}	nano	n
10^{3}	kilo	k	10^{-12}	pico	p
10^{2}	hecto	h	10^{-15}	femto	f
10	deca	da	10^{-18}	atto	a

ARTICLE 6: INTRODUCTION TO METRIC UNITS

For this second edition, the glossary has been broadened to include terms and symbols within the International System of Units (SI). The trend toward worldwide adoption of metric units of measurement means that for some time both the basic inch/pound system and the metric (SI) system will be in use concurrently. The terms and symbols introduced here are those you are most likely to encounter in your technical reading or may want to use in your technical writing.

The acronym "SI" is used in all languages to represent the name "Système International d'Unités." Both the acronym and the name were adopted for universal usage in 1960 by the eleventh Conférence Générale des Poids et Mesures (CGPM), which is the international authority on metrication. Since then, some of the metric terms the conference established have crept into our language. For example, *Hertz,* the unit of frequency measurement, was first introduced as a replacement for *cycles per second* in the early 1960's; now it is used universally, both in the technical world and by the general public. Other terms already gaining recognition are:

	non-SI	*SI*
Temperature:	degrees Fahrenheit	degrees Celsius
Length:	miles, yards, feet, inches	kilometers, meters, millimeters

Weight:	tons, pounds ounces	kilograms, grams, milligrams
Liquid volume:	gallons, quarts	kiloliters, liters

The glossary defines the basic SI units and shows how they should be written, either in full or abbreviated as a symbol, and their multiples and submultiples. It does not attempt to explain how the units are derived, nor does it include many of the less common units. The individual glossary entries are supplemented by the guidelines listed below, many of which are equally applicable to non-SI terms and units.

Guidelines. SI symbols must be written, typed, or printed:

1. In upright type, even if the surrounding type slopes or is in italic letters.
2. In lower case letters, except when the name of the unit is derived from a person's name (e.g. the symbol **F** for *farad* is derived from *Faraday*), in which case the first letter of the symbol is capitalized (e.g. **Wb** for *weber*).
3. With a space between the last numeral and the first letter of the symbol: **355 V, 27 km** (not 355V, 27km).
4. With no "s" added to a plural: **1 g, 236 g.**
5. With no period after the symbol, unless it forms the last word in a sentence.
6. With no space between the multiple or submultiple symbol and the SI symbol: **3.6 kg, 150 mm, 960 kHz**.
7. With a solidus (oblique stroke: /) to represent the word *per*: **cm/s** (centimeters per second). Only one solidus should be used in any one expression.
8. With a dot (·) at midletter height to represent that the symbols are multiplied: **1m•s** (lumen seconds).
9. As a symbol, if a number is used with the SI unit: "... the tank holds **400 L**"; but spelled out when no number is used with the unit: "... capacity is measured in liters," not: "capacity is measured in L."

Spelling: "-er" or "-re"? An anomaly exists concerning the spelling of **liter** and **meter.** SI stipulates the **-re** spelling (**litre** and **metre**), but in the United States we have traditionally used the **-er** spelling. I have continued to use the more familiar traditional spelling in *Technically-Write!* but caution readers that the spelling may change in the United States as metric units of measurement become more widely established.

THE GLOSSARY

General abbreviations used throughout the glossary are:

abbr	abbreviate(d); abbreviation
def	definition
lc	lower case
pl	plural
pref	prefer, preferred, preference
SI	International System of Units

A

a; an use *an* before words that begin with a silent *h* or a vowel; use *a* when the *h* is sounded or if the vowel is sounded as *w* or *y; an hour* but *a hotel, an onion* but *a European*

aberration

ab initio def: from the beginning

abrasion

abscissa

absolute abbr: **abs**

absorb(ent); adsorb(ent) *absorb* means to swallow up completely (as a sponge absorbs moisture); *adsorb* means to hold on the surface, as if by adhesion

abut; abutted; abutting; abutment

accelerate; accelerator; accelerometer

accept; except *accept* means to receive (normally willingly); *he accepted the company's offer of employment; except* generally means exclude: *the night crew completed all the repairs except rewiring of the control panel*

access(ible)

accessory; accessories abbr: **accy**

accidental(ly)

accommodate; accommodation

account abbr: **acct**

accumulate; accumulator

achieve means to conclude successfully, usually after considerable effort; avoid using *achieve* when the intended meaning is simply to reach or to get

acknowledg(e)ment *acknowledgment* pref

acre; acreage

across not *accross*

actually omit this word: it is seldom necessary in technical writing

actuator

A.D. def: Anno Domini

adapter; adaptor *adapter* pref

adaption; adaptation *adaption* pref

addendum pl: *addenda*

adhere to never use *adhere by*

ad hoc def: set up for one occasion

adjective (compound) two or more words that combine to form an adjective are either joined by a hyphen or compounded into a single word; see Article 1

adsorb(ent) see **absorb**

advice; advise use *advice* as a noun and *advise* as a verb: *the engineer's advice was sound; the technician advised the driver to take an alternative route;* spell: **adviser, advisable**

aerate

aerial see **antenna**

aero- a prefix meaning of the air; it combines to form one word: *aerodynamics, aeronautical;* in some instances it has been replaced by *air: airplane, aircraft*

aesthetic; esthetic both are correct; in U.S., *esthetic* pref

affect; effect *affect* is used only as a verb, never as a noun; it means to produce an effect upon or to influence (*the potential difference affects the transit time*); *effect* can be used either as a verb or as a noun; as a verb it means to cause or to accomplish (in *to effect a change);* as a noun it means the consequences or result of an occurrence *(the detrimental effect upon the environment)*, or refers to property, such as *personal effects*

after- as a prefix, usually combines to form one word: *afteracceleration, afterburner, afterglow, afterheat, afterimage*

agenda although plural, *agenda* is generally treated as singular: *the agenda is complete*

aggravate the correct definition of *aggravate* is to increase or intensify (worsen) a situation; try not to use it when the meaning is *annoy*

aggregate

aging; ageing *aging* pref

agree to; agree with to be correct, you should *agree to* a suggestion or proposal, but *agree with* another person

air- as a prefix, normally combines to form one word: *airborne, airfield, airflow;* exceptions: *air-condition(ed) (er) (ing), air-cool(ed) (ing)*

air horsepower abbr: **ahp**

airline; air line an *airline* provides aviation services; an *air line* is a line or pipe that carries air

alkali; alkaline pl: *alkalis* (pref) or *alkalies*

allege(d); alleging

allot(ted)

all ready; already *all ready* means that all (everyone or everything) is ready; *already* means by this time: *the samples are all ready to be tested; the samples have already been tested*

all right def: everything is satisfactory; never use *alright*

all together; altogether *all together* means all collectively, as a group; *altogether* means completely, entirely: *the samples have been gathered all together, ready for testing; the samples are altogether useless*

alphanumeric def: in alphabetical, then numerical sequence

alternate; alternative *alternate(ly)* means by turn and turn about: *the inspector alternated among the four construction sites; alternative(ly)* offers a choice between only two things: *the alternative was to return the samples;* although it is incorrect to say *there are three alternatives,* it is becoming common usage

alternating current abbr: **ac**

alternator

altitude abbr: **alt**

a.m. def: before noon (*ante meridiem*)

amateur

ambient abbr: **amb**

ambiguous; ambiguity

American Wire Gage abbr: **AWG**

among; between use *among* when referring to three or more items; use *between* when referring to only two; *amongst* is seldom used

amount; number use *amount* to refer to general quantity: *the amount of time taken as sick leave has decreased;* use *number* to refer to items that can be counted; *the number of technicians assigned to the project was 30% greater than anticipated*

ampere(s) abbr: **A** (pref) or **amp;** other abbr: **kA, mA, μA, nA, pA, A/m** (ampere per minute)

ampere-hour(s) abbr: **Ah** (pref) or **amp-hr** (more common)

amplitude modulation abbr: **AM**

an see **a**

anaemic; anemic *anemic* pref

anaesthetic; anesthetic *anesthetic* pref

analog; analogous

analyse(r); analyze(r) *analyze(r)* pref

AND-gate

and/or avoid using this term; in most cases it can be replaced by either *and* or *or*

angstrom abbr: **Å**

anion def: negative ion

anneal(ing)

annihilation

antarctic see comment for **arctic**

ante- a prefix that means before; combines to form one word: *antecedent, anteroom*

ante meridiem def: before noon; abbr: **a.m.**; can also be written as *antemeridian,* but never as *antimeridian* (which see)

antenna the proper plural in the technical sense is *antennas;* although *antennae* is sometimes seen, its use should be limited to zoology; *antenna* has generally replaced the obsolescent *aerial*

anti- a prefix meaning opposite or contradictory to; generally combines to form one word: *antiaircraft, antiastigmatism, anticapacitance, anticoincidence, antisymmetric;* if combining word starts with *i* or is a proper noun, insert a hyphen: *anti-icing, anti-American*

antimeridian def: the opposite meridian (of longitude); e.g. the antimeridian of 96° 30′W is 83° 30′E

anxious although *anxious* really implies anxiety, current usage permits it to be used when the meaning is simply keen or eager

anybody; any body *anybody* means any person; *any body* means any object: *anybody can attend; discard the batch if you find any body containing foreign matter*

anyone; any one *anyone* means any person; *any one* means any single item: *you may take anyone with you; you may take any one of the samples*

anyway; any way *anyway* means in any case or in any event; *any way* means in any manner; *the results may not be as good as you expect, but we want to see them anyway; the work may be done in any way you wish*

apparatus(es)

apparent(ly)

appear(s); seems(s) use *appears* to describe a condition that can be seen: *the equipment appears to be new;* use *seems* to describe a condition that cannot be seen: *he seems to be clever*

appendix def: the part of a report that contains supporting data; pl: *appendixes* (pref) or *appendices*

appreciate means to value or to cherish; it should not be used as a synonym for *understand,* as in *we appreciate your difficulty in finding spare parts*

approximate(ly) abbr: **approx;** but *about* is a better word

arbitrary

arc; arced; arcing

architect

arctic capitalize when referring to a specific area: *beyond the Arctic Circle;* otherwise use lc letters: *in the arctic;* don't omit the first *c*

area the SI unit for area is the *hectare* (abbr: **ha**)

areal def: having area

around def: on all sides, surrounding

arrester; arrestor *arrester* pref

artwork

as avoid using when the intended meaning is *since* or *because;* to write *he could not open his desk as he left his keys at home* is incorrect (replace *as* with *because)*

as per avoid using this hackneyed expression, except in specifications

asphalt *asfalt* less pref

assembly; assemblies abbr: **assy**

assure means to state with confidence that something has been or will be made certain; it is sometimes confused with *ensure* and *insure,* which it does not replace; see **ensure**

asymmetrical

asynchronous

athletic not *atheletic*

atmosphere abbr: **atm**

atomic weight abbr: **at. wt**

attenuator

atto def: 10^{-18}; abbr: **a**

audible; audibility

audio frequency abbr: **af** (pref) or **a-f**

audiovisual

aural def: that which is heard; it should not be confused with *oral,* which means that which is spoken

author; writer avoid referring to yourself as *the author* or *the writer;* use *I, me,* or *my*

auto- a prefix meaning self; combines to form one word: *autoalarm, autoconduction, autoionization, autotransformer*

automatic frequency control abbr: **afc** (pref) or **AFC**

automatic volume control abbr: **avc** (pref) or **AVC**

auxiliary abbr: **aux**

average see *mean*

ax; axe **ax** pref; pl: *axes*

axis the plural also is *axes*

azimuth abbr: **az**

B

back- as a prefix normally combines into one word: *backboard, backdate(d), backlog*

balance; remainder use *balance* to describe a state of equilibrium *(discontinuous permafrost is frozen soil delicately balanced between the frozen and unfrozen state),* or as an accounting term; use *remainder* when the meaning is the rest of: *the remainder of the shipment will be delivered next week*

bandwidth

barometer abbr: **bar.**

barrel(l)ed; barrel(l)ing the single *l* is pref; the abbr of *barrel(s)* is **bbl**

barretter

bases this is the plural of both *base* and *basis*

basically

B.C. def: Before Christ

because; for use *because* when the clause it introduces identifies the cause of a result: *he could not open his desk because he left his keys at home;* use *for* when the clause introduces something less tangible: *he failed to complete the project on* *schedule, for reasons he preferred not to divulge*

becquerel def: a unit of activity of radionuclides (SI): abbr: **Bq;** other abbr: **PBq, TBq, GBq, kBq;** in SI, the *becquerel* replaces the *curie*

begin(ning)

benefit; benefited; benefiting

beside, besides *beside* means alongside, at the side of; *besides* means as well as

between see **among**

bi- a prefix meaning two or twice; combines to form one word: *biangular, bidirectional, bifilar, bilateral, bimetallic, bizonal*

biannual(ly); biennial(ly) *biannual(ly)* means twice a year; *biennial(ly)* means every two years; the *bi* of *bimonthly* and *biweekly* means every two

bias; biased; biasing

billion def: 10^9 (U.S.); 10^{12} (Britain)

billion electron volts although the pref abbr is **GeV**, *beV* and *bev* are more commonly used

Bill of Materials abbr: **BOM**

bimonthly def: every two months

binary

binaural

bioelectronics

bionics def: application of biological techniques to electronic design

birdseye (view)

biweekly def: every two weeks

blueprint

blur; blurred; blurring; blurry

board feet abbr: **fbm** (derived from *feet board measure)*

boiling point abbr: **bp**

bona fide def: in good faith, authentic, genuine

borderline

brakedrum; brake lining; brakeshoe

brake horsepower; brake horsepower-hour abbr: **bhp, bhp-hr**

brand-new

break- when used as a prefix to form a compound noun or adjective, *break* combines into one word: *breakfast, breakdown, breakup;* in the verb form it retains its single-word identity; *it was time to break up the meeting*

bridging

Brinell hardness number abbr: **Bhn**

British thermal unit abbr: **Btu**

buoy; buoyant

bureacracy; bureaucrat

burned; burnt *burned* pref

bur(r) *burr* pref

bus(es); bused; busing; bus bar

bypass

by-product

C

calendar; calender; colander a *calendar* is the arrangement of the days in a year; *calender* is the finish on paper, cloth; a *colander* is a sieve

caliber; calibre *caliber* pref

calking; caulking *calking* pref

cal(l)iper *caliper* pref

calorie abbr: **cal**

calorimeter; colorimeter a *calorimeter* measures quantity of heat; a *colorimeter* measures color

cancel(l)ed; cancel(l)ing *canceled, canceling* pref, but always *cancellation*

candela def: unit of luminous intensity (replaces *candle*); abbr: **cd;** recommended abbr for candela per square foot and square meter are **cd/ft²** and **cd/m²**

candlepower; candlehour(s) abbr: **cp, c-hr;** *candle* has been replaced by *candela*

candoluminescence

cannot one word pref; avoid using *can't* in technical writing

capacitor

capacity for never use *capacity to* or *capacity of*

capillary

capital letters abbr: **caps.;** use capital letters as little as possible (see Article 3)

carburet(t)or *carburetor* pref; a third, seldom used spelling is *carburetter*

carcino- as a prefix combines to form one word

caseharden

cassette

caster; castor pref spelling is *caster* when the meaning is to swivel freely; *castor* is used when the reference is to castor oil, etc.

catalog(ue) *catalog, cataloged,* and *cataloging* pref

catalyst

category; categories; categorical

cathode-ray tube abbr: **crt** (pref) or **CRT** (commonly used)

cation def: positive ion

-ceed; -cede; -sede only one word ends in *-sede: supersede;* only three words end in *-ceed: exceed, proceed, succeed;* all others end in *-cede:* e.g. *precede, concede*

centerline abbr: **℄** (pref) or **CL**

Celsius abbr: **C;** see **temperature**

center-to-center abbr: **c-c**

centi- def: 10^{-2}; as a prefix combines to form one word: *centiampere, centigram;* abbr: **c;** other abbr:

centigram	**cg**
centiliter	**cL**
centimeter	**cm**
centimeter-gram-second	**cgs**
centimeter per second	**cm/s**
square centimeter	$\mathbf{cm^2}$

centigrade abbr: **C;** in SI, **centigrade** has been replaced by **Celsius;** see **temperature**

centri- a prefix meaning center; combines to form one word: *centrifugal, centripetal*

chairperson

chamfer

changeable; changeover

channel(l)ed; channel(l)ing single *l* pref

chapter abbr: **chap.**

chassis both singular and plural are spelled the same

check- as a prefix combines to form one word: *checklist, checkpoint, checkup* (noun or adjective)

chisel(l)ed; chisel(l)ing single *l* pref

chrominance

cipher

circuit abbr: **cct**

cite def: to quote; see **site**

climate avoid confusing *climate* with *weather; climate* is the average type of weather, determined over a number of years, experienced at a particular place; *weather* is the state of the atmospheric conditions at a specific place at a specific time

clockwise (turn) abbr: **CW**

co- as a prefix, *co-* generally means jointly or together; it usually combines to form one word: *coexist, coequal, cooperate, coordinate, coplanar (co-worker* is an exception); it is also used as the abbr for *complement of* (an arc or angle): *codeclination, colatitude*

coalesce

coarse; course *coarse* means rough in texture or of poor quality; *course* implies movement or passage of time: *a coarse granular material; the technical writing course*

coaxial abbr: **coax.**

coefficient abbr: **coef**

cologarithm abbr: **colog**

colon when a colon is inserted in the middle of a sentence to introduce an example or short statement, the first word following the colon is not capitalized; when a colon is used at the end of a sentence to introduce subparagraphs that follow, the first word of each subparagraph should be capitalized unless all the subparagraphs are very short (i.e. only one sentence long); a hyphen should not be inserted after the colon

colorimeter see **calorimeter**

column abbr: **col.**

combustible

comma a comma normally need not be used immediately before *and, but,* and *or,* but may be inserted if to do so will increase understanding or avoid ambiguity; also see Article 3

commence in technical writing, replace *commence* with the more direct *begin* or *start*

commit; commitment; committed; committing

committee

compare; comparable; comparison; comparative use *compared to* when suggesting a general likeness; use *compared with* when making a definite comparison

compatible; compatibility

complement; compliment *complement* means the balance required to make up a full quantity or a complete set; to *compliment* means to praise; *in a right angle, the complement of 60° is 30°; Mr. Perchanski complimented him for writing a good report*

composed of; comprising; consists of all three terms mean "made up of" (specific items); if any one of these terms is followed by a list of items, it implies that the list is complete; if the list is not complete, the term should be replaced by *includes* or *including*

compound terms two or more words that combine to form a compound term are joined by a hyphen or are written as one word, depending on accepted usage and whether they form a verb, noun, or adjective; the trend is toward one-word compounds; see Article 1

comprise; comprised; comprising to write *comprised of* is incorrect, because the verb comprise includes the preposition *of;* also see **composed of**

concur; concurred; concurrent; concurring

condenser

conform use *conform to* when the meaning is to abide by; use *conform with* when the meaning is to agree with

conscience

conscious

consensus means a general agreement of opinion; hence to write *consensus of opinion* is incorrect; e.g. *the consensus was that a further series of tests would be necessary*

consistent with never use *consistent of*

consists of; consisting of see **composed of**

contact should not be used as a verb when *write, visit, speak,* or *telephone* better describes the action to be taken

continual; continuous *continual(ly)* means happens frequently but not all the time: *the generator is continually being overloaded* (is frequently overloaded); *continuous(ly)* means goes on and on without stopping: *the noise level is continuously above 100 dB* (it never drops below 100dB)

continue(d) abbr: **cont**

continuous wave abbr: **cw**

contra-rotating

contrast when used as a verb, *contrast* is followed by *with;* when used as a noun, it may be followed by either *to* or *with* (*with* pref)

conversant with never use *conversant of*

corollary

correspond *to correspond to* suggests a resemblance; *to correspond with* means to communicate in writing

cosecant abbr: **csc** (pref) or **cosec**

cosine abbr: **cos**

cotangent abbr: **cot**

coulomb def: a quantity of electricity, electric charge (SI); abbr: **C**; other abbr: **kC, mC, μC, nC, pC, C/m²**

counter- a prefix meaning opposite or reciprocal; combines to form one word: *counteract, counterbalance, counterflow, counterweight*

counterclockwise (turn) abbr: **CCW**

counterelectromotive force abbr: **cemf;** also known as *back emf*

counts per minute abbr: **cpm**

course see **coarse**

criterion pl: *criteria*

criticism; criticize

cross- as a prefix combines erratically: *cross-check, crosshatch, crosstalk, cross-purpose, cross section*

cross-refer(ence) abbr: **x-ref**

cryogenic

crystal abbr: **xtal**

crystalline; crystallize

cubic abbr: **cu** or ³; other abbr:

cubic centimeter(s)	**cm³** (pref); **cc**
cubic decimeter(s)	**dm³**
cubic foot (feet)	**ft³** (pref); **cu ft**
cubic feet per minute	**cfm** (pref); **ft³/min**

cubic feet per second	**cfs** (pref); **ft^3/sec**
cubic inch(es)	**in.^{3}** (pref); **cu in.**
cubic meter(s)	**m^3**
cubic millimeter(s)	**mm^3**
cubic yard(s)	**yd^3** (pref); **cu yd**

curb; kerb *curb* pref

curie abbr: **Ci;** other abbr: **mCi, μCi;** in SI the *curie* is replaced by the *becquerel* (which see)

curriculum pl: *curriculums* (pref) or *curricula*

cursor

cycles per minute abbr: **cpm**

cycles per second abbr: **cps;** although still occasionally used, this term has been replaced by **hertz** (which see)

cylinder; cylindrical abbr: **cyl**

D

daraf def: the unit of elastance

data def: gathered facts; although *data* is plural (derived from the singular *datum*, which is rarely used), it is more often used as a singular noun: *when all the data has been received, the analysis will begin*

date(s) avoid vague statements such as "last month" and "next year" because they soon become indefinite; write as a specific date, using day (in numerals), month (spelled out), and year (in numerals): *January 27, 1982* or *27 January 1982* (the latter form has no punctuation); to abbreviate, reduce month to first three letters and year to last two digits; *Jan 27, 82* or *27 Jan 82*. For SI, use numerals only, in this order: year, month, day: *1982 01 27*

days days of the week are capitalized: *Monday, Tuesday*

de- a prefix that generally combines to form one word: *deaccentuate, deactivate, decentralize, decode, deemphasize, deenergize, deice, derate, destagger;* an exception is *de-ionize*

debug(ging)

decelerate def: to slow down; never use *deaccelerate*

decibel abbr: **dB;** the abbr for decibel referred to 1 mW is **dBm**

decimals for values less than unity (one), place a zero before the decimal point: *0.17, 0.0017*

decimate def: to reduce by one-tenth; can also mean to destroy much of

decimeter abbr: **dm**

declination abbr: **dec**

defective; deficient *defective* means unserviceable or damaged (generally lacking in quality); *deficient* means lacking in quantity (it is derived from *deficit*), and in the military sense incomplete: *a short circuit resulted in a defective transmitter; the installation was completed on schedule except for a deficient rotary coupler*

defense; defence *defense* pref

defer; deferred; deferring; deference

definite; definitive *definite* means exact, precise; *definitive* means conclusive, fully evolved; e.g. *a definite price* is a firm price; *a definitive statement* concerns a topic that has been thoroughly considered and evaluated

degree(s) abbr: **deg** (pref in narrative) or ° (following numerals); see **temperature**

demarcation

demi- a little-used prefix meaning half (generally replaced by *semi-*); combines to form one word: *demivolt*

dependent; dependant *dependent* pref; *dependant* is rarely used

deprecate; depreciate *deprecate* means to disapprove of; *depreciate* means to reduce the value of: *the use of "as per" in technical writing is deprecated; the vehicles depreciated by 50% the first year and 20% the second year*

depth

desiccant; desiccate(d)

desirable

deteriorate

develop not *develope*

device; devise the noun is *device*, the verb is *devise*: *a unique device; he devised a new circuit*

dext(e)rous *dexterous* pref

diagram; diagramed; diagraming; diagrammatic

dial(l)ed; dial(l)ing single *l* pref

dialog(ue) *dialogue* pref

diaphragm

diazo

dielectric

diesel; diesel-electric

dietitian

differ use *differ from* to demonstrate a difference; use *differ with* to describe a difference of opinion

different *different from* is preferred; *different to* is sometimes used; *different than* is better not used

diffraction

diffusion; diffusible

dilemma means to be faced with a choice between two unhappy alternatives; should not be used as a synonym for *difficulty*

diplex; duplex *diplex operation* means the simultaneous transmission of two signals using a single feature, e.g. an antenna; *duplex operation* means that both ends can transmit and receive simultaneously

direct current abbr: **dc**

directly def: immediately; do not use when the meaning is as soon as

disassemble never use *dissemble* when the meaning is to take apart

disassociate see *dissociate*

disc; disk both are correct and commonly used; **disk** pref

discernible

discreet; discrete *discreet* means prudent or discerning: *his answer was discreet; discrete* means individually distinctive and separate: *discrete channels; discretion* is formed from *discreet,* not from *discrete*

disinterested; uninterested *disinterested* means unbiased, impartial; *uninterested* means not interested

dispatch; despatch *dispatch* pref

disseminate

dissimilar

dissipate

dissociate; disassociate *dissociate* pref

distil(l) *distil* pref; but always *distilled, distillate, distillation*

don't; doesn't such contractions should not appear in technical writing

donut; doughnut for electronics/nucleonics, use *donut*

doppler capitalize only when referring to the Doppler principle

down- as a prefix combines into one word: *downrange, downtime, downwind*

dozen abbr: **doz**

drawing(s) abbr: **dwg**

drier; dryer the adjective is always *drier;* the pref noun is *dryer: this material is drier; place the others back in the dryer*

drop; droppable; dropped; dropping

due to an overused expression; *because of* pref

duo- a prefix meaning two; combines to form one word: *duocone, duodiode, duophase*

duplex see **diplex**

E

each abbr: **ea**

east capitalize only if *east* is part of the name of a specific location: *East Africa;* otherwise use lc letters: *the east coast of the U.S.;* abbr: **E**; the abbr for *east-west* (control, movement) is **E-W**; *eastbound* and *eastward* are written as one word

eccentric

echo; echoes

economic; economical use *economical* to describe economy (of funds, effort, time); use *economic* when writing about economics: *an economical operation* (it did not cost much); *an economic disaster* (refers to economics)

effect see **affect**

efficacy; efficiency *efficacy* means effectiveness, ability to do the job intended; *efficiency* is a measurement of capability, the ratio of work done to energy expended: *we hired a consultant to assess the efficacy of our training methods; the powerhouse is to have a high-efficiency boiler*

e.g. pref abbr for *exempli gratia* (the Latin of *for example*); avoid confusing with **i.e.**; no comma is necessary after *e.g.;* may also be abbr **eg**

eighth

electric(al) if in doubt, use *electric;* generally, *electric* means produces or carries electricity, whereas *electrical* means related to the generation or carrying of electricity; abbr: **elec**

electro- a prefix generally meaning pertaining to electricity; it normally combines to form one word: *electro-acoustic, electroanalysis, electrodeposition, electromechanical, electroplate;* if the combining word starts with *o,* insert a hyphen: *electro-optics, electro-osmosis*

electromagnetic units abbr: **emu**

electromotive force abbr: **emf**

electronic(s) use *electronic* as an adjective, *electronics* as a noun: *electronic countermeasures; your career in electronics*

electron volt(s) abbr: **eV** (pref) or **ev**

electrostatic units abbr: **esu**

elevation abbr: **el**

ellipse

embedded

emigrate; immigrate *emigrate* means to go away from; *immigrate* means to come into

emit, emitter, emittance; emission; emissivity

enamel(l)ed; enamel(l)ing single *l* pref

encase; incase *encase* pref

encipher

enclose; inclose *enclose* pref; *inclose* is used mainly as a legal term

endorse, indorse *endorse* pref

engineer; engineering

enquire; inquire *inquire* pref

enrol; enroll both are correct, but *enroll* pref; universal usage prefers *ll* for *enrolled* and *enrolling,* but only a single *l* for *enrolment*

en route def: on the road, on the way; never use *on route*

ensure; insure; assure use *ensure* (or *insure*) when the meaning is to make certain of: *use the new oscilloscope to ensure accurate calibration;* use *insure* when the meaning is to protect against financial loss: *we insured all our drivers;* use *assure* when the meaning is to state with confidence that something has been or will be made certain: *he assured the meeting that production would increase by 8%*

entrust; intrust *entrust* pref

envelop; envelope *envelop* is a verb which means to surround or cover completely; *envelope* is a noun that means a wrapper or a covering

environment(al)

equal *equality* and *equalize* have only one *l; equally* always has *ll; equaled* and *equaling* preferably have only one *l,* but sometimes are seen with *ll*

equi- a prefix that means equality; combines to form one word: *equiphase, equipotential, equisignal*

equilibrium(s)

equip; equipped; equipping; equipment

equivalent abbr: **equiv**

erase; erasable

errata although *errata* is plural (from the singular *erratum,* which is seldom used), it can be used as a singular or plural noun: *the errata is complete* is acceptable usage

erratic

especially; specially *specially* pref when it refers to an adjective (*a specially trained crew*); *especially* should introduce a phrase (*they were all well trained, especially the computer technicians*)

esthetic pref spelling

et al. def: and others

et cetera abbr: **etc**; def: and so forth, and so on; use with care in technical writing: *etc* can create an impression of vagueness or unsureness: *the transmitters, etc, were tested* is much less definite than either *the transmitter, modulator, and power supply were tested,* or (if to restate all the equipment is too repetitious) *the transmitting equipment was tested*

extremely high frequency abbr: **ehf**

everybody; every body *everybody* means every person, or all the persons; *every body* means every single body: *everybody was present; every body was examined for gunpowder scars*

everyone; every one *everyone* means every person, or all the persons; *every* *one* means every single item: *everyone is insured; every one had to be tested in a saline solution*

exa def: 10^{18}; abbr: **E**

exaggerate

except def: to exclude; see **accept**

exhaust

extracurricular

extraordinary

F

Fahrenheit abbr: **F**; see **temperature**

fail-safe

fallout one word in the noun form

familiarize; familiarization

farad def: a unit of electric capacitance; abbr: **F**; other abbr: **μF, nF, pF**

farther; further *farther* means greater distance: *he traveled farther than the other salesmen; further* means a continuation of (as an adjective) or to advance (as a verb): *the promotion was a further step in his career plan,* and *to further his education, he took a part-time course in industrial drafting*

fasten(er)

faultfinder; faultfinding

feasible

feet; foot abbr: **ft;** other abbr:

feet board measure (board feet)	**fbm**
feet per minute	**fpm**
feet per second	**fps**
foot-candle(s)	**fc** (pref); **ft-c**
foot-pound(s)	**fp** (pref); **ft-lb**
foot-pound-second (system)	**fps** system

femto def: 10^{-15}; abbr: **f**; other abbr:

femotampere(s)	**fA**
femtovolt(s)	**fV**

ferri- a prefix meaning contains iron in the ferric state; combines to form one word: *ferricyanide, ferrimagnetic*

ferro- a prefix meaning contains iron in the ferrous state; combines to form one word: *ferroelectric, ferromagnetic, ferrometer*

ferrule; ferule a *ferrule* is a metal cap or lid; a *ferule* is a ruler

fewer; less use *fewer* to refer to items that can be counted: *fewer technicians than we predicted have been assigned to the project;* use *less* to refer to general quantities: *there was less water available than predicted*

fiber; fibrous; Fiberglas *Fiberglas* is a trade name

figure numbers in text, spell out the word *Figure* in full, or abbr it to **Fig**: *the circuit diagram in Figure 26* and *for details, see Fig. 7;* use the abbreviated form beneath an illustration; always use numerals for the figure number

final; finally; finalize

first to write *the first two . . .* (or three, etc.) is better than to write *the two first . . .;* never use *firstly*

fix in technical usage, *fix* means to firm up or establish as a permanent fact; avoid using it when the meaning is repair

flammable def: easily ignited; see **inflammable**

flight usually combines to form two words: *flight control, flight deck, flight plan*

flip-flop

flowchart

fluid abbr: **fl**; the abbr for fluid ounce is **fl oz**

fluorescence

focus; focused; focuses; focusing pl: *focuses* (pref) or *foci*

foot see **feet**

for see **because**

forecast this spelling applies to both present and past tenses

foresee

forestall

foreword; forward a *foreword* is a preface or preamble to a book; *forward* means onward: *the scope is defined in the foreword to the book; he requested that we bring the meeting date forward*

for example abbr: **e.g.** (pref) or **eg**

former; first use *former* to refer to the first of only two things; use *first* if there are more than two

formula pl: *formulas* (pref) or *formulae*

forty def: 40; it is not spelled *fourty*

fourth def: 4th; it is not spelled *forth*

fractions when writing fractions that are less than unity, spell them out in descriptive narrative, but use figures for technical details: *by the end of the heat run, nine-tenths of the installation had been completed; a flat case 15¼ mm square by 7/8 mm deep;* use decimals rather than fractions, except when a quantity is normally stated as a fraction (such as *3/8 in. plywood*)

free from use *free from* rather than *free of: he is free from prejudice*

free on board abbr: **fob** (pref), **f.o.b.** (commonly used), or **FOB**

frequency abbr: **freq**

frequency modulation abbr: **FM**

fulfil(l) the pref spelling is *fulfill, fulfilled, fulfilling,* and *fulfillment; fulfil* and *fulfilment* can also be spelled with a single *l.*

funnel(l)ed; funnel(l)ing single *l* pref

further see **farther**

fuse as a verb, means join together or weld; as a noun, means a circuit protection device

fuze def: a detonation initiation device

G

gage; gauge both spellings are correct; gage is recommended because it is less likely to be misspelled; *gaging* and *gauging* do not retain the *e*

gallon gallons differ between U.S. (231 in.^{3}; 3.785 dm^3) and Britain (277.42 in.^{3}; 4.546 dm^3); abbr: **gal**; other abbr:

gallons per day	**gpd**
gallons per hour	**gph**
gallons per minute	**gpm**
gallons per second	**gps**

gang; ganged; ganging

gas; gases; gassed; gassing; gaseous; gassy

gauge see **gage**

geiger (counter)

gelatin(e) *gelatin* pref

geo- a prefix meaning of the earth; combines to form one word: *geocentric, geodesic, goemagnetic, geophysics*

giga def: 10^9; abbr: **G**; other abbr:

gigabecquerel(s)	**GBq**
gigahertz	**GHz**
gigajoule(s)	**GJ**
gigaohm(s)	**GΩ; Gohm**
gigapascal(s)	**GPa**
gigavolt(s)	**GV**

gimbal

glue; gluing; gluey

glycerin(e) *glycerin* pref

gotten never use this expression in technical writing; use *have got* or simply *have*

government capitalize when referring to a specific government either directly or by implication; use lc if the meaning is government generally: *the U.S. Government; the Government specifications; no government would sanction such restrictions*

gram abbr: **g**; abbr for gram-calorie is **g-cal**

grammar; grammatical(ly)

gray def: absorbed dose of ionizing radiation (SI); abbr: **Gy**; other abbr: **mGy, μGy**; in SI, the *gray* replaces the *rad*

Greenwich mean time abbr: **GMT**

grill(e) def: a loudspeaker covering, or a grating; *grille* pref

ground (electrical) abbr: **gnd**

guage wrongly spelled; the correct spelling is *gauge* or *gage* (pref)

guarantee never *guaranty*

guideline(s)

gyroscope abbr: **gyro**

H

half; halved; halves; halving as a prefix, *half* combines erratically; some common compounds are: *half-hour(ly), half-life, half-month(ly), halftone, halfwave;* for others, consult your dictionary

handful; handfuls

hangar; hanger a *hangar* is a large building for housing aircraft; *a hanger* is a supporting bracket

haversine abbr: **hav**

H-beam

heat- as a prefix, *heat* combines erratically; some typical compounds are: *heat-resistant, heat-run, heatsink, heat-treat;* for others, consult your dictionary

heavy-duty

hectare def: a large unit of area, used in surveying and agriculture; in SI, *hectare* replaces *acre;* abbr: **ha**

height (not *heighth*) abbr: **ht**

heightfinder; heightfinding

helix pl: *helices* (pref) or *helixes*

hemi- a prefix meaning half; combines to form one word: *hemisphere, hemitropic*

henry def: a unit of inductance; abbr: **H;** other abbr: **mH, μH, nH, pH**

here- when used as a prefix, combines to form one word: *hereafter, hereby, herein, herewith*

hertz def: a unit of frequency measurement (similar to *cycle per second,* which it replaces); abbr: **Hz;** other abbr: **THz, GHz, MHz, kHz**

heterodyne

heterogeneous; homogeneous *heterogeneous* means of the opposite kind; *homogeneous* means of the same kind

high fidelity abbr: **hi-fi**

high frequency abbr: **hf**

high-pressure (as an adjective) abbr: **h-p**

high voltage abbr: **hv** (pref) or **HV**

hinge; hinged; hinging

homogeneous see **heterogeneous**

horizontal abbr: **hor**

horsepower abbr: **hp;** the abbr for horse-power-hour is **hp-hr**

hour(s) abbr: **hr** or **h** (SI)

hundred abbr: **C**

hundredweight def: 112 lb; abbr: **cwt**

hybrid

hydro- a prefix meaning of water; combines to form one word: *hydroacoustic, hydroelectric, hydromagnetic, hydrometer*

hyper- a prefix meaning over; combines to form one word: *hyperacidity, hypercritical*

hyperbola the plural is *hyperbolas* (pref) or *hyperbolae*

hyperbole def: an exaggerated statement

hyperbolic cosine; sine; tangent abbr: **cosh, sinh, tanh**

hyphen in compound terms you may omit hyphens unless they need to be inserted to avoid ambiguity or to conform to accepted usage; e.g. *preemptive* is preferred without a hyphen, but *photo-offset* and *re-cover* (when the meaning is *to cover again*) both require one; refer to individual entries and Article 1

hypothesis pl: *hypotheses*

I

I-beam

ibid. def: Latin abbr for *ibidem,* meaning in the same place; used in footnoting

i.e. pref abbr for *id est* (the Latin of *that is*); avoid confusing with **e.g.**; no comma is necessary after *i.e.;* may also be abbr **ie**

if and when avoid using this expression; use either *if* or *when*

ignition abbr: **ign**

illegible

im- see **in-**

imbalance this term should be restricted for use in accounting and medical terminology; use *unbalance* in other technical fields

immalleable

immigrate see **emigrate**

immittance

immovable

impeller

imperceptible

impermeable

impinge; impinging

imply; infer speakers and writers can *imply* something; listeners and readers

imply; infer *(continued)* *infer* it from what they hear or read: *in his closing remarks, Mr. Smith implied that further studies were in order; the technician inferred from the report that his work was better than expected*

impracticable; impractical *impracticable* means not feasible; *impractical* means not practical; a less-preferred alternative for impractical is *unpractical*

in; into *in* is a passive word; *into* implies action: *ride in the car; step into the car*

in-; im-; un- all three prefixes mean not; all combine to form one word: *ineligible, impermeable, unintelligible;* if you are not sure whether you should use in-, im-, or un-, use *not*

inaccessible

inaccuracy

inadvisable, unadvisable *inadvisable* pref

inasmuch as a better word is *since*

inaudible

incandescence; incandescent

incase *encase* pref

inch(es) abbr: **in.**; other abbr:

inches per second	**ips** (pref); **in./s**
inch-pound(s)	**in.-lb**

incidentally in most cases the word *incidentally* is unnecessary

inclose use **enclose**

includes; including abbr: **incl**; when followed by a list of items, *includes* implies that the list is not complete; if the list is complete, use *comprises* or *consists of* (which see)

incomparable

incompatible

incur; incurred; incurring

index pl: *indexes* pref, except in mathematics (where *indices* is common)

indicated horsepower abbr: **ihp**; the abbr for indicated horsepower-hour is **ihp-hr**

indifferent to never use *indifferent of*

indispensable

indorse *endorse* pref

industrywide

inessential; unessential both are correct; *unessential* pref

infer; inferred; inferring; inference also see **imply**

inflammable def: easily ignited (derived from *inflame*); *flammable* is a better word: it prevents readers from mistakenly thinking the *in* of *inflammable* means not

infrared

ingenious; ingenuous *ingenious* means clever, innovative; *ingenuous* means innocent, naive; *ingenuity* is a noun derived from ingenious

inoculate

inoperable not *inoperatable*

inquire; enquire *inquire* pref; also *inquiry*

inseparable

inside diameter abbr: **ID**

in situ def: in the normal position

insofar as

instal(l) *install, installed, installer, installing, installation,* and *installment* pref; a single *l* is acceptable for *instal and instalment*

instantaneous

instrument

insure the pref def is to protect against financial loss; can also mean make certain of; see **ensure**

integer

integral, integrate, integrator

intelligence quotient abbr: **IQ**

intelligible

inter- a prefix meaning among or between; normally combines to form one word: *interact, intercarrier, interdigital, interface, intermodulation, interoffice*

intermediate frequency abbr: **if.** (pref) or **i-f**

intermediate-pressure (as an adjective) abbr: **i-p**

intermittent

internal abbr: **int**

interrupt

into see **in**

intra- a prefix meaning within; normally combines to form one word: *intranuclear;* if combining word starts with *a,* insert a hyphen: *intra-atomic*

intractable

intrigue; intrigued; intriguing

intrust *entrust* pref

irrational

irregardless never use this expression; use *regardless*

irrelevant frequently misspelled as *irrevelant*

irreversible

iso- a prefix meaning the same, of equal size; normally combines to form one word: *isoelectronic, isometric, isotropic;* if combining word starts with *o,* insert a hyphen: *iso-octane*

its; it's *its* means belonging to; *it's* is an abbr for it is: *the transmitter and its modulator; if the fault is not in the remote equipment, then it's most likely in master control;* in technical writing *it's* should seldom be used: replace with *it is*

J

joule def: a unit of energy, work, or quantity of heat (SI); abbr: **J**; other abbr: **TJ, GJ, MJ, kJ, mJ, J/m³, J/K** (joule(s) per kelvin), **J/kg, J/mol** (joule(s) per mole)

judg(e)ment *judgment* pref

juxtaposition

K

kelvin def: the SI unit for thermodynamic temperature; abbr: **K**

kerb; curb *curb* pref

key- as a prefix normally combines to form one word: *keynote, keypunch*

kilo def: 10^3; abbr: **k**; other abbr:

kiloampere(s)	**kA**
kilobecquerel(s)	**kBq**
kilocalorie(s)	**kcal**
kilocoulomb(s)	**kC**
kilogram(s), (see **kilogram**)	**kg**
kilohertz	**kHz**
kilohm(s)	**kΩ; kohm**
kilojoule(s)	**kJ**
kiloliter(s)	**kL**
kilometer(s)	**km**
kilometers per hour	**km/h**
kilomole(s)	**kmol**
kilonewton(s)	**kN**
kilopascal(s)	**kPa**
kilosecond(s)	**ks** (pref); **ksec**
kilosiemens	**kS**
kilovolt(s)	**kV**
kilovolt-ampere(s)	**kVA**
kilovolt-ampere(s), reactive	**kVAr**
kilowatt(s)	**kW**
kilowatthour(s)	**kWh** (pref); **kw-hr**

kilogram def: the SI unit for mass; abbr: **kg**; other typical abbr: **Mg, g, mg, μg;** also:

kilogram-calorie(s)	**kg-cal**
kilogram(s) per meter	**kg/m**
kilogram(s) per square meter	**kg/m²**
kilogram(s) per cubic meter	**kg/m³**
kilogram meter(s) per second	**kg·m/s**

knockout as noun or adjective, one word

knot abbr: **kn**

L

label(l)led; label(l)ing single *l* pref

laboratory abbr: **lab**

lacquer

lambert abbr: **L**; use the abbr **L** with care: it is also the SI abbr for *liter*

lampholder

last, latest, latter *last* means final; *latest* means most recent; *latter* refers to the second of only two things (if more than two, use *last*); it is better to write *the last two* (or three, etc.) than *the two last*

lath; lathe a *lath* is a strip of wood; a *lathe* is a machine

latitude abbr: **lat** or **Φ**

learned; learnt *learned* pref

least common multiple abbr: **lcm**

left-hand(ed) abbr: **LH**

lend; loan use *lend* as a verb, *loan* as a noun; to write or say "*loan* me your iron" is wrong, but "*lend* me your iron" is correct

length the SI unit of length is the *metre* (which see), expressed in multiples and submultiples of *kilometres* (**km**), *metres* (**m**), and *millimetres* (**mm**)

lengthy not *lengthly*

less see **fewer**

letter of intent; letter of transmittal pl: *letters of intent* or *transmittal*

level(l)ed; level(l)ing single *l* pref

liable to means under obligation to; avoid using as a synonym for *apt to* or *likely to*

liaison liaison is a noun; it is sometimes used uncomfortably as a verb: *liaise*

libel(l)ed; libel(l)ous single *l* pref

licence; license *license* pref: *licence* is sometimes used as a noun

lightening; lightning *lightening* means to make lighter; *lightning* is an atmospheric discharge of electricity

likable

linear abbr: **lin**; the abbr for lineal foot is **lin ft**

lines of communication not *line of communications*

liquefy; liquefaction

liquid abbr: **liq**

liter; litre the SI spelling is **litre**, but in U.S. **liter** is more common; abbr: **L**; other abbr: **kL, mL, μL**; the abbr for *liter(s) per day/hour/minute/second* are **L/d, L/h, L/m, L/s**

loan see **lend**

loath; loathe *loath* means reluctant; *loathe* means to dislike intensely

locknut; lockwasher

locus pl: *loci*

logarithm abbr: common—**log**; natural —**ln**

logbook

logistic(s) use *logistic* as an adjective, *logistics* as a noun: *logistic control; the logistics of the move*

longitude abbr: **long.** or **λ**

long-play(ing) (record) abbr: **LP**

lose; loose *lose* is a verb that refers to a loss; *loose* is an adjective or a noun that means free or not secured: *three loose nuts caused us to lose a wheel*

louver; louvre *louver* pref

low frequency abbr: **lf**

low-pressure (as an adjective) abbr: **l-p**

lubricate; lubrication abbr: **lub**

lumen def: a unit of luminous flux (SI); abbr: **lm**; other abbr:

lumen-hour(s)	**lm·h** (pref); **lm-hr**
lumens per square foot	**lm/ft²**
lumens per square meter	**lm/m²**
lumens per watt	**lm/W**
lumen-second(s)	**lm·s**
microlumen(s)	**μ lm**
millilumen(s)	**mlm**

luminance; luminescence; luminosity; luminous

lux def: a unit of illuminance (SI); abbr: **lx**; other abbr: **klx**

M

Mach

macro- a prefix meaning very large; combines to form one word: *macroscopic*

magneto pl: *magnetos;* as a prefix, it normally combines to form one word: *magnetoelectronics, magnetohydrodynamics, magnetostriction*; if combining word starts with *o* or *io*, insert a hyphen: *magneto-optics, magneto-ionization*

magneton; magnetron a *magneton* is a unit of magnetic moment; a *magnetron* is a vacuum tube controlled by an external magnetic field

maintain; maintenance

majority use *majority* mainly to refer to a number, as in *a majority of 27;* avoid using it as a synonym for many or most; do not write *the majority of technicians* when the intended meaning is *most technicians*

malfunction

malleable

man; manned; manning; man-hour(s); manpower

manage(d); manageable; managing

maneuver; manoeuvre *maneuver, maneuvered, maneuvering* pref

manufacturer abbr: **mfr**

marshal(l)ed; marshal(l)ing; marshal(l)-er single *l* pref

marvel(l)ed; marvel(l)ing; marvel(l)ous single *l* pref

mass see **kilogram**

material; materiel *material* is the substance or goods out of which an item is made; when used in the plural, it describes items of a like kind, such as *writing materials; materiel* are all the equipment and supplies necessary to support a project or undertaking (a term commonly used in military operational support)

matrix pl: *matrices*

maximum pl: *maximums* (pref) or *maxima;* abbr: **max**; like *minimize, maximize* can be used as a verb

meager; meagre *meager* pref

mean; median the *mean* is the average of a number of quantities; the *median* is the midpoint of a sequence of numbers; e.g. in the sequence of five numbers 1, 2, 3, 7, 8, the mean is 4.2 and the median is 3

mean effective pressure abbr: **mep**

mean sea level *abbr:* **msl** (pref) or **MSL**

medium when *medium* is used to mean substances, liquids, materials, or the means for accomplishing something (such as advertising), the plural is *media,* in all other senses the plural is *mediums*

mega def: 10^6; abbr: **M**; other abbr:

megacoulomb(s)	**MC**
megaelectronvolt(s)	**MeV**
megahertz	**MHz**
megajoule(s)	**MJ**
meganewton(s)	**MN**
megapascal(s)	**MPa**
megavolt(s)	**MV**
megawatt(s)	**MW**
megohm(s)	**MΩ; Mohm**

memorandum pl: *memorandums* (pref) or *memoranda;* abbr: **memo** (singular) or **memos** (plural)

merit; merited; meriting

metal a single *l* is pref for *metaled* and *metaling* (although *ll* also is acceptable); *metallic* and *metallurgy* always have *ll*

meteorology; metrology *meteorology* pertains to the weather; *metrology* pertains to weights, measures, and calibration

meter; metre def: metric unit of length; the SI spelling is *metre,* but in U.S. *meter* is more common; abbr: **m**; other typical abbr:

square meter(s)	$\mathbf{m^2}$
cubic meter(s)	$\mathbf{m^3}$
meters per second	**m/s**
newton meter(s)	**N·m**
newtons per square meter	$\mathbf{N/m^2}$
kilogram(s) per cubic meter	$\mathbf{kg/m^3}$

micro def: 10^{-6}; abbr: **μ** (pref) or **u**; other abbr:

microampere(s)	**μA**
microcoulomb(s)	**μC**
microfarad(s)	**μF**
microgram(s)	**μg**
microgray(s)	**μGy**
microhenry(s)	**μH**
microhm(s)	**μΩ; μohm**
microlumen(s)	**μlm**
micromho(s)	**μmho**
micrometer(s)	**μm**
micromole(s)	**μmol**
micronewton(s)	**μN**
micropascal(s)	**μPa**
microsecond(s)	**μs** (pref); **μsec**
microsiemens	**μS**
microtesla(s)	**μT**
microvolt(s)	**μV**
microwatt(s)	**μW**

as a prefix meaning very small, **micro-** normally combines to form one word: *microameter, micrometer, microorganism, microprocessor, microswitch, microwave;* the term *micromicro-* (10^{-12}) has been replaced by **pico** (which see)

microphone abbr: **MIC** (pref) or **mike**

mid- a prefix that means in the middle of; generally combines into one word: *midday, midpoint, midweek;* if used with a proper noun, insert a hyphen: *mid-Atlantic*

mile the word mile is generally understood to mean a statute mile of 5280 ft (1609 m), so the statement *I drove 326 miles* implies statute miles; when referring to the *nautical mile* (6080 ft; 1853 m), always identify it as such: *the flight*

mile *(continued)*
distance was 4210 nautical miles (or *4210 nmi*); abbr:

statute mile(s)	**mi**
nautical mile(s)	**nmi** (pref); **n.m.**
miles per gallon	**mpg**
miles per hour	**mph**

mileage; milage *mileage* pref

milli def: 10^{-3}; abbr: **m**; other abbr:

milliampere(s)	**mA**
millicoulomb(s)	**mC**
millicurie(s)	**mCi**
millifarad(s)	**mF**
milligram(s)	**mg**
milligray(s)	**mGy**
millihenry(s)	**mH**
millijoule(s)	**mJ**
millikelvin(s)	**mK**
milliliter(s)	**mL**
millilumen(s)	**mlm**
millimeter(s)	**mm**
millimho(s)	**mmho**
millimole(s)	**mmol**
milliohm(s)	**mΩ; mohm**
millinewton(s)	**mN**
millipascal(s)	**mPa**
milliroentgen(s)	**mR**
millisecond(s)	**ms** (pref); **msec**
millisiemens	**mS**
millitesla(s)	**mT**
millivolt(s)	**mV**
milliwatt(s)	**mW**
milliweber(s)	**mWb**

as a prefix, **milli-** combines to form one word: *milliammeter, milligram, millimicron*

millibar def: a unit of pressure (= 100 Pa); abbr: **mbar**

mini- as a prefix combines to form one word: *minicomputer, minireport*

miniature; miniaturization

minimum pl: *minimums* (pref) or *minima;* abbr: **min**

minority use mainly to refer to a number: *a minority by 2;* avoid using it as a synonym for several or a few; to write *a minority of the technicians* is incorrect when the intended meaning is *a few technicians*

minute abbr:

time	**min**
angular measure	′

mis- a prefix meaning wrong(ly) or bad(ly); combines to form one word: *misalign, misfired, mismatched, misshapen*

miscellaneous

miscible

misspelled; misspelt *misspelled* pref

miter(ed); mitre(d) *miter(ed)* pref

mnemonic

model(l)ed; model(l)er; model(l)ing single *l* pref

mold; mould *mold* pref

mole def: the SI unit for amount of substance; abbr: **mol** other abbr: **kmol, mmol, μmol, mol/m^3**

mono- a prefix meaning one or single; combines to form one word: *monopulse, monorail, monoscope*

monotonous

months the months of the year are always capitalized: *January, February,* etc; if abbr, use only the first three letters: *Jan, Feb,* etc; the abbr for *month* is **mo**

mortice; mortise *mortise* pref

mosaic

most never use as a short form for *almost;* to say *most everyone is here* is incorrect

movable; moveable *movable* pref

Mr.; Ms. address men as *Mr.* and women as *Ms.*; use *Miss* or *Mrs.* only if you know the person prefers to be so addressed; the period (punctuation) may be omitted after *Mr* and *Ms*

multi- a prefix meaning many; combines to form one word: *multiaddress, multicavity, multielectrode, multistage*

municipal; municipality

N

nano def: 10^{-9}; abbr: **n**; other abbr:

nanoampere(s)	**nA**
nanocoulomb(s)	**nC**
nanofarad(s)	**nF**
nanohenry(s)	**nH**
nanometer(s)	**nm**
nanosecond(s)	**ns** (pref); **nsec**

nanotesla(s)	**nT**
nanovolt(s)	**nV**
nanowatt(s)	**nW**

naphtha(lene)

nationwide

nautical mile def: 6080 ft (1853 m); abbr: **nmi** (pref) or **n.m.**; see **mile**

NB means note well, and is the abbr for *nota bene;* it's more common to use the word *NOTE*

NC abbr for *normally closed* (contacts)

nebula pl: *nebulas* (pref) or *nebulae*

negative abbr: **neg**

negligible

nevertheless

newton def: a unit of force (SI); abbr: **N**; other abbr: **MN, kN, mN, μN, N·m** (newton meter), **N/m** (newtons per meter)

next it is better to write *the next two* (or next three, etc) than *the two next* (etc.)

night never use *nite;* write *nighttime* as one word

nineteen; ninety; ninth all three are frequently misspelled

NO abbr for *normally open* (contacts)

No. abbr for **number** (which see)

noise-cancel(l)ing single *l* pref

nomenclature

nomogram; nomograph *nomogram* pref

non- as a prefix meaning not or negative, normally combines to form one word: *nonconductor, nondirectional, nonnegotiable, nonlinear, nonstop;* if combining word is a proper noun, insert a hyphen: *non-American;* avoid forming a new word with *non-* when a similar word that serves the same purpose already exists (i.e. you should not form *nonaudible* because *inaudible* already exists)

none when the meaning is "not one," treat as singular; when the meaning is "not any," treat as plural: *none* (not one) *was satisfactory; none* (not any) *of the receivers were repaired*

nonplus(s)ed *nonplused* pref

no one two words

norm def: the average or normal (situation or condition)

normalize

normally closed; normally open (contacts) abbr: **NC, NO**

normal to def: at right angles to

north abbr: **N**; other abbr:
northeast **NE**
northwest **NW**
north-south **N-S** (control, movement)
northbound and *northward* are written as one word; for rule on capitalization, see **east**

not applicable abbr: **N/A**

note well abbr: **NB** (derived from *nota bene),* but *NOTE* is more common

NOT-gate

noticeable

not to exceed an overworked phrase that should be used only in specifications; in all other cases use *not more than*

nth (harmonic, etc.)

nucleus the plural is *nuclei*

null

number although **no.** would appear to be the most logical abbr for number (and is pref), **No.** is much more common (the symbol # is not an abbr for number); the abbr *no.* or *No.* must always be followed by a quantity in numerals; it is incorrect to write *we have received a No. of shipments;* for the difference in usage between *amount* and *number,* see **amount**

numbers (in narrative) as a general rule, spell out up to and including nine, and use numerals for 10 and above; for specific rules, see Article 4

O

oblique; obliquity

oblivious def: unaware or forgetful; oblivious should be followed by *of,* not *to: absorbed in his work, he was oblivious of the disturbance caused by the installation crew*

obstacle

obtain; secure use *obtain* when the meaning is simply to get; use *secure* when the meaning is to make safe or to take possession of (possibly after some difficulty): *we obtained four additional samples; we secured space in the prime display area*

occasional(ly)

occur; occurred; occurrence; occurring

o'clock avoid using; see **time**

of avoid using in place of *have;* write *we should have measured,* not *we should of measured*

off- as a prefix either combines into one word, or a hyphen is inserted: *offset, off-center(ed), off-scale, off-the-shelf*

off of an awkward construction; omit the word *of*

ohm def: a unit of electric resistance; abbr: **Ω** or **ohm**; other abbr: **GΩ, Gohm, MΩ, Mohm, kΩ, kohm, mΩ, mohm, μΩ, uohm**; abbr for ohm-centimeter(s) is **ohm-cm;** *ohmmeter* has two *m*'s

oil-filled

OK; okay these are slang expressions which should never appear in technical writing

omit; omitted; omission

omni- a prefix meaning all or in all ways; combines to form one word: *omnibearing, omnidirectional, omnirange*

once-over

onward(s) *onward* pref

op. cit. def: Latin abbr for *opere citato,* meaning in the work cited; used in footnoting

operate; operator; operable not *operatable*

optimum pl: *optima* (pref), and sometimes *optimums*

oral def: spoken; see **aural**

orbit; orbital; orbited; orbiting

OR-gate

orientation this is the noun; the verb form is *orient, oriented, orienting*

orifice

origin; original; originally

oscillate

oscilloscope slang abbr: **scope**

ounce(s) abbr: **oz**; other abbr:

ounce-foot	**oz-ft**
ounce-inch	**oz-in.**

out- as a prefix normally combines to form one word: *outbreak, outcome, outdistance;* but when *out-* is followed by *of,* insert hyphens: *out-of-date, out-of-phase*

outside diameter abbr: **OD**

outward(s) *outward* pref

over- as a prefix meaning above or beyond, normally combines to form one word: *overbunching, overcurrent, overdriven, overexcited, overrun;* avoid using as a synonym for more than, particularly when referring to quantities: *more than 17 were serviceable* is better than *over 17 were serviceable*

overage means either too many or too old

overall an overworked word; as an adjective it often gives unnecessary additional emphasis (as in *overall impression*) and should be deleted; avoid using as a synonym for *altogether, average, general,* or *total*

oxidize *oxidation* is better than *oxidization*

oxyacetylene

P

page; pages abbr: **p; pp**

paid not *payed,* when the meaning is spent

pair(s) abbr: **pr**

pamphlet

panel(l)ed; panel(l)ing single *l* pref

paperwork

parabola(s); parabolic; paraboloid

paragraph(s) abbr: **para**

parallax

parallel; paralleled; paralleling; parallelism; parallelogram both *parallel to* and *parallel with* are correct

parenthesis this is the singular form; pl: *parentheses*

particles

partly; partially use *partly* when the meaning is "a part" or "in part"; use *partially* when the meaning is "to a certain extent," or when preference or bias is implied.

parts per million abbr: **ppm**

part-time

pascal def: a unit of pressure (SI); abbr: **Pa**; other abbr: **Gpa, Mpa, kPa, mPa, μPa, pPa, Pa·s** (pascal second)

passed; past as a general rule, use *passed* as a verb and *past* as an adjective or a noun: *the test equipment has been* *passed by quality control; past experience has demonstrated a tendency to fail at low temperature; in the past . . .*

pencil(l)ed; pencil(l)ing single *l* pref

pendulum pl: *pendulums*

people; persons *people* pref: *all the people were present;* use *persons* to refer only to small numbers of people: *three persons were interviewed* (and even here, *people* could be used)

per in technical writing it is acceptable to use *per* to mean either by or a(n), as in *per diem* (by the day) and *miles per hour;* in literary writing, take care not to use *per* in place of *a* or *an;* avoid using *as per* in all writing

percent; percentage the abbr for percent is %; avoid using the expression *a percentage of* as a synonym for *a part of* or *a small part*

perceptible

permeable; permeameter; permeance

permissible

permit; permitted; permitting; permittivity

perpendicular abbr: **perp**

perseverance

persistent; persistence; persistency

personal; personnel *personal* means concerning one person; *personnel* means the members of a group, or the staff: *a personal affair;* the *personnel in the powerhouse;* for *person(s)* see **people**

peta def: 10^{15}; abbr: **P;** other abbr: **PBq** (petabecquerel)

pharmacy; pharmacist; pharmaceutical

phase in the nonelectric sense, *phase* means a stage of transmission or development; it should not be used as a synonym for aspect; it is used correctly in *the second phase called for a detailed cost breakdown*

phase-in; phaseout but use two words for the verb forms: *to phase in, to phase out*

phenolic

phenomenon pl: *phenomena*

photo- as a prefix, normally combines to form one word: *photoelectric, photogrammetry, photoionization, photomultiplier;* if combining word starts with *o*, insert a hyphen: *photo-offset*

pico def: 10^{-12}; abbr: **p;** other abbr:

picoampere(s)	**pA**
picocoulomb(s)	**pC**
picofarad(s)	**pF**
picohenry(s)	**pH**
picosecond(s)	**ps** (pref); **psec**
picowatt(s)	**pW**

piezoelectric; piezo-oscillator

pint abbr: **pt**

pipeline

plateau pl: *plateaus* (pref) or *plateaux*

plug; plugged; plugging

plumbbob; plumb line

p.m. def: after noon (post meridiem)

pneumatic

polarize; polarization; polarizing

poly- a prefix meaning many; combines to form one word: *polydirectional, polyethylene, polyphase*

polyvinyl chloride abbr: **pvc**

positive abbr: **pos**

post- a prefix meaning after or behind; combines to form one word: *postacceleration, postgraduate*

post meridiem def: after noon; abbr: **p.m.** can also be written as *postmeridian* (less pref)

post office

postpaid

potentiometer abbr: **pot.**

pound(s) (weight) abbr: **lb;** other abbr:

pound-foot	**lb-ft**
pound-inch(es)	**lb-in.**
pounds per square foot	**psf** (pref); **lb/ft²**
pounds per square inch	**psi** (pref); **lb/in.²**
pounds per square inch, absolute	**psia**

power factor abbr: **pf** or spell out

powerhouse; power line; powerpack

practicable; practical these words have similar meanings but different applications that sometimes are hard to identify; *practicable* means feasible to do: *it was difficult to find a practicable solution*

practicable; practical *(continued)* (one that could reasonably be implemented); *practical* means handy, suitable, able to be carried out in practice: *a practical solution would be to combine the two departments*

practice; practise the noun always is *practice;* the verb is *practice* (pref), but can also be *practise*

pre- a prefix meaning before or prior; normally combines to form one word: *preamplifier, predetermined, preemphasis, preignite, preset;* if combining word is a proper noun, insert a hyphen: *pre-Roman*

precede; proceed *precede* means go before; *proceed* generally means carry on or continue: *the dinner was preceded by a brief business meeting; after dinner, we proceeded with the annual presentation of awards;* see **proceed**

precedence; precedent *precedence* means priority (of position, time, etc.): *the pressure test has precedence* (it must be done first); a *precedent* is an example that is or will be followed by others: *we may set a precedent if we grant his request* (others will expect similar treatment)

prefer; preferred; preference; preferable avoid overstating *preferable* (which states its meaning quite clearly on its own) by using it to make a comparison; e.g. it is wrong to write *more preferable* or *highly preferable*

presently use *presently* only to mean soon or shortly; never use it to mean *now* (use *at present* instead)

pressure-sensitive

pretense; pretence *pretense* pref

preventive; preventative *preventive* pref

previous def: earlier, that which went before; avoid writing *previous to* (use *before*); see **prior**

principal; principle as a noun, *principal* means (1) the first one in importance, the leader; or (2) a sum of money on which interest is paid: *one of the firm's principals is Mr. H. Winman; the invested principal of $10,000 earned $650 in interest last year;* as an adjective, *principal* means *most important* or *chief; principle* means a strong guiding rule, a code of conduct, a fundamental or primary source (of information, etc): *his principles prevented him from taking advantage of the error*

prior; previous use only as adjectives meaning earlier: *he had a prior appointment,* or *a previous commitment prevented Mr. Perchanski from attending the meeting;* write *before* rather than *prior to* or *previous to*

privilege

proceed; proceeding; procedure use *proceed to* when the meaning is to start something new; use *proceed with* when the meaning is to continue something that was started previously

program; program(m)ed; program(m)-ing; program(m)er use of single and double *m* varies widely; double *m* pref

prohibit use *prohibit from;* never *prohibit to*

prominent; prominence

propel; propelled; propelling; propellant (noun); **propellent** (adjective); **propeller, propellor**: *er* pref

prophecy; prophesy use *prophecy* only as a noun, *prophesy* only as a verb

proportion avoid writing *a proportion of* or *a large proportion of* when *some, many,* or a specific quantity would be simpler or more direct

proposition in its proper sense, *proposition* means a suggestion put forward for argument; it should not be used as a synonym for *plan, project,* or *proposal*

pro rata def: assign proportionally; sometimes used in the verb form as *prorate: I want you to prorate the cost over two years' operations*

protein

proved; proven use *proven* only as an adjective or in the legal sense; otherwise use *proved: he has been proven guilty; he proved his case*

psycho- as a prefix normally combines to form one word: *psychoanalysis, psychopathic, psychosis;* if combining word starts with *o,* insert a hyphen: *psycho-organic*

purge; purging

Q

quality control abbr: **QC**

quantity; quantitative the abbr of quantity is **qty**

quart abbr: **qt**

quasi- a prefix meaning seemingly or almost; insert a hyphen between the prefix and the combining word: *quasi-active, quasi-bistable, quasi-linear*

question mark insert a question mark after a direct question: *how many booklets will you require?;* omit the question mark when the question posed is really a demand: *may I have your decision by noon on Monday*

questionnaire

quiescent; quiescence

R

rack-mounted

racon def: a radar beacon

radian def: a unit of angular measurement; abbr: **rad**

radio- as a prefix, combines to form one word: *radioactive, radiobiology, radioisotope, radioluminescence;* if combining word starts with *o*, omit one of the *o*'s: *radiology, radiopaque;* in other instances *radio* may be either combined or treated as a separate word, depending on accepted usage; typical examples are *radio compass, radio countermeasures, radio direction-finder, radio frequency* (as a noun), *radio-frequency* (as an adjective), *radio range, radiosonde, radiotelephone*

radio frequency abbr: **rf**

radius pl: *radii*

radix pl: *radices* (pref) or *radixes*

range; ranging; rangefinder; range marker

rare; rarity; rarefy; rarefaction

ratemeter

ratio; ratios

re def: a Latin word meaning in the case of; avoid using *re* in technical writing, particularly as an abbr for *regarding, concerning, with reference to*

re- a prefix meaning to do again, to repeat; normally combines to form one word: *reactivate, rediscover, reemphasize, reentrant, reignition, rerun, reset;* if the compound term forms an existing word that has a different meaning, insert a hyphen to identify it as compound: *re-cover* (to cover again)

reaction use *reaction* to describe chemical or mechanical processes, not as a synonym for *opinion* or *impression*

reactive kilovolt-ampere; reactive volt-ampere see **kilo** or **volt**

readability

readout

recede

receive; receiver; receiving; receivable

rechargeable

recommend

reconcile; reconcilable

reconnaissance

recur; recurred; recurring; recurrence these are the correct spellings; never use *reoccur* (etc)

reducible

reenforce; reinforce *reenforce* means to enforce again; *reinforce* means to strengthen: *Rick Davis reenforced his original instructions by circulating a second memorandum; the Artmo Building required 34,750 tons of reinforced concrete*

refer; referred; referring; referral; referee; reference

referendum pl: *referendums* (pref); *referenda* (less common)

reoccur(rence) never use; see **recur**

repairable; reparable both words mean in need of repair and capable of being repaired; *reparable* also implies that the cost to repair the item has been taken into account and it is economically worthwhile to effect repairs

repellant; repellent use *repellant* as a noun, *repellent* as an adjective; **repeller**

replaceable

reservoir

reset; resetting; resettability

resin; rosin these words have become almost synonymous, with a preference for *resin;* use *resin* to describe a gluey substance used in adhesives, and *rosin* to describe a solder flux-core

respective(ly) this overworked word is not really needed in sentences that differentiate between two or more items; it should be deleted from sentences such as: *pins 4, 5, and 7 are marked R, S, and V respectively*

retro- a prefix meaning to take place before, or backward; normally combines to form one word: *retroactive, retrofit, retrogression;* if combining word starts with *o,* insert a hyphen: *retro-operative*

reverse; reverser; reversal; reversible

revolutions per minute; revolutions per second abbr: **rpm; rps**

rheostat

rhombus pl: *rhombuses* (pref) or *rhombi*

rhythm

ricochet; richocheted; ricocheting

right-handed(ed) abbr: **RH**

rivet; riveted; riveter; riveting

roentgen abbr: **R**

role; roll a *role* is a person's function or the part that he plays (in an organization, project, or play); a *roll,* as a technical noun, is a cylinder; as a verb, it means to rotate: *the technicians' role was to make the samples roll toward the magnet*

root mean square abbr: **rms**

rosin see **resin**

rotate; rotator; rotatable; rotary

ruggedize

rustproof; rust-resistant

S

salable; saleable *salable* pref

same avoid using *same* as a pronoun; to write *we have repaired your receiver and tested same* is awkward; a better version is *we have repaired and tested your receiver*

sapphire

satellite

saturate; saturation; saturable

sawtooth; saw-toothed

scalar; scaler *scalar* is a quantity that has magnitude only; *scaler* is a measuring device

scarce; scarcity

sceptic(al); skeptic(al) *skeptic(al)* pref

schedule

schematic although really an adjective (as in *schematic diagram),* in technical terminology *schematic* can be used as a noun (meaning *a schematic drawing)*

science; scientific(ally); scientist

screwdriver; screw-driven

seamweld

seasonal; seasonable *seasonal* means affected by or dependent on the season; *seasonable* means appropriate or suited to the time of year: *a seasonal activity; seasonable weather*

seasons the seasons are not capitalized: *spring, summer, autumn* or *fall, winter*

secant abbr: **sec**

second abbr:
time **sec**
angular measure ″

secure see **obtain**

-sede *supersede* is the only word to end with *-sede;* others end with *-cede* or *-ceed* (which see)

seems(s) see **appear(s)**

self- insert a hyphen when used as a prefix to form a compound term: *self-absorption, self-bias, self-excited, self-locking, self-resetting;* but there are exceptions: *selfless, selfsame*

semi- a prefix meaning half; normally combines to form one word: *semiactive, semiannually (*every six months), *semiconductor, semimonthly* (half-monthly), *semiremote, semiweekly* (half weekly); if combining word starts with *i,* insert a hyphen: *semi-idle, semi-immersed*

separate; separable; separator; separation

sequence; sequential

serial number abbr: **ser no.** or **S/N**

series-parallel

serrated

serviceable; serviceman, servicewoman, serviceperson (pref)

servo- as a prefix, combines to form one word: *servoamplifier, servocontrol, servosystem;* as a noun, *servo* is an abbreviation for *servomotor* or *servomechanism*

sewage; sewerage *sewage* is waste matter; *sewerage* is the drainage system that carries away the waste matter

short- as a prefix, most often combines with a hyphen: *short-circuit, short-form* (report), *short-lived, short-term;* in some cases it combines into one word: *shorthand* (writing), *shorthanded, shortcoming, shortsighted*

shrivel(l)ed; shrivel(l)ing single *l* pref

siemens def: a unit of electric conductance (SI); abbr: **S**; other abbr: **kS, mS, μS**

sight def: the ability to see; see **site**

signal(l)ed; signal(l)ing; signal(l)er single *l* pref

signal-to-noise (ratio)

silverplate; silver-plate use *silverplate as* a noun or adjective, *silver-plate* as a verb

similar not *similiar*

sine abbr: **sin**

singe; singeing the *e* must be retained to avoid confusion with *singing*

singlehanded

siphon not *syphon*

sirup; syrup *syrup* pref

site, sight, cite three words that often are misspelled; a *site* is a location: *the construction site; sight* implies the ability to see: *mud up to the axles became a familiar sight; cite* means quote: *I cite the May 17 progress report as a typical example of good writing*

siz(e)able *sizable* pref

skeptic(al); sceptic(al) *skeptic(al)* pref

skil(l)ful *skillful* pref; note that general usage dictates that *ll* is pref, contrary to the preference in most *l* and *ll* situations in this glossary

slip- usually combines into a single word: *slippage, slipshod, slipstream;* but *slip ring(s)*

smelled; smelt *smelled* pref

smo(u)lder *smolder* pref

solder

someone; some one *someone* is correct when the meaning is any one person; *some one* is seldom used

some time; sometimes *some time* means an indefinite time: *some time ago; sometimes* means occasionally: *he sometimes works until after midnight*

sound combines irregularly: *sound-absorbent, sound-absorbing, sound-powered, soundproof, sound track, sound wave*

south abbr: **S**; other abbr:

southeast	**SE**
southwest	**SW**

southbound and *southward* are written as one word; for rule on capitalization, see **east**

space- as a prefix normally combines to form one word: *spacecraft, spaceflight*

spare(s) as a noun, frequently means spare part(s)

specially see **especially**

specific gravity abbr: **sp gr**

specific heat abbr: **sp ht**

spectro- as a prefix, combines to form one word: *spectrometer, spectroscope;* if combining word starts with *o,* omit one *o*: *spectrology*

spectrum pl: *spectra*

spelled; spelt *spelled* pref

spilled; spilt *spilled* pref

spiral(l)ed; spiral(l)ing one *l* pref

split infinitive to split an infinitive is permissible when not to split it would result in awkward construction or ambiguity, or require extensive rewriting

spoiled; spoilt *spoiled* pref

spotweld

square abbr: **sq** or ²; other abbr:

square foot	**ft²** (pref); **sq ft**
square inch	**in.²** (pref); **sq in.**
square meter	**m²**
square centimeter	**cm²**
square millimeter	**mm²**
curies per square meter	**Ci/m²**
milliwatts per square meter	**mW/m²**

standby; standoff; standstill all combine into one word when used as a noun or an adjective

standing-wave ratio *abbr:* **swr**

state-of-the-art

stationary; stationery *stationary* means not moving: *the vehicle was stationary when the accident occurred; stationery* refers to writing materials: *the main item in the October stationery requisition was an order for one thousand writing pads*

statute mile def: 5280 ft (1609 m); see **mile**

statutory

stencil(l)ed; stencil(l)ing single *l* pref

stereo- as a prefix, combines to form one word: *stereometric, stereoscopic; stereo* can be used alone as a noun meaning multi-channel system

stimulus pl: *stimuli*

stock; stockholder; stocklist; stock market; stockpile

stop- as a prefix usually combines to form one word: *stopgap; stopnut; stopover* (when used as noun or adjective); *stoppage; stop watch*

strato- a prefix that combines to form one word: *stratocumulus, stratosphere*

stratum pl: *strata*

structural

stylus pl: *styluses* (pref) or *styli*

sub- a prefix generally meaning below, beneath, under; combines to form one word: *subassembly, subcarrier, subcommittee, subnormal, subpoint*

subparagraph abbr: **subpara**; abbr for *subsubparagraph* is **subsubpara**

subtle; subtlely, subtly *subtly* pref

sufficient in technical writing, *enough* is a better word than *sufficient*

sulfur; sulphur *sulfur* pref; as a prefix, *sulf-* combines to form one word: *sulfanilamide*

super- a prefix meaning greater or over; combines to form one word: *superconductivity, superregeneration*

superhigh frequency abbr: **shf**

superimpose; superpose *superimpose* means to place or impose one thing on top of another; *superpose* means to lay or place exactly on top of, so as to be coincident with

supersede see **-sede**

supra- a prefix meaning above; normally combines to form one word: *supramolecular;* if combining word starts with *a*, insert a hyphen: *supra-auditory*

surveillance

susceptible

switch- *switchboard; switchbox; switchgear*

swivel(l)ed; swivel(l)ing single *l* pref

syllabus **pl:** *syllabuses* (pref) or *syllabi*

symmetry; symmetrical

symposium pl: *symposia* (pref) or *symposiums*

synchro- as a prefix combines to form one word: *synchromesh, synchronize, synchroscope; synchro* can also be used alone as a noun meaning synchronous motor

synonymous use *synonymous with, not synonymous to*

synopsis pl: *synopses*

synthesis pl: *syntheses*

synthetic

syphon *siphon* pref

syrup(y)

systemwide

T

tablespoon use *tablespoonfuls* rather than *tablespoonsful;* abbr: **tbsp**

tail- as a prefix normally combines to form one word: *tailboard, tailless, tailwind;* but *tail end, tail fin*

take- *takeoff; takeover; takeup;* as nouns and adjectives these terms all combine into a single word

tangent abbr: **tan**

teaspoon use *teaspoonfuls* rather than *teaspoonsful;* abbr: **tsp**

technician

tele- a prefix that means at a distance; combines to form one word: *teleammeter, telemetry, telephony, teletype (writer)*

telecom; telecon *telecom* is the abbr for *telecommunication(s); telecon* is the abbr for *telephone conversation*

television abbr: **TV**

temperature *abbr:* **temp**; combinations are *temperature-compensating* and *temperature-controlled;* when recording temperatures, the abbr for *degree* (deg or °) may be omitted: *an operating temperature of 85C; the water boils at 100C or 212F;* the pref (SI) unit for temperature is the degree Celsius (°**C**)

tempered

template; templet both spellings are correct; *template* pref; def: a pattern or guide

temporary; temporarily

tensile strength abbr: **ts**

tera def: 10^{12}; abbr: **T**; other abbr:

terabecquerel(s)	**TBq**
terahertz	**THz**
terajoule(s)	**TJ**
terawatt(s)	**TW**

terminus pl: *termini* (pref) or *terminuses;* note that this plural is contrary to the pref plurals for most *-us* endings

tesla def: a unit of magnetic flux density, magnetic inductance (SI); abbr:**T**; other abbr: **mT, μT, nT**

that is abbr: **i.e.** (pref) or **ie**

their; there; they're the first two of these words are frequently misspelled, more through carelessness than as an outright error; *their* is a possessive meaning belonging to them: *the staff took their holidays earlier than normal; there* means in that place: *there were 18 desks in the room,* or *put it there; they're* is a contraction of *they are* and should not appear in technical or business writing

there- as a prefix combines to form one word: *thereafter, thereby, therein, thereupon*

therefor(e) *therefore* pref

thermo- a prefix generally meaning heat; combines to form one word: *thermoammeter, thermocouple, thermoelectric, thermoplastic*

thermodynamic temperature the SI unit is the kelvin (abbr: **K**), expressed in degrees Celsius (°C)

thesis pl: *theses*

thousand abbr:

thousand foot-pound(s)	**kip-ft**
thousand pound(s)	**kip**

three- when used as a prefix, a hyphen normally is inserted between the combining words: *three-dimensional, three-phase, three-ply, three-wire;* exceptions are *threefold* and *threesome*

threshold

through never use *thru*

tieing; tying *tying* pref

timber; timbre *timber* is wood; *timbre* means tonal quality

time always write time in numerals, if possible using the 24-hour clock: *08:17* or *8:17 a.m., 15:30* or *3:30 p.m.* 24-hour times may be written as *20:45* (pref), *20:45 hr,* or *20:45 hours;* never use the term "o'clock" in technical writing: write *15:00* or *3 p.m.* rather than *3 o'clock*

time- typical combinations are *time base; time-card; time clock; time constant; time-consuming, time lag; timesaving; timetable; time-wasting*

tinplate; tin-plate use *tinplate* as a noun, *tin-plate* as a verb or adjective

to; too; two frequently misspelled, most often through carelessness; *to* is a preposition that can mean in the direction of, against, before, or until; *too* means as well; *two* is the quantity "2"

today; tonight; tomorrow never use *tonite*

tolerance abbr: **tol**

ton; tonne the U.S. ton is 2000 lb and is known as a *short ton;* the British ton is 2240 lb and is known as a *long ton;* the metric ton is 1000 kg (2204.6 lb) and is known as a *tonne* (abbr: **t**); other terms: *tonmile* and *tonnage*

top- *top-heavy; top-loaded; top-up*

torque(d); torquing

total(l)ed; total(l)ing single l pref

toward(s) *toward pref*

traceable

trade- *trademark, tradeoff*

trans- a prefix meaning over, across, or through; it normally combines to form

trans- *(continued)* one word: *transadmittance, transcontinental, transship;* if combining word is a proper noun, insert a hyphen: *trans-Canada* (an exception is *transatlantic,* which through common usage has dropped the capital A and combined into one word); *transonic* has only one *s*

transceiver def: a transmitter-receiver

transfer; transferred; transferring; transferable; transference

transmit; transmitted; transmitting; transmittal; transmitter; transmission

transverse; traverse *transverse* means to lie across; *traverse* means to track horizontally

travel(l)ed; travel(l)ing; travel(l)er single *l* pref

tri- a prefix meaning three or every third; combines to form one word: *triangulation, tricolor, trilateral, tristimulus, triweekly*

trouble-free; troubleshoot(ing)

tune; tunable; tuneup (noun or adjective)

tunnel(l)ed; tunnel(l)ing single *l* pref

turbo- a prefix meaning turbine-powered; combines to form one word: *turboelectric, turboprop*

turbulence

turn- *turnaround* and *turnover* form one word when used as adjectives or nouns; *turnstile* and *turntable* always form one word; *turns-ratio* is hyphenated

two- when used as a prefix to form a compound term, a hyphen normally is inserted: *two-address, two-phase, two-ply, two-position, two-wire;* an exception is *twofold*

type- as a prefix normally combines into one word: *typeface; typeset(ting); typewriter; typewriting*

U

ultimatum *ultimatums* (pref) or *ultimata* (seldom used)

ultra- a prefix meaning exceedingly; normally combines to form one word: *ultrasonic, ultraviolet;* if combining word starts with *a,* insert a hyphen: *ultra-audible, ultra-audion*

ultrahigh frequency abbr: **uhf**

un- a prefix that generally means not or negative; normally combines to form one word: *uncontrolled, undamped, unethical, unnecessary;* if combining word is a proper noun, insert a hyphen: *un-American;* if term combines to form an existing word that has a different meaning, insert a hyphen: *un-ionized* (meaning not ionized) to avoid confusion with *unionized* (to belong to a union); if uncertain whether to use *un-, in-,* or *im-,* try using *not*

unadvisable; inadvisable both are correct; *inadvisable* pref

unbalance; imbalance for technical writing, *unbalance* pref

unbias(s)ed *unbiased* pref

under- a prefix meaning below or lower; combines to form one word: *underbunching, undercurrent, underexposed, underrated, undershoot, undersigned*

underage means a shortage or deficit, or too young

unequal(l)ed *unequaled* pref; see **equal**

unessential; inessential *unessential* pref

uni- a prefix meaning single or one only; combines to form one word: *uniaxial, unidirectional, unifilar, univalent*

uninterested def: not interested; avoid confusing with *disinterested* (which see)

unique def: the one and only, without equal, incomparable; use with great care and never in any sense where a comparison is implied; you cannot write *this is the most unique design;* rewrite as *this design is unique,* or (if a comparison must be made) *this is the most unusual design*

unparalleled

unpractical *impractical* pref

unsanitary *insanitary* pref

unserviceable abbr: **u/s**

unstable *instability* is better than *unstability*

up- as a prefix combines to form one word: *update, upend, upgrade, uprange, upswing*

upper case def: capital letters; abbr: **uc**

uppermost

up-to-date

use; usable; usage; using; useful

V

vacuum

valve-grind(ing)

vari- as a prefix meaning varied, combines into one word: *varicolored, variform*

variance write *at variance with;* never write *at variance from*

varimeter; varmeter def: a meter for measuring reactive power; *varimeter* pref

vehicle; vehicular

vender; vendor *vendor* pref

versed sine abbr: **vers**

versus def: against; abbr: **vs**

vertex def: top; pl: *vertexes* (pref) or *vertices;* avoid confusing with *vortex*

very high frequency abbr: **vhf**

vice versa def: in reverse order

video- *videocast; videotape; video mapper*

video frequency abbr: **vf**

viewfinder; viewpoint

vise def: a clamp; in U.S. spelled *vise;* elsewhere: *vice*

visor; vizor *visor* pref

viz def: namely; this term is seldom used in technical writing

volt def: electric potential or potential difference; abbr: **V;** other abbr: **MV, kV, mV, μV, nV;** also:

volt-ampere(s)	**VA**
volt-ampere(s), reactive	**VAr**
volts, alternating current	**Vac**
volts, direct current	**Vdc**
volts, direct current, working	**Vdcw**
volts per meter	**V/m**

volt- combines into one word: *voltammeter, voltohmyst*

volume abbr: **vol**

vortex def: spiral; pl: *vortexes* (pref) or vortices; avoid confusing with *vertex*

VU-meter

W

walkie-talkie

war- as a prefix, combines to form one word: *warfare, wartime*

warranty

waste, wastage

water- combines irregularly: *water-cool(ed); water cooler; waterflow; water level; waterline; waterproof; water-soluble; watertight*

watt def: a unit of power, or radiant flux (SI); abbr: **W;** other abbr: **TW, GW, MW, kW, mW, μW, nW, pW, W/m²;** the abbr for *watt-hour(s)* is **Wh** (pref) or **W-hr;** as a prefix *watt-* forms *watthourmeter* and *wattmeter*

wave- normally combines to form one word: *waveband, waveform, wavefront, waveguide, wavemeter, waveshape;* exceptions are *wave angle* and *wave-swept*

wavelength abbr: **λ**

weather use only as a noun; never write *weather conditions;* avoid confusing with *climate* and *whether* (which see)

weber def: a unit of magnetic flux (SI); abbr: **Wb;** other abbr: **mWb**

week(s) abbr: **wk**

weight abbr: **wt**

west abbr: **W;** *westbound* and *westward* are written as one word; for rule on capitalization, see **east**

where- as a prefix, combines to form one word: *whereas, wherefore, wherein;* when combining word starts with *e,* omit one *e: wherever*

whether; weather *whether* means if; *weather* has to do with rain, snow, sunshine, etc

while; whilst *while* pref

whoever

wide; width abbr: **wd**

wideband; widespread

wirecutter(s); wire-cutting; wirewound

withheld; withhold

words per minute abbr: **wpm**

work- as a prefix usually combines to form one word: *workbench, workflow, workload, workshop*

working volts, dc abbr: **Vdcw**

worldwide

wrap; wrapped; wrapping; wraparound

writeoff; writeup both combine into one word when used as noun or adjective

writer see **author**

writing only one *t*

X

x- *x-axis; X-band; x-particle; x-radiation; x-ray*

Xerox

X-Y recorder

Y

y- *Y-antenna; y-axis; Y-connected; Y-network; Y-signal*

yard(s) abbr: **yd**

yardstick

year(s) abbr: **yr**; typical combinations are *year-end* and *year-round*

your; you're *your* means belonging to or originating from you: *I have examined your prototype analyzer; you're* is a contraction of *you are* and should not appear in technical or business writing

Z

Z-axis

zero pl: *zeros* (pref) or *zeroes;* typical combinations are *zero-access, zero-adjust, zero-beat, zero-hour, zero level, zero-set, zero reader*

zip code

zoology; zoological

INDEX

MARKING CONTROL CHART

This control chart will show you which aspects of your writing need attention and, as time progresses, whether you have successfully corrected your most predominant faults. As each assignment is returned to you, count up the errors indicated as marginal notations by your instructor and enter them on the chart. For instance, if on assignment 1 your instructor enters "F" and "U" once, and "S" three times, in the margin, you are being told you have used the wrong format (F), your work is untidy (U), and you have three spelling errors (S). In column 1 of the chart enter "1" in the squares opposite F and U, and "3" in the square opposite "S".

ASSIGNMENT NUMBER

1	2	3	4	5	6	7	8	9	10	11	12	13	14	15	
															A – Awkward construction
															B – Brevity overdone; too few details
															C – Continuity weak; paragraph lacks coherence/unity
															D – Development inadequate; support your argument
															E – Error! Check your facts, data, information
															F – Format incorrect
															G – Grammar fault
															H – Heavy going; dull; uninteresting
															I – Illogical or irrelevant (correct or omit)
															J – Jumpy – too many short sentences, reads like primary reader
															K – King-size paragraphs, sentences or words (shorten them)
															L – Low Information Content words or phrase (delete)
															M – Missing words or information
															N – No! Never do this; never use slang, contractions, unexplained abbreviations, etc.
															O – Organization poor
															P – Punctuation error, or punctuation missing
															Q – Query: what does this mean? not understood; can't read your writing
															R – Repetition
															S – Spelling error
															T – Tone wrong
															U – Untidy, messy, or careless work (improve "presentation")
															V – Vague; ambiguous; not clear enough
															W – Wishy-washy; weak argument; unconvincing
															X – X-out (delete, omit) this unnecessary statement
															Y – Yak! Yak! Yak! – too wordy; too many generalities
															Z – Lacks continuity; needs better transitions
															# – Numbers wrongly presented
															// – Use parallel construction